AF411413

Physical and Chemical Interactions between Insects and Plants

Physical and Chemical Interactions between Insects and Plants

Editors

Gianandrea Salerno
Manuela Rebora
Stanislav N. Gorb

Basel • Beijing • Wuhan • Barcelona • Belgrade • Novi Sad • Cluj • Manchester

Editors

Gianandrea Salerno
Department of Chemistry,
Biology and Biotechnology
University of Perugia
Perugia
Italy

Manuela Rebora
Department of Chemistry,
Biology and Biotechnology
University of Perugia
Perugia
Italy

Stanislav N. Gorb
Department of Functional
Morphology and Biomechanics
Kiel University
Kiel
Germany

Editorial Office
MDPI
St. Alban-Anlage 66
4052 Basel, Switzerland

This is a reprint of articles from the Special Issue published online in the open access journal *Insects* (ISSN 2075-4450) (available at: www.mdpi.com/journal/insects/special_issues/interactions_insects_plants).

For citation purposes, cite each article independently as indicated on the article page online and as indicated below:

Lastname, A.A.; Lastname, B.B. Article Title. *Journal Name* **Year**, *Volume Number*, Page Range.

ISBN 978-3-0365-8709-7 (Hbk)
ISBN 978-3-0365-8708-0 (PDF)
doi.org/10.3390/books978-3-0365-8708-0

Cover image courtesy of Stanislav N. Gorb

Contents

About the Editors

Gianandrea Salerno

Gianandrea Salerno is Professor of Entomology at the Department of Agricultural, Food and Environmental Sciences (DSA3), University of Perugia (Italy). He received his Ph.D. degree in Agricultural Entomology from the Department of Arboriculture and Plant Protection, University of Perugia (Italy). His research focuses on insect–plant mechanical and chemical interactions as well as the behavior and chemical ecology of parasitic Hymenoptera and agricultural pests. Since 2013, he has been a Board Member of Professors of the Ph.D. course in "Agricultural, Food and Environmental Science and Biotechnology", University of Perugia (Italy).

Manuela Rebora

Manuela Rebora is a Professor in the Department of Chemistry, Biology and Biotechnology within the University of Perugia, Italy. She received her PhD degree in agricultural entomology from the Department of Arboricolture and Plant Protection, University of Perugia (Italy). Rebora's research focuses on the insect functional morphology (attachment devices and sensory systems) involved in insect–plant interactions. Since 2021, she has been the Coordinator of the PhD course Biological and Naturalistic Sciences of the University of Perugia.

Stanislav N. Gorb

Stanislav Gorb is a Professor and Director at the Zoological Institute of the Kiel University, Germany. He received his PhD degree in Zoology and Entomology from the Schmalhausen Institute of Zoology of the Ukrainian Academy of Sciences in Kiev (Ukraine). Gorb's research focuses on the morphology, structure, biomechanics, and evolution of surface-related functional systems in animals and plants, as well as the development of biologically inspired technological surfaces and systems. In 2018, he received the Karl-Ritter-von-Frisch Medal of the German Zoological Society. Gorb is a Corresponding Member of the Academy of Science and Literature Mainz and a Member of the National Academy of Sciences Leopoldina, Germany.

Preface

Insects and plants have been interacting for more than 350 million years, often resulting in species variability and radiation. Studying insect–plant chemical and mechanical interactions at different levels can help shed light on the complex factors driving the evolutionarily successful relationship between these two groups. This Special Issue focuses on the chemical and physical interactions between insects and plants to understand the functional significance of plant chemistry, plant surface structures and their relationships with a broad range of ecological groups of insects, including pollinators, herbivores and predators. The results of the investigations presented in this Special Issue can help develop effective management strategies to successfully control insect pest infestations or pollination strategies for diverse cropping environments.

Gianandrea Salerno, Manuela Rebora, and Stanislav N. Gorb
Editors

Editorial

Mechanoecology and Chemoecology: Physical and Chemical Interactions between Insects and Plants

Gianandrea Salerno [1,*], Manuela Rebora [2,*] and Stanislav Gorb [3,*]

1   Dipartimento di Scienze Agrarie, Alimentari e Ambientali, University of Perugia, Borgo XX Giugno, 06121 Perugia, Italy
2   Dipartimento di Chimica, Biologia e Biotecnologie, University of Perugia, Via Elce di Sotto 8, 06121 Perugia, Italy
3   Department of Functional Morphology and Biomechanics, Zoological Institute, Kiel University, Am Botanischen Garten 9, 24098 Kiel, Germany
*   Correspondence: gianandrea.salerno@unipg.it (G.S.); manuela.rebora@unipg.it (M.R.); sgorb@zoologie.uni-kiel.de (S.G.)

Citation: Salerno, G.; Rebora, M.; Gorb, S. Mechanoecology and Chemoecology: Physical and Chemical Interactions between Insects and Plants. *Insects* 2023, 14, 657. https://doi.org/10.3390/insects14070657

Received: 11 July 2023
Accepted: 17 July 2023
Published: 23 July 2023

Plants and herbivorous insects, as well as their natural enemies such as predatory and parasitoid insects, are united by intricate relationships. During the long period of co-evolution with insects, plants developed a wide diversity of chemical and mechanical features for defense against herbivores, and attracting pollinators and natural herbivore enemies. The chemical basis of insect–plant interactions has been established, and in many examples the feeding and oviposition site selection of phytophagous insects are dependent on plant secondary metabolites. Volatile organic compound (VOC) emission by plants, influenced by insect feeding or oviposition, can repel herbivores and attract natural enemies. In this context, phytophagous and entomophagous insects evolved a finely tuned sensory system for the detection of plant cues.

Despite being often overlooked, mechanical interactions between insects and plants can be rather crucial in the process of host plant selection by phytophagous insects. The evolution of plant surfaces and insect adhesive pads is an interesting example of competition between insect attachment systems and plant anti-attachment surfaces. To achieve sufficient attachment that enables locomotion on widely diverse plant surfaces, insects have evolved various types of leg attachment devices, allowing them to overcome physical plant defenses such as cuticular microfolds, various kinds of trichomes and crystalline wax coverage. Additionally, insect mouthpart mechanics are adapted to certain mechanical properties of the plant surface, which, in combination, may influence insect–plant interactions.

This Special Issue focuses on the chemical and physical interactions between insects and plants in order to understand the functional significance of plant chemistry, plant surface structures and their relationships with a broad range of ecological groups of insects, including pollinators, herbivores and predators.

Dealing with the mechanical interaction between oligophagous phytophagous insect pests and host plant mechanical barriers, Rebora et al. [1] show that olive fruit fly adhesion is reduced by epicuticular waxes on the olive surface, and that the female shows a different ability to attach to the olive surface of different cultivars of *Olea europaea*, in relation to different values of olive surface wettability. On the other hand, Saitta et al. [2] reveal that Cucurbitaceae glandular trichomes do not affect insect attachment ability of the melon ladybird at adult and larval stages, suggesting some adaptation of this insect species to its host plants; moreover, the authors demonstrate that non-glandular trichomes heavily reduce the attachment ability of adults and larvae only when they are dense, short and flexible. In an insect group, such as Phasmatodea, which is highly associated with plants through herbivory and camouflage, Burack et al. [3] reveal that wax crystal-covered plant substrates with fine roughness cause the lowest attachment ability; whereas, strongly structured natural substrates show the highest attachment ability. Gorb and Gorb [4] highlight the contribution of wax coverage on flower stems to impeding the locomotion of the generalist ant species, *Lasius niger*, and provide further evidence for the hypothesis that when having a diversity of plant stems in the field, generalist ants prefer substrates where their locomotion is less hindered by obstacles and/or surface slipperiness. In another study on the interaction between the mechanical features of flowers and the attachment ability of generalist insect pollinators [5], the same authors observe that insect adhesion is surprisingly reduced for petals, where the color intensity is enhanced due to papillate epidermal cells covered by cuticular folds which contribute to adhesion reduction in generalist insect pollinators.

The influence of pollen properties, insect and floral surfaces on the adhesion forces that mediate pollen transfer has been poorly studied thus far; the paper by Huth et al. [6] makes a contribution to this topic, by reporting on the adhesive properties of pollen related to its aging time.

An important aspect of insect–plant interactions is the multitrophic relationships between plants, pests, and natural enemies. In this regard, Farina et al. [7] analyze the damage to plants caused by the cotton whitefly, *Bemisia tabaci*, in presence of a predator which feeds on both insect prey and plant tissue, while Liu et al. [8] study the effects of herbivore-induced rice volatiles and reveal their positive effects on spider attraction and predation ability, with beneficial effects in improving the control of rice pests.

Plant responses to phytophagous insects are extremely complex because of the presence of various types of interactions between plants and pests. In this context, Gao et al. [9] investigate the interplay between insect venom and wounding stress, and the specific expression genes and transcription factors of the Mongolian pine *Pinus sylvestris* var. mongolica. The authors find that insect venom induces a series of physiological changes in the host that weaken the host's defense response, and hence contribute to the growth of insect symbiotic fungus and the development of eggs.

The use of resistant cultivars is an efficient management strategy against insect pests. Mortazavi Malekshah et al. [10] show that the physicochemical properties of sugarcane cultivars significantly affect *Sesamia nonagrioides* oviposition behavior, life history and population parameters, and recommend a resistant cultivar to reduce damage caused by this pest.

Jakubska-Busse et al. [11] analyze the VOCs emitted by the flowers of an invasive alien plant species and show a list of potential pollinators, in order to shed light on factors enabling the species to rapidly expand.

Changes during leaf ontogeny affect the palatability of insect herbivores, and Lirette et al. [12] summarize the literature describing how chemical defenses of foliage change during the growing season in white spruce, an economically important conifer tree attacked by the eastern spruce budworm *Choristoneura fumiferana*, a specialist growing conifer leaf and bud feeder.

In conclusion, insects and plants have been interacting for more than 350 million years, often resulting in species variability and radiation. Studying insect–plant chemical and mechanical interactions at different levels can help shed light on the complex factors driving the evolutionarily successful relationship between these two groups. Moreover, the results of such investigations can help develop helpful management strategies to successfully control insect pest infestations in cropping environments.

Conflicts of Interest: The authors declare no conflict of interest.

References

1. Rebora, M.; Salerno, G.; Piersanti, S.; Gorb, E.; Gorb, S. Role of Fruit Epicuticular Waxes in Preventing *Bactrocera oleae* (Diptera: Tephritidae) Attachment in Different Cultivars of *Olea europaea*. *Insects* **2020**, *11*, 189. [CrossRef] [PubMed]
2. Saitta, V.; Rebora, M.; Piersanti, S.; Gorb, E.; Gorb, S.; Salerno, G. Effect of Leaf Trichomes in Different Species of Cucurbitaceae on Attachment Ability of the Melon Ladybird Beetle *Chnootriba elaterii*. *Insects* **2022**, *13*, 1123. [CrossRef] [PubMed]
3. Burack, J.; Gorb, S.N.; Büscher, T.H. Attachment Performance of Stick Insects (Phasmatodea) on Plant Leaves with Different Surface Characteristics. *Insects* **2022**, *13*, 952. [CrossRef] [PubMed]

4. Gorb, E.V.; Gorb, S.N. Combined Effect of Different Flower Stem Features on the Visiting Frequency of the Generalist Ant *Lasius niger*: An Experimental Study. *Insects* **2021**, *12*, 1026. [CrossRef] [PubMed]
5. Gorb, E.V.; Gorb, S.N. Petals Reduce Attachment of Insect Pollinators: A Case Study of the Plant *Dahlia pinnata* and the Fly *Eristalis tenax*. *Insects* **2023**, *14*, 285. [CrossRef] [PubMed]
6. Huth, S.; Schwarz, L.-M.; Gorb, S.N. Quantifying the Influence of Pollen Aging on the Adhesive Properties of *Hypochaeris radicata* Pollen. *Insects* **2022**, *13*, 811. [CrossRef] [PubMed]
7. Farina, A.; Massimino Cocuzza, G.E.; Suma, P.; Rapisarda, C. Can *Macrolophus pygmaeus* (Hemiptera: Miridae) Mitigate the Damage Caused to Plants by *Bemisia tabaci* (Hemiptera: Aleyrodidae)? *Insects* **2023**, *14*, 164. [CrossRef] [PubMed]
8. Liu, J.; Sun, L.; Fu, D.; Zhu, J.; Liu, M.; Xiao, F.; Xiao, R. Herbivore-Induced Rice Volatiles Attract and Affect the Predation Ability of the Wolf Spiders, *Pirata subpiraticus* and *Pardosa pseudoannulata*. *Insects* **2022**, *13*, 90. [CrossRef] [PubMed]
9. Gao, C.; Ren, L.; Wang, M.; Wang, Z.; Fu, N.; Wang, H.; Shi, J. Full-Length Transcriptome Sequencing-Based Analysis of *Pinus sylvestris* var. mongolica in Response to *Sirex noctilio* Venom. *Insects* **2022**, *13*, 338. [CrossRef] [PubMed]
10. Mortazavi Malekshah, S.A.; Naseri, B.; Ranjbar Aghdam, H.; Razmjou, J.; Fathi, S.A.A.; Ebadollahi, A.; Changbunjong, T. Physico-chemical Properties of Sugarcane Cultivars Affected Life History and Population Growth Parameters of *Sesamia nonagrioides* (Lefebvre) (Lepidoptera: Noctuidae). *Insects* **2022**, *13*, 901. [CrossRef] [PubMed]
11. Jakubska-Busse, A.; Dziadas, M.; Gruss, I.; Kobyłka, M.J. Floral Volatile Organic Compounds and a List of Pollinators of *Fallopia baldschuanica* (Polygonaceae). *Insects* **2022**, *13*, 904. [CrossRef] [PubMed]
12. Lirette, A.-O.; Despland, E. Defensive Traits during White Spruce (*Picea glauca*) Leaf Ontogeny. *Insects* **2021**, *12*, 644. [CrossRef] [PubMed]

Article

Petals Reduce Attachment of Insect Pollinators: A Case Study of the Plant *Dahlia pinnata* and the Fly *Eristalis tenax*

Elena V. Gorb * and Stanislav N. Gorb 

Department of Functional Morphology and Biomechanics, Zoological Institute, Kiel University, Am Botanischen Garten 9, 24098 Kiel, Germany
* Correspondence: egorb@zoologie.uni-kiel.de; Tel.: +49-431-8804509

Simple Summary: One of the important aspects in the relationship between plants and insects is pollination including attraction of pollinators, which is related to mainly optical effects. Since insects pollinating the majority of plants are in contact with a petal surface for a certain time, it is plausible to hypothesize that such petals should also enable or even promote attachment of insects to their surface. The aim of this study was to understand whether the petal surface in "cafeteria"-type flowers, which offer their nectar and pollen to insect pollinators in an open way, is adapted to a stronger attachment of pollinators. We selected the garden dahlia plant *Dahlia pinnata* and the hovering fly *Eristalis tenax*, examined leaves, petals, and flower stems using cryo scanning electron microscopy, and performed force measurements of fly attachment to surfaces of these plant organs. Our experimental results showed that insect attachment on the petal surface was significantly weaker compared to that on smooth leaf and smooth reference glass. In our opinion, these "cafeteria"-type flowers have the petals, where the colour intensity is enhanced due to papillate/conical epidermal cells covered by micro- and nanoscopic cuticular folds, and exactly these latter structures mainly contribute to attachment reduction in insect pollinators.

Citation: Gorb, E.V.; Gorb, S.N.
Petals Reduce Attachment of Insect
Pollinators: A Case Study of the Plant
Dahlia pinnata and the Fly *Eristalis
tenax*. *Insects* **2023**, *14*, 285. https://
doi.org/10.3390/insects14030285

Academic Editor: Peter H. Adler

Received: 30 January 2023
Revised: 9 March 2023
Accepted: 10 March 2023
Published: 14 March 2023

Abstract: In order to understand whether the petal surface in "cafeteria"-type flowers, which offer their nectar and pollen to insect pollinators in an open way, is adapted to a stronger attachment of insect pollinators, we selected the plant *Dahlia pinnata* and the hovering fly *Eristalis tenax*, both being generalist species according to their pollinator's spectrum and diet, respectively. We combined cryo scanning electron microscopy examination of leaves, petals, and flower stems with force measurements of fly attachment to surfaces of these plant organs. Our results clearly distinguished two groups among tested surfaces: (1) the smooth leaf and reference smooth glass ensured a rather high attachment force of the fly; (2) the flower stem and petal significantly reduced it. The attachment force reduction on flower stems and petals is caused by different structural effects. In the first case, it is a combination of ridged topography and three-dimensional wax projections, whereas the papillate petal surface is supplemented by cuticular folds. In our opinion, these "cafeteria"-type flowers have the petals, where the colour intensity is enhanced due to papillate epidermal cells covered by cuticular folds at the micro- and nanoscale, and exactly these latter structures mainly contribute to adhesion reduction in generalist insect pollinators.

Keywords: attachment force; "cafeteria"-type flowers; cuticular folds; epicuticular wax projections; flower stems; leaves; pollination; papillae

1. Introduction

One of the evolutionary pressures that has led to rise and development of a great variety of surface micro-and nanostructures in plants is a co-evolution between plants and insects [1–3]. Thus, herbivory has resulted in an appearance of features and adaptations associated with plant defence against herbivorous insects. An important role of plant trichomes (hair-like protuberances extending from the epidermis of aerial plant tissues), covering leaves against herbivores, is well known [1,2,4]. For example, hooked trichomes of *Phaseolus* plants (Fabaceae) are able to entrap and kill a number of insect species belonging to different insect orders, such as Hemiptera and Diptera [5,6]. Effects of plant pubescence differ according to the insect herbivore species [1,7] as well as the trichome type (non-glandular or glandular), size, and distribution [8]. Glandular trichomes may additionally produce sticky (e.g.,*Datura, Lycopersicon, Nicotiana,* and *Solanum* species from the Solanaceae family) [9] or even poisonous secretions, which provide resistance against insect pests [10,11]. Additionally, microscopic surface features, such as cuticular folds (cuticle sculpturing of the plant surface usually caused by folding of the cuticle over the outer cell wall of epidermal cells), can deter herbivores, presumably due to reduction in insect attachment caused by minimization of the real contact area between insect attachment organs and microrough plant surface. This effect has been experimentally shown for petals and leaves of several plant species and the beetle *Leptinotarsa decemlineata* (Coleoptera: Chrysomelidae) [12–14]. Three-dimensional projections of epicuticular wax (a complex mixture of long-chain aliphatic and cyclic hydrocarbons, fatty acids, aldehydes, b-diketones, primary and secondary alcohols deposited onto the plant surface) composing a pruinose microrough coverage on aerial primary surfaces of higher plants also greatly contribute to plant protection against herbivorous and phytophagous insects. The protective function has been repeatedly reported and experimentally supported for numerous plants, for example *Eucalyptus* (Myrtaceae), *Pisum* (Fabaceae), and *Brassica* (Brassicaceae) species (see review by Gorb and Gorb [15]). These micro-/nanoscopic surface structures reduce the attachment ability of insects and impair their locomotion performance, possibly due to (1) the reduction in the real contact area between the tips of insect attachment organs (pads) and the plant surface (the roughness hypothesis); (2) the contamination of insect adhesive organs by plant wax (the contamination hypothesis); (3) the absorption of fluid from the insect adhesive pads due to the high capillarity of the wax coverage (the fluid absorption hypothesis); (4) hydroplaning caused by the appearance of a thick layer of liquid caused by the dissolving of the wax in insect adhesive fluid (the wax dissolving hypothesis); and (5) the formation of a separation layer between the plant substrate and insect attachment organs [16,17].

On the other hand, carnivorous plants, which rely on insect prey, have evolved specialized trapping organs bearing various surface structures, in order to capture and retain insects and other small animals for additional nutrition. For instance, in representatives of the plant genera *Gensilea* (Lentibulariaceae) with lobster-pot traps and *Cephalotus, Darlingtonia, Heliamphora,* and *Sarracenia* (all Sarraceniaceae) with pitfall traps, long, often sharp, downward- or inward-pointing trichomes allow an animal's progress into the trap, but prevent its moving into the opposite direction [2,18]. Movable glandular trichomes covering leaves in *Drosera* species (Droseraceae) (flypaper traps) use sticky mucilage to impede and immobilize prey animals [18]. Three-dimensional epicuticular wax coverages on the inner surface of pitchers in carnivorous plants from the genera *Nepenthes* (Nepentheceae), *Sarracenia,* and *Darlingtonia,* and in certain insect-trapping Bromeliaceae, contribute to prey trapping and retention [18–22].

Another important aspect of relationships between plants and insects is pollination, including attraction of pollinators, which is related to mainly optical effects, primarily generation and amplification of colour intensity or chemical attributes [22–24], or to other specialized mechanisms developed in kettle trap flowers. For pollination purposes, pitfall traps have been developed in representatives of the plant families Araceae, Aristolochiaceae, Apocynaceae s.l., Hydnoraceae, and Orchidaceae [25]. In plants from the genera *Aristolochia* (Aristolochiaceae) and *Ceropegia* (Apocynaceae), flowers are equipped with specialized trapping trichomes showing similar effects on pollinating insects during capture and retention stages as those in carnivorous pitfall traps [25–27]. Certain plants with kettle trap flowers, for example *Aristolochia* and *Arisaema* (Araceae) species, bear slippery wax-covered surfaces contributing to temporary capture of their pollinators [25,28,29].

Since insects pollinating the majority of plants are in contact with a petal surface for a certain time, it is plausible to hypothesize that such petals should enable or even promote attachment of insects to their surface. It is known that petals of many insect-pollinating flowers have papillate or cone-shaped epidermal cells/structures [30–35] mainly serving colour intensity enhancements in different directions that result in an increase in flower perceptibility both from a long distance and under different angles of view. The mechanical effects of such petal surface characteristics on the attachment of insect-pollinators still remain rather poorly understood. Thus, the contribution of petal surfaces to their mechanical stability that ensures the interlocking of flower visitors (insects) with well-developed claws to the conical epidermal cells and/or papillae has been reported on [31–33,35]. On the contrary, viscous or oily coverage as well as epicuticular wax projections, present on particular petal sites in some flowers, may prevent attachments of any insect [24,36]. Experimental study on attachment ability of the honeybee *Apis mellifera carnica* Pollmann (Hymenoptera, Apidae) and greenbottle fly *Lucilia caesar* L. (Diptera, Calliphoridae) to petals of ten plant species differing in their surface texture showed that both insect pollinators had a good foothold on rough surfaces including conical and papillate epidermal cells, while flat epidermal cells with microstructures such as cuticular folds and wax projections significantly reduced insect attachment [37].

However, taking into account the diversity of petal microstructures shown in the above studies, it is difficult to judge the particular surface adaptations of petals without comparing them to other surfaces in the same plant. For this study, we selected an insect-pollinating species *Dahlia pinnata* Cav. (Asteraceae) having distinctly different surfaces of leaves vs. petals vs. stems and a hovering fly pollinator *Eristalis tenax* (L.) (Diptera: Syrphidae). Both the fly and plant are generalist (not specialist) species in terms of their diet and pollinators spectrum, respectively. This plant is usually visited by representatives of various insect groups including bees, bumblebees, flies, etc. Adults of *E. tenax* as well as those in the majority of syrphid flies feed on pollen of numerous plants from different families (Asteraceae, Apiaceae, Adoxaceae, etc.), which offer their nectar and pollen in an open way ("cafeteria"-type of flowers). The study is based on a combination of cryo scanning electron microscopy (cryo-SEM) examination of related plant and insect surfaces and force measurements of the attachment of *E. tenax* to surfaces of different plant organs. Our aim was to find out whether the petal surface is adapted to a stronger attachment of the insect. For comparison, we used leaf and flower stem surfaces of *D. pinnata*.

2. Materials and Methods

2.1. Plants and Insects

The garden dahlia *D. pinnata* (Figure 1a) is a 30–200 cm high perennial herbaceous plant with erect stems branching only in the inflorescence, usually simple ovate leaves, and 2–8 flower stems later called "stems" (5–15 cm long) each bearing a flower head (capitulum) with both tubular and ovate strap-shaped florets (ray flowers or ligulate flowers). As a ligula represents fused petals, we called it "petal" throughout the text. Being native to Mexico, Central America and Colombia, *D. pinnata* is now extensively cultivated worldwide as a decorative plant [38]. Plant samples (stems, upper leaves, and petals) for cryo-SEM examination and experiments (Figure 1) were collected from garden plants (Jagotyn, Kiev District, Ukraine, 50°17′ N 31°46′ E).

The common drone fly *E. tenax* mimicks a stock bee [39] that feeds on the nectar and pollen of flowers [40,41]. It has a cosmopolitan distribution and is the most widely distributed syrphid species in the world [42]. Insects were captured in August 2005 near Jagotyn and used for structural examination and experiments. The attachment system of *E. tenax* is composed of paired claws and paired setose pulvilli on each foot (Figure 2a,b) (for details see [43]). Microscopic tenent setae covering the pulvilli have spatula-shaped terminal elements (Figure 2d) responsible for building intimate contact with a substrate (Figure 2b–e). Additionally, a pad fluid is secreted into a contact zone (Figure 2c–e).

Figure 1. The plant *Dahlia pinnata* (**a**) and the fly *Eristalis tenax* in the experiments on different test substrates (**b**–**e**): glass (**b**), the adaxial (upper) side of the leaf (**c**), the adaxial side of the petal (**d**), and the stem (**e**). In (**e**), arrow points to a hair attached dorsally to the fly thorax using a droplet of molten beewax (asterisk). Abbreviations: fs, flower stem; lf, upper leaf; pt, petal (actually, ray flower).

2.2. Microscopy

Plant samples (stem, upper leaf, and petal) (Figure 1a) for cryo-SEM examination were cut off from living plants and kept for 24 h inside small plastic vials containing wet paper in order to prevent desiccation of the plant material. Next, small (1 cm × 1 cm) portions were cut out from the middle regions of the leaf and petal and from the stem, glued with polyvinyl alcohol Tissue-Teck O.C.T.TM Compound (Sakura Finetek Europe B.V., Zoeterwoude, The Netherlands) to a metal holder, and frozen in a cryo stage preparation chamber at −140 °C (Gatan ALTO 2500 cryo preparation system, Gatan Inc., Abingdon, UK). Frozen samples of the upper (adaxial) side, in both the leaf and petal, and the stem surface were sputter coated with gold–palladium (6 nm thickness) and studied in a frozen condition in a cryo-SEM Hitachi S-4800 (Hitachi High-Technologies Corporation, Tokyo, Japan) at

3 kV accelerating voltage and −120 °C temperature. Microscopy was performed at the Max Planck Institute for Metals Research (Stuttgart, Germany).

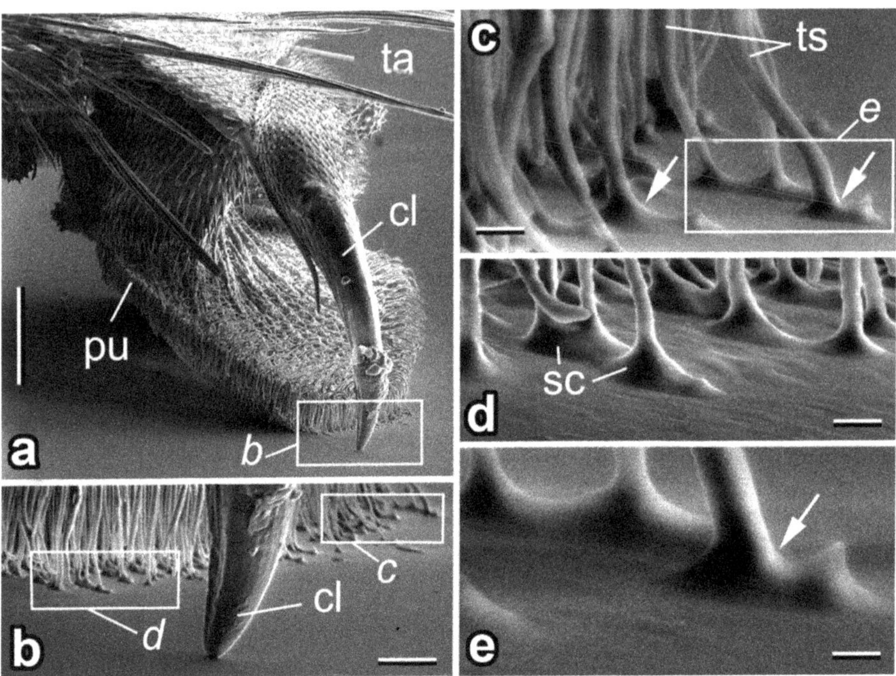

Figure 2. The attachment system of the fly *Eristalis tenax* (cryo-SEM). (**a**) Dorso-lateral view of the distal part of the tarsus (ta) with pretarsus bearing claws (cl) and pulvilli (pu). (**b**) The claw (cl) and distal portion of the pulvillus in contact with a smooth substrate. (**c–e**) Tenent setae (ts) covering the pulvillus in contact with the substrate. Note the pad secretion (sc) delivered into a contact zone. Arrows in (**c,e**) show examples of the setal terminal elements, spatulae, in contact. Scale bars: 100 μm (**a**), 10 μm (**b**), 2 μm (**c,d**), 1 μm (**e**).

Types of wax projections were identified according to [44]. Morphometrical variables of plant surface features were measured from digital images using the software SigmaScan Pro 5 (SPSS Inc., Chicago, IL, USA). These data are given in the text as mean ± SD for n = 10.

2.3. Experiments

Force measurements were carried out using a load cell force sensor FORT-10 (10 g capacity; World Precision Instruments Inc., Sarasota, FL, USA) connected to a force transducer MP 100 (Biopac Systems Ltd., Santa Barbara, CA, USA) [45,46]. Prior to the experiments, test insects were made incapable of flying by cutting off their wings. After a certain recovery time, the fly was attached to the force sensor by means of a 10–15 cm long hair glued to the dorsal side of its thorax with a droplet of molten beeswax (Figure 1e). The force produced by the insect walking horizontally on the test substrates and pulling the hair (Figure 1b–e) for approximately 5–30 s was measured. Force–time curves were used to estimate the maximal friction (traction) force produced by insects. Since the flies were constrained to pulling parallel (not at an angle) to the measurement axis of the sensor, the registered force corresponded to the total traction force. Three types of fresh plant samples were used as substrates: (i) the adaxial side of the upper leaf; (ii) the adaxial side of the petal; and (iii) the stem, where insects move towards the apical stem direction (Figure 1c–e). A smooth hydrophilic glass surface served as a control (Figure 1b). For each insect individual, the test substrates were used in the following sequence: (1) glass as a reference, (2 or 3) leaf, (3 or 2) petal, and (4) stem. Leaf and petal samples were randomized in the consecutive test series. Taking into account that the stem surface is covered with three-dimensional epicuticular wax projections [47], we performed experiments on the stem surface at the end of a test series, in order to avoid the effect of a possible contamination of insect feet by wax particles on the experimental results.

Force tests were carried out at 22–25 °C and ca. 60–75% humidity. On each substrate type, experiments with 10 individual flies (10 repetitions per insect) were performed. In all, 400 force tests were conducted.

Force values were analysed using two-way analysis of variance (ANOVA), considering the insect individual and the substrate as factors (software SigmaPlot 11.0, SPSS Inc., Chicago, IL, USA).

3. Results

3.1. Micromorphology of Dahlia pinnata Surfaces

The adaxial leaf side was slightly uneven because of epidermal cells showing a somewhat convex shape of the outer walls and of the sparsely dispersed (abundance: ca. 0.15 mm^{-2}) stomata, having slightly sunk guard cells (Figure 3a,b). The surface had a smooth appearance (Figure 3b).

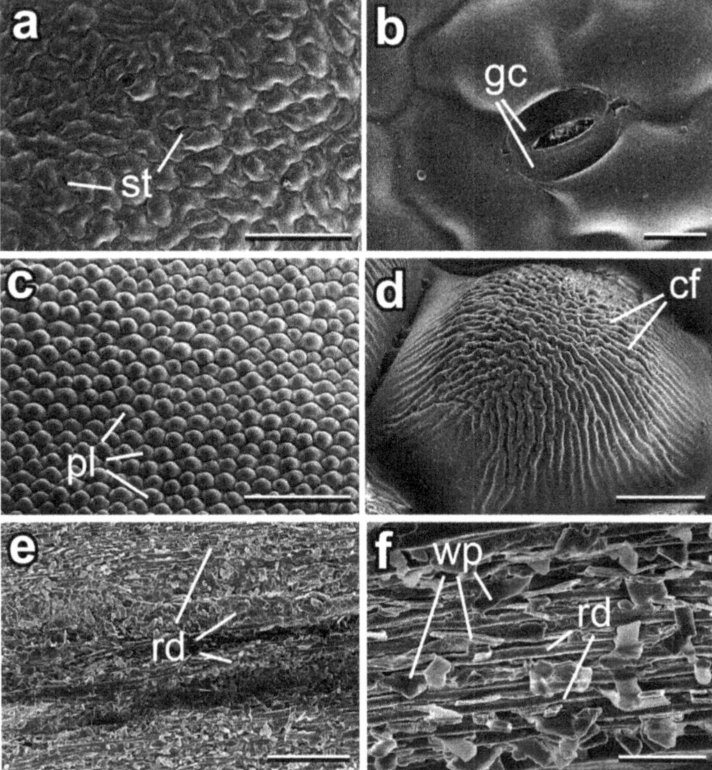

Figure 3. Surfaces of different organs in the plant *Dahlia pinnata* (cryo-SEM). (**a,b**) Adaxial leaf side. (**c,d**) Adaxial petal side. (**e,f**). Stem. Abbreviations: cf, cuticular folds; gc, guard cells of the stoma; pl, papillae; rd, ridges; st, stomata; wp, epicuticular wax plates. Scale bars: 200 µm (**a,c**), 20 µm (**b,e**), 10 µm (**d**), 5 µm (**f**).

The adaxial petal surface was strongly uneven (Figure 3c). It was characterized by the papillate, polygonal epidermal cells, which were noticeably structured with rather uniform cuticular folds (width: 1.68 ± 0.34 µm; height: 0.68 ± 0.16 µm) (Figure 3c,d). Originating at the cell periphery, at first the folds were straight and rather low, becoming more and more wavy, higher (height: 0.69 ± 0.11 µm), and densely spaced (distance between folds: 0.51 ± 0.08 µm) to the centre of the papillae (Figure 3d).

Stems of *D. pinnata* had a hierarchically arranged system of ridges and grooves (Figure 3e,f). The surface bore a three-dimensional epicuticular coverage consisting of separate wax plates (abundance: ca. 0.3–0.5 plates per 1 µm) (Figure 3f). The plates varied both in dimensions (length: 2.58 ± 0.67 µm; width: 1.46 ± 0.38 µm; thickness: 0.19 ± 0.06 µm) and shapes, however, being more often rhomb shaped. Their edges were usually rather distinct, but also projections with slightly irregular edges occurred. The plates showed no preferred orientation. They protruded from the cuticle surface at various angles and were attached to the latter either by their narrow or larger sides. Such a variation in plate connections to the underlying surface was possibly caused by the particular (hierarchical ridges/grooves) geometry of the stem.

3.2. Traction Forces on Different Test Substrates

Traction force values of 10 individuals of *E. tenax* obtained on glass and different plant surfaces are presented in Figure 4a. Since the force generated by different insect individuals varied greatly even on glass (1.18 ± 0.38 mN, n = 100), we analysed the obtained data using two-way ANOVA, considering both the insect individual and the substrate as factors. The flies showed significantly different forces ($p < 0.001$, two-way ANOVA). Both factors, insect individuals and substrates, significantly influenced the force values (individuals: $F_{9,391} = 37.565$, substrates: $F_{3,397} = 213.416$, both $p < 0.001$). The similarly higher forces were registered on reference glass and the leaf surface (Figure 4). These force values significantly differed from much lower ones obtained on petals and stems, which did not differ significantly from each other (Figure 4).

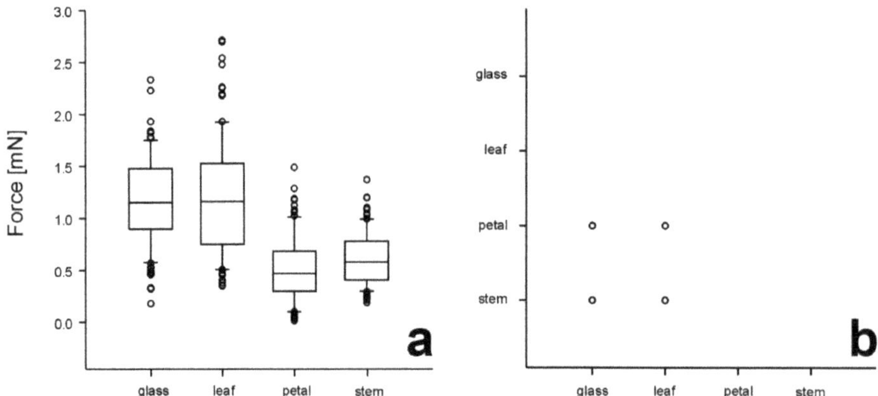

Figure 4. Traction forces produced by flies *Eristalis tenax* on different substrates (glass, adaxial sides of the upper leaf and the petal, and the stem in the plant *Dahlia pinnata*) (**a**) and results of the statistical comparison of the force values obtained on different substrates (**b**). In (**a**), boxplots show the interquartile range and the medians, whiskers indicate the 1.5 × interquartile range, and "°" are outliers. In (**b**), circles mean significant differences in force values between substrates at $p < 0.05$ (Tukey post hoc test, two-way ANOVA).

4. Discussion

Our experimental results clearly distinguished two groups among tested surfaces. Reference glass and the upper leaf surface ensured rather high attachment force of the *E. tenax* fly, whereas both the flower stem and petal significantly reduced it. Strong attachment on smooth, both artificial and natural, surfaces has been observed in many insect species for a rather long time [16,46,48–51]. However, the attachment force reduction on flower stems and petals is caused by different structural effects. In the case of the flower stem, it is the combination of ridged topography and three-dimensional wax projections [47], whereas the papillate petal surface is supplemented by cuticular folds. The effect of the three-dimensional wax coverage on insect attachment has been repeatedly reported in numerous studies (reviewed in [15]) and several contributing mechanisms, such as surface roughness, insect pad contamination, pad fluid absorption, hydroplaning, and separation layer, have been proposed [16,17] and, in part, experimentally proved [21,46,52–54]. The role of cuticular folds in a decrease in insect attachment force has also been recently revealed by several experimental studies [12–14].

The biological reason of insect attachment force reduction on the flower stem is rather clear and is associated with a greasy pole syndrome, i.e., a plant defence mechanism preventing ants from visiting flowers and robbing nectar [55–57]. It has been described in *Salix* (Salicaceae*)*, *Hypenia*, and *Eriope* (both Lamiaceae) plant species and is based on the combined effect of several stem features, such as slender elongate erect stems, rigid spreading trichomes on lower internodes, three-dimensional wax coverage, and, often, swellings in upper internodes, which hamper access of ants to apically located plant reproductive organs [57]. Recently, this syndrome was experimentally studied in several other plants [47,58,59], *D. pinnata* among them [47], where main attention was given to contribution of the wax coverage on flower stems to impeding the locomotion of the generalist ant species *Lasius niger* (Hymenoptera: Formicidae).

However, the reason behind insect attachment force reduction on the petals is less evident. It is plausible to suggest that optimal conditions for attachment of insect pollinators on petals would be advantageous from a plant perspective. That is why the obtained (opposite) result was rather surprising. One possible explanation is related to a need of plants, relying on generalist insect pollinators, to attract latter from a certain distance to their "cafeteria"-type flowers. These individual flowers are usually rather large or, if small, then combined into a rather large inflorescence, as it is the case in the plant species studied here. Many such complex inflorescences (anthodiums) in plants from the family Asteraceae have brightly coloured petals of marginal (ray) flowers. The colour intensity is usually enhanced due to conical or papilla-shaped epidermal cells often supplemented by micro- and nanostructures, such as cuticular folds. It seems to be a result of evolutionary optimization that the enhancement of one function (optical in this particular case), relying on the surface microstructuring, leads to the reduction in the other (mechanical) one. It can be further hypothesized that the function of pollinator attraction is more dominant for "cafeteria"-flowers than providing optimal attachment conditions for pollinators. Since the petals were not particularly large in comparison to the leg span of the syrphid fly studied here, and also to other typical pollinators, such as bees or calliphorid flies, these insects can presumably secure good attachment to the petals by using mechanical interlocking of their tarsal claws to the petal margins without contacting the petal surface with their adhesive pads. Another way, how the insects may avoid problems of low adhesion on *D. pinnata* petals is their tendency to land in the central area of the inflorescence, where the tubular flowers lacking prominent petals are situated.

Additionally, the combination of conical surface structures with cuticular folds found in a number of plant species may be responsible for so called "rose petal effect", which is based on the ability of certain rough surfaces (with hierarchically organized surface micro- and nanostructures) to have a high contact angle with water, simultaneously with strong adhesion of water (high contact angle hysteresis) [60,61]. This is well-known for petals of plants from, for example, the Rosaceae and Asteraceae families (see Figure 1a) and was even biomimetically implemented in certain technical surfaces [62–64]. In plants, this kind of surfaces prevents water from wetting the petals and from rolling drops off the surface into the direction of the flower/inflorescence centre. The surfaces having low wetting ability by water, but stably holding water drops, provides optimal conditions for a quick evaporation of the fluid. Although a thin water film has a high evaporation rate due to its large surface area, breaking such a film in numerous small round droplets, having punctual contacts with the substrate, leads to an even larger overall area that can enhance evaporation rate [60–64]. Moreover, in contrast to the thin water film, this effect supports water evaporation in an almost non-contact state with the petal surface.

Thus, based on the combination of cryo-SEM examination of different surfaces (leaf, flower stem, petal) in the plant *D. pinnata* and the force measurements of the fly *E. tenax* attachment to these surfaces, we showed that the petal surface is not adapted to stronger attachment of generalist insect pollinators. In our opinion, these "cafeteria"-type flowers have the petals, where the colour intensity is enhanced due to papillate epidermal cells covered by cuticular folds at the micro- and nanoscale, and exactly these latter structures mainly contribute to adhesion reduction in insect pollinators. Obtained results may be potentially interesting for fabrication of technical coatings with anti-adhesive properties for insects [52,65–69].

Author Contributions: All authors have equally contributed to the work reported (conceptualization, methodology, investigation, data curation, writing, and visualization). All authors have read and agreed to the published version of the manuscript.

Funding: This research received no external funding.

Data Availability Statement: The data presented in this study are available on request from the corresponding author.

Acknowledgments: We are grateful to the Max Planck Institute for Metals Research (Stuttgart, Germany) for providing microscopy equipment for this study and the four anonymous reviewers for their useful comments on the initial version of this paper. Mohsen Jafarpour kindly provided linguistic corrections.

Conflicts of Interest: The authors declare no conflict of interest.

References

1. Southwood, S. Plant Surfaces and Insects—An Overview. In *Insects and the Plant Surface*; Juniper, B., Southwood, R., Eds.; Edward Arnold Publishers: London, UK, 1986; pp. 1–22.
2. Jeffree, C.F. The Cuticle, Epicuticular Waxes and Trichomes of Plants, with Reference to Their Structure, Function and Evolution. In *Insects and the Plant Surface*; Juniper, B., Southwood, R., Eds.; Edward Arnold Publishers: London, UK, 1986; pp. 23–63.
3. Gorb, E.V.; Gorb, S.N. Anti-Adhesive Surfaces in Plants and Their Biomimetic Potential. In *Materials Design Inspired by Nature: Function Through Inner Architecture*; Fratzl, P., Dunlop, J.W.C., Weinkamer, R., Eds.; RSC Publishing: Cambridge, UK, 2013; pp. 282–309.
4. Levin, D.A. The role of trichomes in plant defense. *Q. Rev. Biol.* **1973**, *48*, 3–15. [CrossRef]
5. Johnson, B. The injurious effects of the hooked epidermal hairs of French beans (*Phaseolus vulgaris* L.) on *Aphis craccivora* Koch. *Bull. Entomol. Res.* **1953**, *44*, 779–788. [CrossRef]
6. Rebora, M.; Salerno, G.; Piersanti, S.; Gorb, E.; Gorb, S.N. Entrapment of *Bradysia paupera* (Diptera: Sciaridae) by *Phaseolus vulgaris* (Fabaceae) plant leaf. *Arthropod. Plant. Interact.* **2020**, *14*, 499–509. [CrossRef]
7. Hare, J.; Elle, E. Variable impact of diverse insect herbivores on dimorphic *Datura wrightii*. *Ecology* **2002**, *83*, 2711–2720. [CrossRef]
8. Andres, M.R.; Connor, E.F. The community-wide and guild-specific effects of pubescence on the folivorous insects of manzanitas *Arctostaphylos* spp. *Ecol. Entomol.* **2003**, *28*, 383–396. [CrossRef]
9. Gibson, R.W. Glandular hairs providing resistance to aphids in certain wild potato species. *Ann. Appl. Bio.* **1971**, *68*, 113–119. [CrossRef]
10. Dufey, S.S. Plant Glandular Trichomes: Their Partial Role in Defense Against Insects. In *Insects and the Plant Surface*; Juniper, B., Southwood, R., Eds.; Edward Arnold Publishers: London, UK, 1986; pp. 151–172.
11. Voigt, D.; Gorb, E.; Gorb, S. Plant surface–bug interactions: *Dicyphus errans* stalking along trichomes. *Arthropod Plant Interact.* **2007**, *1*, 221–243. [CrossRef]
12. Prüm, B.; Seidel, R.; Bohn, H.F.; Speck, T. Plant surfaces with cuticular folds are slippery for beetles. *J. R. Soc. Interface* **2011**, *9*, 127–135. [CrossRef]
13. Prüm, B.; Bohn, H.F.; Seidel, R.; Rubach, S.; Speck, T. Plant surfaces with cuticular folds and their replicas: Influence of microstructuring and surface chemistry on the attachment of a leaf beetle. *Acta Biomater.* **2013**, *9*, 6360–6368. [CrossRef]

14. Surapaneni, V.A.; Aust, T.; Speck, T.; Thielen, M. Polarity in cuticular ridge development and insect attachment on leaf surfaces of *Schismatoglottis calyptrata* (Araceae). *Beilstein J. Nanotechnol.* **2021**, *12*, 1326–1338. [CrossRef]
15. Gorb, E.V.; Gorb, S.N. Anti-adhesive effects of plant wax coverage on insect attachment. *J. Exp. Bot.* **2017**, *68*, 5323–5337. [CrossRef] [PubMed]
16. Gorb, E.V.; Gorb, S.N. Attachment ability of the beetle *Chrysolina fastuosa* on various plant surfaces. *Entomol. Exp. Appl.* **2002**, *105*, 13–28. [CrossRef]
17. Gorb, E.V.; Purtov, J.; Gorb, S.N. Adhesion force measurements on the two wax layers of the waxy zone in *Nepenthes alata* pitchers. *Sci. Rep.* **2014**, *4*, 5154. [CrossRef] [PubMed]
18. Juniper, B.E.; Robins, R.J.; Joel, D.M. *The Carnivorous Plants*; Academic Press: London, UK, 1989; 353p.
19. Gaume, L.; Gorb, S.; Rowe, N. Function of epidermal surfaces in the trapping efficiency of *Nepenthes alata* pitchers. *New Phytol.* **2002**, *156*, 479–489. [CrossRef]
20. Gorb, E.; Haas, K.; Henrich, A.; Enders, S.; Barbakadze, N.; Gorb, S. Composite structure of the crystalline epicuticular wax layer of the slippery zone in the pitchers of the carnivorous plant *Nepenthes alata* and its effect on insect attachment. *J. Exp. Biol.* **2005**, *208*, 4651–4662. [CrossRef] [PubMed]
21. Scholz, I.; Bückins, M.; Dolge, L.; Erlinghagen, T.; Weth, A.; Hischen, F.; Mayer, J.; Hoffmann, S.; Riederer, M.; Riedel, M.; et al. Slippery surfaces of pitcher plants: *Nepenthes* wax crystals minimize insect attachment via microscopic surface roughness. *J. Exp. Biol.* **2010**, *213*, 1115–1125. [CrossRef]
22. Daumer, K. Blumenfarben, wie sie die Bienen sehen. *Z. Vergl. Physiol.* **1958**, *41*, 49–110. [CrossRef]
23. Kugler, H. *Blütenokologie*; Gustav Fischer Verlag: Stuttgart, Germany, 1970; 345p.
24. Willmer, P. *Pollination and Floral Ecology*; Princeton University Press: Princeton, NJ, USA, 2011; 832p.
25. Oelschlägel, B.; Gorb, S.; Wanke, S.; Neinhuis, C. Structure and biomechanics of trapping flower trichomes and their role in the pollination biology of *Aristolochia* plants (Aristolochiaceae). *New Phytol.* **2009**, *184*, 988–1002. [CrossRef]
26. Müller, L. Zur biologischen Anatomie der Blüte von *Ceropegia woodii* Schlechter. *Biol. Gen.* **1926**, *2*, 799–814.
27. Vogel, S. Die Bestäubung der Kesselfallen-Blüten von *Ceropegia*. *Beiträge Zur Biol. Der Pflanz.* **1961**, *36*, 159–237.
28. Vogel, S.; Martens, J. A survey of the lethal kettle traps of *Arisaema* (Araceae), with records of pollinating fungus gnats from Nepal. *Bot. J. Linnean Soc.* **2000**, *133*, 61–100. [CrossRef]
29. Poppinga, S.; Koch, K.; Bohn, H.F.; Barthlott, W. Comparative and functional morphology of hierarchically structured anti-adhesive surfaces in carnivorous plants and kettle trap flowers. *Funct. Plant Biol.* **2010**, *37*, 952–961. [CrossRef]
30. Barthlott, W.; Ehler, N. *Raster-Elektronenmikroskopie der Epidermis-Oberflächen von Spermatophyten (Tropische und Subtropische Pflanzenwelt)*; Franz Steiner Verlag: Stuttgart, Germany, 1977; 105p.
31. Christensen, K.; Hansen, H. SEM-studies of epidermal patterns of petals in the angiosperms. *Opera Bot.* **1998**, *135*, 1–91.
32. Whitney, H.M.; Chittka, L.; Glover, B.J. Conical epidermal cells allow bees to grip flowers and increase foraging efficiency. *Curr. Biol.* **2009**, *19*, 948–953. [CrossRef]
33. Whitney, H.M.; Bennett, K.M.; Dorling, M.; Sandbach, L.; Prince, D.; Chittka, L.; Glover, B.J. Why do so many petals have conical epidermal cells. *Ann. Bot.* **2011**, *108*, 609–616. [CrossRef]
34. Prüm, B.; Seidel, R.; Bohn, H.F.; Speck, T. Impact of cell shape in hierarchically structured plant surfaces on the attachment of male Colorado potato beetles (*Leptinotarsa decemlineata*). *Beilstein J. Nanotechnol.* **2012**, *3*, 57–64. [CrossRef] [PubMed]
35. Papiorek, S.; Junker, R.R.; Lunau, K. Gloss, colour and grip: Multifunctional epidermal cell shapes in bee- and bird-pollinated flowers. *PLoS ONE* **2014**, *9*, e112013. [CrossRef]
36. Knoll, F. Insekten und Blumen: Experimentelle Arbeiten zur Vertiefung unserer Kenntnisse über die Wechselbeziehungen zwischen Pflanzen und Tieren. *Abbandl. D Zool Botan. Gesellsch. Wien* **1921**, *12*, 1–646. [CrossRef]
37. Bräuer, P.; Neinhuis, C.; Voigt, D. Attachment of honeybees and greenbottle flies to petal surfaces. *Arthropod Plant Interact* **2017**, *11*, 171–192. [CrossRef]
38. Jäger, E.J.; Ebel, F.; Hanelt, P.; Müller, G.K. *Rothmaler Exkursionsflora von Deutschland. 5: Krautige Zier- und Nutzpflanzen*; Spektrum: Heidelberg, Germany, 2008; 879p.
39. Heal, J.R. Colour patterns of syrphidae: IV. Mimicry and variation in natural populations of *Eristalis tenax*. *Heredity* **1982**, *49*, 95–109. [CrossRef]
40. Buckton, G.B. *The Natural History of Eristalis Tenax or the Drone-Fly*; Macmillan: London, UK; New York, NY, USA, 1895; 140p.
41. Holloway, B.A. Pollen-feeding in hover-flies (Diptera: Syrphidae). *N. Z. J. Zool.* **1979**, *3*, 339–350. [CrossRef]
42. Skevington, J.H.; Locke, M.M.; Young, A.D.; Moran, K.; Crins, W.J.; Marshall, S.A. *Field Guide to the Flower Flies of Northeastern North America*; Princeton University Press: Princeton, NJ, USA, 2019; 512p.
43. Gorb, S.; Gorb, E.; Kastner, V. Scale effects on the attachment pads and friction forces in syrphid flies (Diptera, Syrphydae). *J. Exp. Biol.* **2001**, *204*, 1421–1431. [CrossRef]
44. Barthlott, W.; Neinhuis, C.; Cutler, D.; Ditsch, F.; Meusel, I.; Theisen, I.; Wilhelmi, H. Classification and terminology of plant epicuticular waxes. *Bot. J. Linn. Soc.* **1998**, *126*, 237–260. [CrossRef]
45. Gorb, S.N.; Popov, V.L. Probabilistic fasteners with parabolic elements: Biological system artificial model and theoretical considerations. *Phil. Trans. R. Soc. Lond. A* **2002**, *360*, 211–225. [CrossRef]
46. Gorb, E.; Hosoda, N.; Miksch, C.; Gorb, S. Slippery pores: Anti-adhesive effect of nanoporous substrates on the beetle attachment system. *J. R. Soc. Interface* **2010**, *7*, 1571–1579. [CrossRef]

47. Gorb, E.; Gorb, S. How a lack of choice can force ants to climb up waxy plant stems. *Arthropod Plant Interact* **2011**, *5*, 297–306. [CrossRef]
48. Stork, N.E. Experimental analysis of adhesion of *Chrysolina polita* (Chrysomelidae, Coleoptera) on a variety of surfaces. *J. Exp. Biol.* **1980**, *88*, 91–107. [CrossRef]
49. Edwards, P.B. Do waxes of juvenile *Eucalyptus* leaves provide protection from grazing insects. *Aust. J. Ecol.* **1982**, *7*, 347–352. [CrossRef]
50. Federle, W.; Maschwitz, U.; Fiala, B.; Riederer, M.; Hölldobler, B. Slippery ant-plants and skilful climbers: Selection and protection of specific ant partners by epicuticular wax blooms in *Macaranga* (Euphorbiaceae). *Oecologia* **1997**, *112*, 217–224. [CrossRef]
51. Eigenbrode, S.D.; White, C.; Rohde, M.; Simon, C.J. Behavior and effectiveness of adult *Hyppodamia convergens* (Coleoptera: Coccinellidae) as a predator of *Acyrthosiphon pisum* on a glossy-wax mutant of *Pisum sativum*. *Environ. Entomol.* **1998**, *91*, 902–909. [CrossRef]
52. Gorb, E.; Gorb, S. Effects of surface topography and chemistry of *Rumex obtusifolius* leaves on the attachment of the beetle *Gastrophysa viridula*. *Entomol. Exp. Appl.* **2009**, *130*, 222–228. [CrossRef]
53. Gorb, E.V.; Gorb, S.N. 2006. Do Plant Waxes Make Insect Attachment Structures Dirty? Experimental Evidence for the Contamination Hypothesis. In *Ecology and Biomechanics: A Mechanical Approach to the Ecology of Animals and Plants*; Herrel, A., Speck, T., Rowe, N.P., Eds.; CRC Press: Boca Raton, FL, USA, 2006; pp. 147–162.
54. Gorb, E.V.; Lemke, W.; Gorb, S.N. Porous substrate affects a subsequent attachment ability of the beetle *Harmonia axyridis* (Coleoptera, Coccinellidae). *J. R. Soc. Interface* **2019**, *16*, 20180696. [CrossRef]
55. Kerner von Marilaun, A. *Flowers and Their Unbidden Guests*; C. K. Paul & Co.: London, UK, 1878.
56. Harley, R.M. Evolution and Distribution of Eriope (Labiatae) and its Relatives in Brazil. In *Proceedings of a Workshop on Neotropical Distributions*; Academia Brasileira de Ciencias: Rio de Janeiro, Brazil, 1988; pp. 71–120.
57. Harley, R. The Greasy Pole Syndrome. In *Ant-Plant Interactions*; Huxley, C.R., Cutler, D.E., Eds.; Oxford University Press: Oxford, UK, 1991; pp. 430–433.
58. Gorb, S.N.; Gorb, E.V. Frequency of plant visits by the generalist ant *Lasius niger* depends on the surface microstructure of plant stems. *Arthropod Plant Interact* **2019**, *13*, 311–320. [CrossRef]
59. Gorb, S.N.; Gorb, E.V. Combined effect of different flower stem features on the visiting frequency of the generalist ant *Lasius niger*: An experimental study. *Insects* **2021**, *12*, 1026. [CrossRef]
60. Feng, L.; Zhang, Y.; Xi, J.; Zhu, Y.; Wang, N.; Xia, F.; Jiang, L. Petal effect: A superhydrophobic state with high adhesive force. *Langmuir* **2008**, *24*, 4114–4119. [CrossRef]
61. Bhushan, B.; Nosonovsky, M. The rose petal effect and the modes of superhydrophobicity. *Philos. Trans. Royal. Soc. A* **2010**, *368*, 4713–4728. [CrossRef]
62. Jin, M.H.; Feng, X.L.; Feng, L.; Sun, T.L.; Zhai, J.; Li, T.J.; Jiang, L. Superhydrophobic aligned polystyrene nanotube films with high adhesive force. *Adv. Mater.* **2005**, *17*, 1977–1981. [CrossRef]
63. Bormashenko, E.; Stein, T.; Pogreb, R.; Aurbach, D. "Petal Effect" on surfaces based on lycopodium: High-stick surfaces demonstrating high apparent contact angles. *J. Phys. Chem. C* **2009**, *113*, 5568–5572. [CrossRef]
64. Park, Y.M.; Gang, M.; Seo, Y.H.; Kim, B.H. Artificial petal surface based on hierarchical micro- and nanostructures. *Thin Solid Films* **2011**, *520*, 362–367. [CrossRef]
65. Féat, A.; Federle, W.; Kamperman, M.; van der Gucht, J. Coatings preventing insect adhesion: An overview. *Progr. Org. Coat.* **2019**, *134*, 349–359. [CrossRef]
66. Zhou, Y.; Robinson, A.; Steiner, U.; Federle, W. 2014 Insect adhesion on rough surfaces: Analysis of adhesive contact of smooth and hairy pads on transparent microstructured substrates. *J. R. Soc. Interface* **2014**, *11*, 20140499. [CrossRef]
67. Gorb, E.; Böhm, S.; Jacky, N.; Maier, L.P.; Dening, K.; Pechook, S.; Pokroy, B.; Gorb, S. Insect attachment on crystalline bioinspired wax surfaces formed by alkanes of varying chain lengths. *Beilstein J. Nanotechnol.* **2014**, *5*, 1031–1041. [CrossRef] [PubMed]
68. Eichler-Volf, A.; Xue, L.; Dornberg, G.; Chen, H.; Kovalev, A.; Enke, D.; Wang, Y.; Gorb, E.V.; Gorb, S.N.; Steinhart, M. The influence of surface topography and surface chemistry on the anti-adhesive performance of nanoporous monoliths. *ACS Appl. Mater. Interfaces* **2016**, *8*, 22593–22604. [CrossRef] [PubMed]
69. Eichler-Volf, A.; Kovalev, A.; Wedeking, T.; Gorb, E.V.; Xue, L.; You, C.; Piehler, J.; Gorb, S.N.; Steinhart, M. Bioinspired monolithic polymer microsphere arrays as generically anti-adhesive surfaces. *Bioinspir. Biomim.* **2016**, *11*, 025002. [CrossRef] [PubMed]

Article

Can *Macrolophus pygmaeus* (Hemiptera: Miridae) Mitigate the Damage Caused to Plants by *Bemisia tabaci* (Hemiptera: Aleyrodidae)?

Alessia Farina *, Giuseppe Eros Massimino Cocuzza, Pompeo Suma and Carmelo Rapisarda

Department of Agriculture, Food and Environment (Di3A), Applied Entomology Section, University of Catania, 95123 Catania, Italy
* Correspondence: alessia.farina@phd.unict.it

Simple Summary: The whitefly *Bemisia tabaci* is an invasive pest that causes extensive damage to many vegetable crops and ornamental plants. To control this pest, the release of natural enemies has become increasingly important as an ecologically safe and effective method of biological control. Some species in the family Miridae are effective at controlling whitefly populations, but because they feed on both insect prey and plant tissue, their overall effect on plant performance is not well understood. In this study, the impact of the mirid predator *Macrolophus pygmaeus* on the morphological and physiological traits of *Solanum melongena* in the presence of *Bemisia tabaci* was evaluated. Overall, the results show how the presence of this natural enemy mitigates the damage caused by whitefly infestations. They also help to clarify the multitrophic relationships between plant, pest, and natural enemy, enabling the prediction of plant development in the presence of both pest and predator.

Abstract: Nowadays, in protected vegetable crops, pest management based mainly on biological control represents the most sustainable alternative to pesticide use. The cotton whitefly, *Bemisia tabaci*, is one of the key pests that negatively impact the yield and quality of such crops in many agricultural systems. The predatory bug *Macrolophus pygmaeus* is one of the main natural enemies of the whitefly and is widely used for its control. However, the mirid can sometimes behave as a pest itself, causing damage to crops. In this study, we investigated the impact of *M. pygmaeus* as a plant feeder, by analyzing the combined impact of the whitefly pest and the predator bug on the morphology and physiology of potted eggplants under laboratory conditions. Our results showed no statistical differences between the heights of plants infested by the whitefly or by both insects compared with noninfested control plants. However, indirect chlorophyll content, photosynthetic performance, leaf area, and shoot dry weight were all greatly reduced in plants infested only by *B. tabaci*, compared with those infested by both pest and predator or with noninfested control plants. Contrarily, root area and dry weight values were more reduced in plants exposed to both of the insect species, compared with those infested only by the whitefly or compared with noninfested control plants, where the latter showed the highest values. These results show how the predator can significantly reduce the negative effects of *B. tabaci* infestation, limiting the damage it causes to host plants, though the effect of the mirid bug on the underground parts of the eggplant remains unclear. This information might be useful for a better understanding of the role that *M. pygmaeus* plays in plant growth, as well as for the development of management strategies to successfully control infestations by *B. tabaci* in cropping environments.

Keywords: whitefly; predator; zoophytophagy; trophic interactions; plant morphology; plant physiology

Citation: Farina, A.; Massimino Cocuzza G.E.; Suma, P.; Rapisarda, C. Can *Macrolophus pygmaeus* (Hemiptera: Miridae) Mitigate the Damage Caused to Plants by *Bemisia tabaci* (Hemiptera: Aleyrodidae)? *Insects* **2023**, *14*, 164. https://doi.org/10.3390/insects14020164

Academic Editors: Gianandrea Salerno, Stanislav N. Gorb and Manuela Rebora

Received: 21 December 2022
Revised: 2 February 2023
Accepted: 6 February 2023
Published: 8 February 2023

1. Introduction

Multitrophic interactions, understood as relationships between organisms across different trophic levels of a food web [1,2], are gaining growing interest in ecological studies. Especially in the agricultural sector, which is increasingly oriented to achieve a reduction in chemical inputs, complex interactions that involve the binding of different organisms, living both above and below the ground, with the cultivated plants are now recognized [3–5]. Improving our understanding of these relationships may help in planning a rational management of the populations involved, and it may lead to a reduction in pest infestations and their negative effects [6–8].

Bemisia tabaci (Gennadius) is a notable insect pest that affects vegetable crops and many ornamental plants. It causes damage directly by the piercing of leaves, the suction of sap, and the production of honeydew on which sooty molds develop [9] and also indirectly through its ability to transmit phytopathogenic viruses to numerous host plants [10]. Over 100 virus species belonging to the Begomovirus, Carlavirus, Crinivirus, Ipomovirus, and Torradovirus groups are presently known to be transmitted by *B. tabaci* [11,12], causing worldwide economic damage, the value of which is difficult to estimate. For a long time, *B. tabaci* was considered as a single species, but variability among its populations has led scholars to consider *B. tabaci* as a complex of more than 40 species that are morphologically indistinguishable but that differ in their biology (host range, development performance, suitable environmental conditions, virus transmission capability, etc.) and geographic distribution [13–16].

Numerous studies have examined the interactions of *B. tabaci* with other components of its food web and how these influence the population levels of the insect, as well as the negative impact of the insect on the host plants. For instance, regarding the interactions of *B. tabaci* with other pests, Tan [17] showed how the infestation of tomato plants by the green peach aphid *Myzus persicae* (Sulzer) had a negative impact on *B. tabaci* development, indicating that the latter clearly prefers to settle on plant leaves not infested by the aphid. A deeper interaction between these two hemipteran species has also been demonstrated by the effect of preinfesting tomato plants with *M. persicae* on the feeding dynamics of *B. tabaci* and on the acquisition and transmission mechanisms of TYLCV (tomato yellow leaf curl virus) by this vector—showing a clear and significant influence of aphid preinfestation on the tomato–whitefly–TYLCV system [18].

At a higher trophic level, whitefly feeding induces plant defense responses, which affect more-complex interactions with natural enemies. Thus, in the whiteflies–aphids system on tomato, a preinfestation with *B. tabaci* MEAM1 impacts the predation ability of the ladybird *Propylea japonica* (Thunberg) on *M. persicae* [19]. At the top of the feeding pyramid, interactions between whiteflies and natural enemies can lead to cases of intraguild predation, which can lower their effectiveness. For instance, the overall predation on the whitefly was reduced when the two mirid bugs, *Macrolophus pygmaeus* (Rambur) and *Nesidiocoris tenuis* Reuter, occurred together on the same plant [20]. On the other hand, when feeding on plants, the two zoophytophagous mirids stimulate plants to release volatile organic compounds (VOCs), which repel pests, such as *B. tabaci* and *Frankliniella occidentalis* Pergande, but attract whitefly parasitoids, such as *Encarsia formosa* Gahan [21].

The complex interactions between plants and phytophagous insects, from which physiological, morphological, or behavioral plasticity derives in both hosts and herbivores, have been the focus of numerous studies [22]. For *B. tabaci*, interactions with plants are extremely important, and they are based on the attractiveness of plants, which is communicated to whiteflies by both visual and (to a lesser extent) biochemical cues. The nutrient composition of host plants therefore impacts whitefly performance, such that nutrient changes or stresses in plants affect nutrition in whiteflies [23]. In turn, whitefly feeding alters the physiology and morphology of plants, causing changes in physiological (e.g., gas exchange, chlorophyll fluorescence, indirect chlorophyll content), biochemical (e.g., enzymes, phenols, flavonoids), or morphological (e.g., plant height, leaf area, shoot dry weight, root dry weight) parameters [24–32]. Most of these phenomena remain poorly understood and therefore need to be more deeply investigated in order to improve sustainable whitefly management.

In one previous work, the impact of *B. tabaci* MED on eggplants and tomatoes was investigated. This study considered the principal morphological and physiological traits (e.g., plant height, dry plant biomass, chlorophyll content, etc.) [33] and found that eggplant and tomato plants infested by the whitefly showed strong and significant reductions in height, shoot dry weight, leaf area, and indirect chlorophyll content, though with different levels of intensity among the two plant species. Starting from the above results, and in order to widen our knowledge of multitrophic interaction mechanisms related to the impact of whiteflies on plant biology, a further trophic level in the analysis was added. In the Mediterranean basin, *M. pygmaeus* spontaneously colonizes tomato crops when pesticide applications are reduced [34,35]. However, as a zoophytophagous insect, it can also feed on the mesophyll of leaves, the tissues of stems, inflorescences, and fruits [36,37]; the suitability of this predator for establishment also varies depending on the species of the host plant and the part of the plant on which the predator lives [38]. Because of this, and bearing in mind the considerable diffusion of this mirid bug in horticultural areas of the Mediterranean basin, we sought to investigate the effects on the morphology and physiology of eggplants through the combined action of *B. tabaci* and *M. pygmaeus*, and we evaluated the modifications to various morphological and physiological parameters of host plants following infestation by *B. tabaci* MED and by a combined presence of this whitefly with its predator *M. pygmaeus*, compared with totally noninfested plants.

2. Materials and Methods

2.1. Insects and Plants

A colony of *Bemisia tabaci* was collected from an eggplant crop grown under greenhouse conditions in southeast Sicily (Vittoria, province of Ragusa, 36.97134 lat.; 14.424505 lon.). The specimens were then transferred and maintained on eggplant plants in the laboratory under controlled environmental conditions (T = 25 ± 2 °C, RH = 65 ± 5%, and photoperiod of 14L:10D h).

Macrolophus pygmaeus individuals came from commercial sources (MIRICAL; Koppert Biological Systems, S.L., Águilas, Murcia, Spain). These were maintained in the laboratory, under the same environmental conditions as those for *B. tabaci*, on eggplants infested by the whitefly. These were also fed every 3 days with eggs of *Ephestia kuehniella* Zeller (Koppert B.V., Berkel en Rodenrijs, BE, The Netherlands).

Host plants (*Solanum melongena* L. cv. Gloria) were grown from seeds germinated and raised in polystyrene planting trays in a nursery. The seedlings were then individually transplanted into black plastic pots (10 cm × 10 cm × 12 cm), using a professional potting soil specific for vegetable sowing, and maintained under controlled environmental conditions in the laboratory (T = 25 ± 2 °C; R.H. = 65 ± 5%, and photoperiod of 14L:10D h) throughout the experiment.

2.2. Experimental Design and Sampling

The study was carried out at the laboratories of the Applied Entomology section, Department of Agriculture, Food and Environment (Di3A), University of Catania, Italy, in the period October 2021–January 2022.

The species identity of *B. tabaci* was genetically attained on about 30 whiteflies collected from the rearing described above, before running the test. Using molecular biological methods [39,40], all tested individuals were identified as *B. tabaci* MED, Q2 subclade.

The impact of *B. tabaci* and *M. pygmaeus* on the host plants was assessed on a total of 36 eggplant plants with 6 fully expanded leaves. The trial was set up using a completely randomized design with 12 replicates under each of the following three evaluated conditions (hereafter treatments): noninfested control plants (C); plants infested by *B. tabaci* (PIB); and plants infested by *B. tabaci* where *M. pygmaeus* was also released (PIBM). In order to infest the plants representing the PIB and the PIBM treatments, 4 weeks after transplanting groups of three plants were isolated in netted cages (L × W × H: 60 cm × 60 cm × 60 cm), and 60 unsexed newly emerged (<24 h old) whitefly adults (i.e., 20/plant) collected from the insectary were released on the floor in the center of each cage. The whitefly adults were allowed to lay eggs for 3 days before being removed from the cages by a mouth aspirator (John W. Hock Company, Gainesville, FL, USA).

Next, to assess whether oviposition had occurred, the number of eggs laid was counted on three leaves of each plant, using a stereomicroscope (Olympus Optical Co., Ltd., Tokyo, Japan, SZX-ILLK200). To verify the progress of the infestation, the nymphs fixed on the lower surface of each of the three previously selected leaves were checked 2 weeks after removal of the adults. In line with procedures described in the literature [41,42], 3 weeks after oviposition by *B. tabaci*, 24 unsexed newly emerged *M. pygmaeus* adults (<24 h old) were released on the floor in the center of each cage to test the "pest + zoophytophagous predator" condition (PIBM) (i.e., 8/plant). The mirid adults were allowed to lay eggs for 6 days [43] before being removed from the cages. The monitoring of nymphs' emergence started 8 days after their release [44] and continued daily for 5 weeks. All newly emerged specimens were also fed with eggs of *E. kuehniella*, glued on a paper strip and provided every 3 days. By following the method described by Sanchez [45], these specimens were counted visually on all entire eggplant plants and were removed from the cages using the mouth aspirator at the end of the experiment (i.e., after about 40 days from the introduction of *M. pygmaeus*). All plants were watered twice a week.

To assess the combined effects of both insects on *S. melongena* development, the height of the plants (PH), the indirect chlorophyll content (ICC), the chlorophyll fluorescence (CF), the dry plant biomass (roots and shoots: RDW and SDW), and the leaf and root areas (LA and RA) were all measured at the end of the experiment. Plant height, expressed in centimeters (cm), was measured from stem base to apex [46] with a ruler. To obtain values for RDW and SDW, expressed in grams, the fresh hypogeal and epigeal biomass was oven-dried (Thermo Fisher Scientific, Langenselbold, Germany, Heratherm OGS100) at 65 °C for 3 days and finally weighed with a high-precision balance (ORMA BC 1000, Orma srl, Milan, Italy; resolution 0.1 g). To calculate the amount of chlorophyll present in the leaf [47]; ICC measurements were taken using a Soil Plant Analysis Development (SPAD-502, Minolta, Sakai, Osaka, Japan) chlorophyll meter on three leaves per plant, which were at principal growth stage 1, according to the BBCH scale [48]. To measure plant stress, the CF data were collected using an OS1-FL Chlorophyll Fluorometer (Opti-Sciences Inc., Hudson, NH, USA). Initially, a leaf in the middle third of each plant was dark-adapted for at least 20 min with detachable leaf clips. Next, the device emitted a saturation pulse through a beam of light, which was read by the device when reflected [49]. The parameter considered in our experiment was F_V/F_M, which is the ratio of the variable to the maximum fluorescence after dark-adaptation, which provides information on the functioning of photosystem II (PSII), representing the maximum quantum yield of PSII [50]. In other words, it is a sensitive indicator of plant photosynthetic performance that enables the comparison of plant samples of the same known dark-adapted state using a normalized ratio [51]. The LA and RA of plants were expressed in square centimeters (cm^2) and determined by ImageJ software (Wayne Rasband—Services Branch, National Institute of Mental Health, Bethesda, MD, USA), which processed the pictures taken by a digital camera (Nikon D850 45.4 megapixel). To monitor the stress state of the plants, ICC and CF were performed monthly for the duration of the test [49].

2.3. Data Analysis

Data of parameters relating to the responses of different plants to the three tested conditions were analyzed using a one-way ANOVA. Where significant differences were detected, the means were separated by Tukey's HSD test ($p < 0.05$). Statistical analysis was carried out using the program Statistica (StatSoft, TIBCO Software Inc., Tulsa, OK, USA).

3. Results

The mean number of eggs laid by the whitefly adults on the lower surface of each of the three selected leaves per plant after 3 days of exposition was 46.6 ± 6.94 (average: 2.2 eggs/cm^2). Further, 17 days after the adults were released, the average number of nymphs was 44.1 ± 9.98 (mean: 2.1 nymphs/cm^2). These results confirmed that oviposition and the progress of infestation evenly occurred in all plants under both PIB and PIBM conditions (Figure 1a).

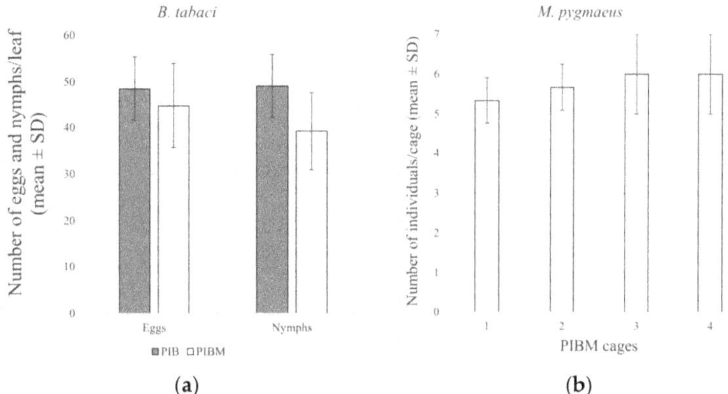

Figure 1. Mean number (±SE) of (**a**) *B. tabaci* eggs laid and nymphs fixed on the lower surface of each of the three selected leaves per plant under both PIB and PIBM conditions and (**b**) *M. pygmaeus* specimens per PIBM cage.

The average total density of *M. pygmaeus*, expressed as the number of specimens per plant 40 days after release, was 5.7 ± 0.7 insects/plant (Figure 1b). About 2 weeks after the release of *M. pygmaeus*, its first-instar nymphs appeared, and these became adults by the end of the experiment.

The stress state of plants during the test was indicated by the first nondestructive measurement taken 1 month after the beginning of the experiment. This showed no statistical differences in the ICC values (Table 1) between C, PIB, and PIBM modalities ($F_{2,33} = 0.32$; $p < 0.727$) (Figure 2a). In contrast, CF values recorded in the same period (Table 1) showed significant differences between PIB and the other two conditions, C and PIBM ($F_{2,33} = 4.89$; $p < 0.0138$) (Figure 2b). Starting from the second measurement carried out 2 months after the beginning of the test, the ICC parameter was influenced by the presence of the two insects (Table 1), revealing statistical differences among the three conditions ($F_{2,33} = 86.31$; $p < 0.001$) (Figure 2a). However, with respect to the CF parameter (Table 1), in the second nondestructive measurement, the statistical differences found in the first measurement were maintained ($F_{2,33} = 91.9$; $p < 0.001$), with the lowest mean values recorded in plants infested by *B. tabaci* (Figure 2b).

Table 1. Stress state of plants recorded during the first and second nondestructive measurements, as indicated by the calculation of the amount of chlorophyll present in the leaf (ICC) and by calculation of the plant's photosynthetic performance (CF).

Treatment	1st ICC (SPAD Unit ± SE)	2nd ICC (SPAD Unit ± SE)	1st CF (F_V/F_M ± SE)	2nd CF (F_V/F_M ± SE)
C	37.48 ± 0.36 a	35.68 ± 0.34 a	0.77 ± 0.004 a	0.77 ± 0.001 a
PIB	37.07 ± 0.33 a	29.68 ± 0.35 c	0.76 ± 0.001 b	0.73 ± 0.002 b
PIBM	37.23 ± 0.39 a	32.61 ± 0.27 b	0.77 ± 0.002 a	0.77 ± 0.002 a
F; df; p	0.32; 2, 33; <0.727	86.31; 2, 33; <0.001	4.89; 2, 33; <0.0138	91.9; 2, 33; <0.001

ANOVA parameters are reported for each test condition. Within each column, data followed by different letters are significantly different ($p < 0.05$, Tukey's HSD test).

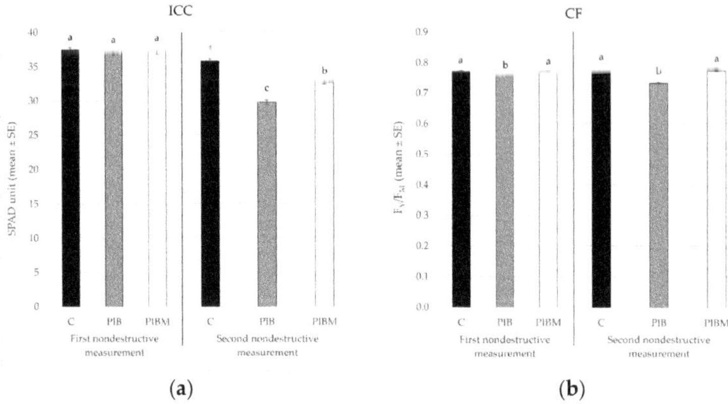

Figure 2. Mean values (±SE) of (**a**) the indirect chlorophyll content (ICC) and (**b**) the chlorophyll fluorescence (CF) detected in the first and second nondestructive measurements performed on noninfested control plants (C), eggplants infested by *B. tabaci* (PIB), and host plants in the presence of both *B. tabaci* and *M. pygmaeus* (PIBM). Different letters indicate statistically significant differences at $p < 0.05$.

At the end of the experiment, the values for all considered plant physiology and morphology parameters showed statistical differences between the conditions examined, as reported in Table 2. Overall, plant height (PH) was negatively affected ($F_{2,33} = 16.297$; $p < 0.001$) by the presence of both the insect species, with higher mean values recorded in the noninfested plants (Figure 3a).

Table 2. Impact of the biological activity of *Bemisia tabaci* (PIB), either alone or in association with *Macrolophus pygmaeus* (PIBM), on the main morphological and physiological parameters of eggplants. (PH—plant height; ICC—indirect chlorophyll content; CF—chlorophyll fluorescence; LA—leaf area; RA—root area; SDW—dry shoot biomass; RDW—dry root biomass).

Treatment	PH (cm ± SE)	ICC (SPAD Unit ± SE)	CF (F_V/F_M ± SE)	LA (cm² ± SE)	SDW (g ± SE)	RA (cm² ± SE)	RDW (g ± SE)
C	26.07 ± 0.57 a	35.60 ± 0.64 a	0.78 ± 0.003 a	1141.49 ± 35.67 a	0.74 ± 0.03 a	214.90 ± 13.69 a	0.13 ± 0.01 a
PIB	20.35 ± 1.03 b	26.23 ± 1.23 c	0.74 ± 0.007 b	765.17 ± 74.21 c	0.40 ± 0.06 c	178.91 ± 11.38 b	0.11 ± 0.01 b
PIBM	21.73 ± 0.51 b	31.91 ± 0.56 b	0.77 ± 0.051 a	939.17 ± 31.81 b	0.57 ± 0.03 b	149.83 ± 9.73 b	0.08 ± 0.001 b
F; df; p	16.297; 2, 33; <0.001	29.728; 2, 33; <0.001	16.159; 2, 33; <0.001	13.45; 2, 33; <0.001	17.12; 2, 33; <0.001	7.74; 2, 33; <0.0017	6.42; 2, 33; <0.0044

ANOVA parameters are reported for each test condition. Within each column, data followed by different letters are significantly different ($p < 0.05$, Tukey's HSD test).

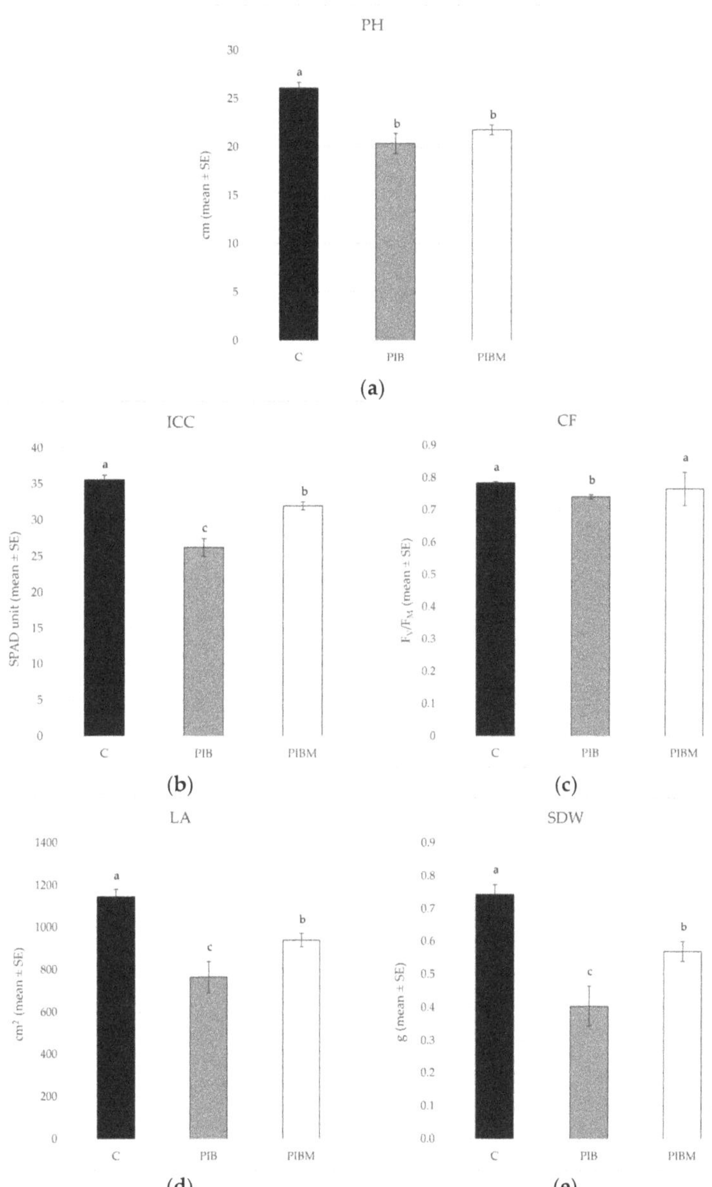

Figure 3. *Cont.*

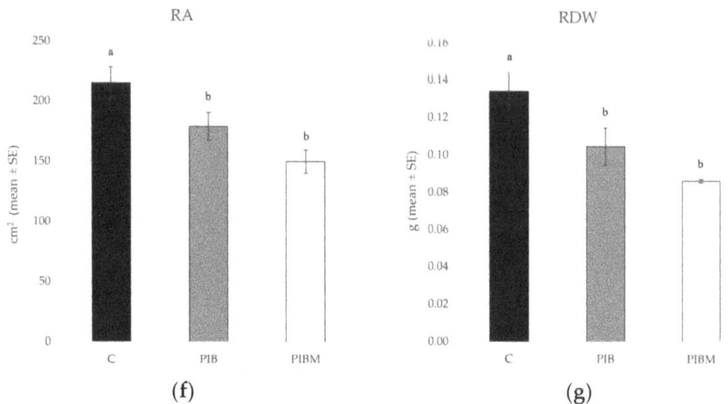

Figure 3. Mean values (±SE) of (**a**) plant height (PH), (**b**) indirect chlorophyll content (ICC), (**c**) chlorophyll fluorescence (CF), (**d**) leaf area (LA), (**e**) shoot dry weight (SDW), (**f**) root area (RA), and (**g**) root dry weight (RDW) of noninfested control plants (C), host plants infested by *B. tabaci* (PIB), and eggplants in the presence of both whitefly and *M. pygmaeus* (PIBM). Different letters indicate statistically significant differences at *p* < 0.01.

The release of *B. tabaci* (PIB) caused a clear detrimental effect on indirect chlorophyll content (ICC) in the leaves of infested plants, which was statistically different when compared with the other two tested modalities ($F_{2,33}$ = 29.728; p < 0.001); indeed, in the presence of the predators (PIBM), the plants exhibited a slightly higher chlorophyll content, though it was still less than that of the noninfested plants (C) (Figure 3b).

Similarly, the CF values indicating the plant's photosynthetic performances followed a broadly similar trend; in this case, a data analysis revealed a statistically significant decrease in the CF values ($F_{2,33}$ = 16.159; p < 0.001) of the PIB treatment in comparison with PIBM treatment (i.e., when the predator was released in the cage) that was not statistically different from the noninfested plants of the C condition (Figure 3c).

Even in the case of the leaf area (LA) of plants, there was a statistical difference between the three tested conditions ($F_{2,33}$ = 13.45; p < 0.001); in particular, the lowest mean values of the PIB condition suggested that the presence of the predator limited the damage caused by the whitefly (Figure 3d).

Similarly, in the case of the SDW parameter, the analysis revealed a statistical difference among plants under the three conditions ($F_{2,33}$ = 17.12; p < 0.001) (Figure 3e), where those of the PIB condition, in the absence of *M. pygmaeus*, once again exhibited the lowest values.

As was the case with plant height, the root area and root dry weight (i.e., RA and RDW) were also negatively influenced by the presence of both insect species. In particular, the highest mean values were recorded in the noninfested plants, with statistically significant differences in comparison with the plants of the other two treatments ($F_{2,33}$ = 7.74; p < 0.0017 and $F_{2,33}$ = 6.42; p < 0.0044—for RA and RDW, respectively) (Figure 3f,g).

4. Discussion

Integrated pest management aims to guide agriculture strategies by controlling arthropod infestations through the optimal selection of host plants [52] and management of the activity of natural enemies [53]. In this context, the responses of host plants to the presence of pests have often been studied, but it remains unclear how plants respond to the zoophytophagy of predatory omnivorous insects [54].

The present research indicates that the mirid predator *M. pygmaeus* exerts a significant influence on some physiological and morphological traits of *S. melongena* (e.g., indirect chlorophyll content, chlorophyll fluorescence, root area and dry weight, etc.) that have been poorly investigated so far.

It is well known that infestation by *B. tabaci* affects the quantity and quality of yields in many varieties of vegetable crops [55]. The means by which whitefly infestations exert a negative effect on horticultural species (eggplant and tomato) have also been confirmed in a recent study [33]. The findings reported here show that the presence of *M. pygmaeus* reduces the negative effects of the whitefly on plants [38,46], resulting in significantly higher values of indirect chlorophyll content, chlorophyll fluorescence, leaf area, and shoot dry weight in the PIBM condition compared with the PIB. In this regard, according to Pappas [54], zoophytophagous hemipterans, such as *Orius insidiosus* (Say) and pentatomids, feed on plants mainly to acquire water from the xylem and also potentially to obtain nutrients from the mesophyll or pollen, most likely causing only some minor cell wounding on leaves. Feeding by mirid bugs therefore depends on the plant sap and not only on prey. This explains the greater suitability of the eggplant compared with other vegetable crops, which results in a relatively longer survival of the predator even in the absence of prey [55]. However, it is also known that plant sap may influence the taste of prey individuals, making them either more or less attractive and desirable to the predators, so that eggplants might generate better prey than other vegetable crops do [56]. Furthermore, the presence of the eggs of *E. kuehniella*, another optimal food source for *M. pygmaeus*, may have helped the predator to successfully establish and increase a stable population [57].

However, as observed by Bresch [46] in a study of tomato and tobacco plants infested by *Trialeurodes vaporariuorum* (Westwood), the mirid predator *M. pygmaeus* cannot always significantly reduce the negative impact of the pest. Indeed, the results of the present study reveal that *M. pygmaeus* was unable to limit the effect of *B. tabaci* on PH and on the characteristics of the roots (i.e., RA and RDW) that were not significantly different from those obtained in the PIB condition.

Insect pests can trigger the production of deterrents or toxic secondary metabolites that reduce the suitability of plant tissues for further insect colonization and may compromise the ability of the plant to activate certain resistance-related pathways [58–60]. This could be the case with *B. tabaci*, which induces the activation of salicylic acid (SA) [61,62], a phytohormone able to suppress the activation of jasmonic acid (JA) [53,63]. However, the latter phytohormone can be induced by mirid insect activities (e.g., oviposition by adult females and feeding by *M. pygmaeus* nymphs) [54] and functions in the mediation of plant responses. Specifically, the root stimulation of JA responses, following shoot damage, is completely dependent on hormone translocation [64]; thus, the phloem transport system is crucial to allocate resources among plant tissues and organs and move the jasmonates, which accumulate in vascular bundles after wounding [65–67].

Our findings may be seen in line with the results obtained by Schulze [64], who reported that even if, following wounding, shoot-produced jasmonates on *Arabidopsis thaliana* (L.) Heynh. move downward into the root through the phloem, the wounding on shoots is not always able to trigger the expression of JA marker genes in roots of all genotypes. Because JA may produce a local and nonsystemic effect [68] and because our eggplants (PIBM) were exposed to *M. pygmaeus* for mating and oviposition for only 3 days, it is probable that compounds related to plant defense were not transferred up to the roots during the experimental period. As indicated by Zhang [68], a longer exposure of host plants to mirids, or exposure to a greater density, could result in more-evident differences in phytohormone concentrations in various parts of plants, with consequent increases in root area and root dry weight, as was the case in our roots.

Although our results offer a preliminary insight into multitrophic interactions and mechanisms among host plant, pest, and predator, the specific response of whole eggplants to the presence of *M. pygmaeus* remains unclear. The wound-induced responses of plants are often modified by the perception of herbivore-specific elicitors, and this may be the case for zoophytophagous omnivores, especially within the Miridae, as they produce many different salivary enzymes [69,70]. Further investigations are needed to more deeply study how phytophagy by *M. pygmaeus* directly and indirectly affects host plants and whether *B. tabaci* is able to suppress or resist plant defenses stimulated by the predator.

Author Contributions: Conceptualization, C.R., A.F. and P.S.; methodology, A.F., P.S., G.E.M.C. and C.R.; formal analysis, A.F. and P.S.; investigation, A.F.; writing—original draft preparation, A.F. and C.R.; writing—review and editing, C.R., A.F., P.S. and G.E.M.C. All authors have read and agreed to the published version of the manuscript

Funding: This research has received funding from the European Union's Horizon 2020 research and innovation program under grant agreement number 101000570.

Institutional Review Board Statement: Not applicable.

Informed Consent Statement: Not applicable.

Data Availability Statement: The data presented in this study are available on request from the corresponding author.

Conflicts of Interest: The authors declare no conflict of interest.

References

1. Gange, A.C.; Brown, V.K. *Multitrophic Interactions in Terrestrial Systems*; Blackwell Science Ltd.: Oxford, UK, 1997; p. 458.
2. Tscharntke, T.; Hawkins, B.A. *Multitrophic Level Interactions*; Cambridge University Press: Cambridge, UK, 2002; pp. 1–7.
3. Ohgushi, T. Indirect interaction webs: Herbivore-induced effects through trait change in plants. *Annu. Rev. Ecol. Evol. Syst.* **2005**, *36*, 81–105. [CrossRef]
4. Shikano, I. Evolutionary ecology of multitrophic interactions between plants, insect herbivores and entomopathogens. *J. Chem. Ecol.* **2017**, *43*, 586–598. [CrossRef] [PubMed]
5. Jacobsen, D.J.; Raguso, R.A. Lingering effects of herbivory and plant defenses on pollinators. *Curr. Biol.* **2018**, *28*, R1164–R1169. [CrossRef] [PubMed]
6. Jaber, L.R.; Araj, S.E. Interactions among endophytic fungal entomopathogens (Ascomycota: Hypocreales), the green peach aphid *Myzus persicae* Sulzer (Homoptera: Aphididae), and the aphid endoparasitoid *Aphidius colemani* Viereck (Hymenoptera: Braconidae). *Biol. Control* **2018**, *116*, 53–61. [CrossRef]
7. Zytynska, S.E.; Meyer, S.T. Effects of biodiversity in agricultural landscapes on the protective microbiome of insects—A review. *Entomol. Exp. Et Appl.* **2019**, *167*, 2–13.
8. Schifani, E.; Castracani, C.; Giannetti, D.; Spotti, F.A.; Reggiani, R.; Leonardi, S.; Mori, A.; Grasso, D.A. New tools for conservation biological control: Testing ant-attracting artificial Nectaries to employ ants as plant defenders. *Insects* **2020**, *11*, 129. [CrossRef]
9. Gerling, D. *Whiteflies: Their Bionomics, Pest Status and Management*; Intercept Ltd. Press: Andover, UK, 1990; p. 348.
10. Martin, J.H.; Mifsud, D.; Rapisarda, C. The whiteflies (Hemiptera: Aleyrodidae) of Europe and the Mediterranean basin. *Bull. Entomol. Res.* **2000**, *90*, 407–448. [CrossRef]
11. Navas-Castillo, J.; Fiallo-Olivé, E.; Sánchez-Campos, S. Emerging virus diseases transmitted by whiteflies. *Annu. Rev. Phytopathol.* **2011**, *49*, 219–248.
12. Fiallo-Olivé, E.; Pan, L.L.; Liu, S.S.; Navas-Castillo, J. Transmission of begomoviruses and other whitefly-borne viruses: Dependence on the vector species. *Phytopathology* **2020**, *110*, 10–17. [CrossRef]
13. De Barro, P.J.; Liu, S.S.; Boykin, L.M.; Dinsdale, A.B. *Bemisia tabaci*: A statement of species status. *Annu. Rev. Phytopathol.* **2011**, *56*, 1–19.
14. Liu, S.S.; Colvin, J.; De Barro, P.J. Species concepts as applied to the whitefly *Bemisia tabaci* systematics: How many species are there? *J. Integr. Agric.* **2012**, *11*, 176–186. [CrossRef]
15. Lee, W.; Park, J.; Lee, G.S.; Lee, S.; Akimoto, S.I. Taxonomic status of the *Bemisia tabaci* complex (Hemiptera: Aleyrodidae) and reassessment of the number of its constituent species. *PLoS ONE* **2013**, *8*, e63817. [CrossRef] [PubMed]
16. Boykin, L.M.; Savill, A.; De Barro, P. Updated mtCOI reference dataset for the *Bemisia tabaci* species complex. *F1000Research* **2017**, *6*, 1835. [CrossRef]
17. Tan, X.L.; Wang, S.; Ridsdill-Smith, J.; Liu, T.X. Direct and indirect impacts of infestation of tomato plant by *Myzus persicae* (Hemiptera: Aphididae) on *Bemisia tabaci* (Hemiptera: Aleyrodidae). *PLoS ONE* **2014**, *9*, e94310.
18. Tan, X.L.; Chen, J.L.; Benelli, G.; Desneux, N.; Yang, X.Q.; Liu, T.X.; Ge, F. Pre-infestation of tomato plants by aphids modulates transmission-acquisition relationship among whiteflies, tomato yellow leaf curl virus (TYLCV) and plants. *Front. Plant Sci.* **2017**, *8*, 1597.
19. Yang, F.; Tan, X.L.; Liu, F.H.; Wang, S.; Chen, J.L.; Yi, T.Y. A functional response evaluation of pre-infestation with *Bemisia tabaci* cryptic species MEAM1 on predation by *Propylea japonica* of *Myzus persicae* on host plant tomatoes. *Arthropod Plant Interact.* **2017**, *11*, 825–832.
20. Moreno-Ripoll, R.; Gabarra, R.; Symondson, W.O.C.; King, R.A.; Agustí, N. Do the interactions among natural enemies compromise the biological control of the whitefly *Bemisia tabaci*? *J. Pest Sci.* **2014**, *87*, 133–141. [CrossRef]
21. Pérez-Hedo, M.; Bouagga, S.; Zhang, N.X.; Moerkens, R.; Messelink, G.; Jaques, J.A.; Flors, V.; Broufas, G.; Urbaneja, A.; Pappas, M.L. Induction of plant defenses: The added value of zoophytophagous predators. *J. Pest Sci.* **2022**, *95*, 1501–1517. [CrossRef]
22. Ashra, H.; Nair, S. Trait plasticity during plant-insect interactions: From molecular mechanisms to impact on community dynamics. *Plant Sci.* **2022**, *317*, 111188.

23. Dinant, S.; Kehr, J. Sampling and analysis of phloem sap. In *Plant Mineral Nutrients*; Humana Press: Totowa, NJ, USA, 2013; pp. 185–194.
24. De Lima Toledo, C.A.; da Silva Ponce, F.; Oliveira, M.D.; Aires, E.S.; Seabra Júnior, S.; Lima, G.P.P.; de Oliveira, R.C. Change in the Physiological and Biochemical Aspects of Tomato Caused by Infestation by Cryptic Species of *Bemisia tabaci* MED and MEAM1. *Insects* **2021**, *12*, 1105.
25. Buntin, D.G.; Gilbertz, D.A.; Oetting, R.D. Chlorophyll loss and gas exchange in tomato leaves after feeding injury by *Bemisia tabaci* (Homoptera: Aleyrodidae). *J. Econ. Entomol.* **1993**, *86*, 517–522.
26. Islam, M.T.; Qiu, B.; Ren, S. Host preference and influence of the sweetpotato whitefly, *Bemisia tabaci* (Homoptera: Aleyrodidae) on eggplant (*Solanum melongena* L.). *Acta Agric. Scand. Sect. B Soil Plant Sci.* **2010**, *60*, 320–325. [CrossRef]
27. Li, Q.; Tan, W.; Xue, M.; Zhao, H.; Wang, C. Dynamic changes in photosynthesis and chlorophyll fluorescence in *Nicotiana tabacum* infested by *Bemisia tabaci* (Middle East–Asia Minor 1) nymphs. *Arthropod Plant Interact.* **2013**, *7*, 431–443. [CrossRef]
28. Chand, R.; Jokhan, A.; Prakash, R. Egg deposition by Spiralling whiteflies (*Aleurodicus dispersus*) reduces the stomatal conductance of cassava (*Manihot esculenta*). *Wētā* **2018**, *52*, 55–60.
29. Yee, W.L.; Toscano, N.C.; Chu, C.C.; Henneberry, T.J.; Nichols, R.L. Bemisia argentifolii (Homoptera: Aleyrodidae) action thresholds and cotton photosynthesis. *Environ. Entomol.* **1996**, *25*, 1267–1273.
30. Lin, T.B.; Wolf, S.; Schwartz, A.; Saranga, Y. Silverleaf whitefly stress impairs sugar export from cotton source leaves. *Physiol. Plant.* **2000**, *109*, 291–297.
31. Lin, T.B.; Schwartz, A.; Saranga, Y. Photosynthesis and productivity of cotton under silverleaf whitefly stress. *Crop Sci.* **1999**, *39*, 174–184. [CrossRef]
32. Islam, T.; Shunxiang, R. Effect of sweetpotato whitefly, *Bemisia tabaci* (Homoptera: Aleyrodidae) infestation on eggplant (*Solanum melongena* L.) leaf. *J. Pest Sci.* **2009**, *82*, 211–215. [CrossRef]
33. Farina, A.; Barbera, A.C.; Leonardi, G.; Massimino Cocuzza, G.E.; Suma, P.; Rapisarda, C. Bemisia tabaci (Hemiptera: Aleyrodidae): What Relationships with and Morpho-Physiological Effects on the Plants It Develops on? *Insects* **2022**, *13*, 351. [CrossRef]
34. Urbaneja, A.; González-Cabrera, J.; Arno, J.; Gabarra, R. Prospects for the biological control of Tuta absoluta in tomatoes of the Mediterranean basin. *Pest Manag. Sci.* **2012**, *68*, 1215–1222.
35. De Backer, L.; Megido, R.C.; Haubruge, É.; Verheggen, F.J. *Macrolophus pygmaeus* (Rambur) as an efficient predator of the tomato leafminer Tuta absoluta (Meyrick) in Europe. A review. *Biotechnol. Agron. Soc. Environ.* **2014**, *18*, 536–543.
36. Wheeler, A.G. *Biology of the Plant Bugs (Hemiptera: Miridae): Pests, Predators, Opportunists*; Cornell University Press: Ithaca, NY, USA, 2001.
37. Ingegno, B.L.; Pansa, M.G.; Tavella, L. Plant preference in the zoophytophagous generalist predator *Macrolophus pygmaeus* (Heteroptera: Miridae). *Biol. Control* **2011**, *58*, 174–181. [CrossRef]
38. Sanchez, J.A.; Lopez-Gallego, E.; Pérez-Marcos, M.; Perera-Fernandez, L.G.; Ramirez-Soria, M.J. How safe is it to rely on *Macrolophus pygmaeus* (Hemiptera: Miridae) as a biocontrol agent in tomato crops? *Front. Ecol. Evol.* **2018**, *6*, 132.
39. Walsh, P.S.; Metzger, D.A.; Higuchi, R. Chelex 100 as a medium for simple extraction of DNA for PCR-based typing from forensic material. *Biotechniques* **1991**, *10*, 506–513. [PubMed]
40. De Barro, P.J.; Driver, F. Use of RAPD PCR to distinguish the B biotype from other biotypes of *Bemisia tabaci* (Gennadius) (Hemiptera: Aleyrodidae). *Aust. J. Entomol.* **1997**, *36*, 149–152. [CrossRef]
41. Calvo, J.; Urbaneja, A. *Nesidiocoris tenuis*, un aliado para el control biológico de la mosca blanca. *Hortic. Int.* **2004**, *44*, 20–25.
42. Alomar, O.; Riudavets, J.; Castañé, C. *Macrolophus caliginosus* in the biological control of *Bemisia tabaci* on greenhouse melons. *Biol. Control* **2006**, *36*, 154–162.
43. Margaritopoulos, J.T.; Tsitsipis, J.A.; Perdikis, D.C. Biological characteristics of the mirids *Macrolophus costalis* and *Macrolophus pygmaeus* preying on the tobacco form of *Myzus persicae* (Hemiptera: Aphididae). *Bull. Entomol. Res.* **2003**, *93*, 39–45.
44. Moerkens, R.; Berckmoes, E.; Van Damme, V.; Wittemans, L.; Tirry, L.; Casteels, H.; De Clercq, P.; De Vis, R. Inoculative release strategies of *Macrolophus pygmaeus* Rambur (Hemiptera: Miridae) in tomato crops: Population dynamics and dispersal. *J. Plant Dis. Prot.* **2017**, *124*, 295–303. [CrossRef]
45. Sanchez, J.A.; López-Gallego, E.; Pérez-Marcos, M.; Perera-Fernández, L. The effect of banker plants and pre-plant release on the establishment and pest control of *Macrolophus pygmaeus* in tomato greenhouses. *J. Pest Sci.* **2021**, *94*, 297–307. [CrossRef]
46. Bresch, C.; Ottenwalder, L.; Poncet, C.; Parolin, P. Tobacco as banker plant for *Macrolophus pygmaeus* to control *Trialeurodes vaporariorum* in tomato crops. *Univers. J. Agric. Res.* **2014**, *2*, 297–304. [CrossRef]
47. Ling, Q.; Huang, W.; Jarvis, P. Use of a SPAD-502 meter to measure leaf chlorophyll concentration in *Arabidopsis thaliana*. *Photosynth. Res.* **2011**, *107*, 209–214. [CrossRef] [PubMed]
48. Meier, U. *Growth Stages of Mono-and Dicotyledonous Plants*; Blackwell Wissenschafts-Verlag: Berlin, Germany, 1997.
49. Schutze, I.X.; Yamamoto, P.T.; Malaquias, J.B.; Herritt, M.; Thompson, A.; Merten, P.; Naranjo, S.E. Correlation-Based Network Analysis of the Influence of *Bemisia tabaci* Feeding on Photosynthesis and Foliar Sugar and Starch Composition in Soybean. *Insects* **2022**, *13*, 56. [CrossRef] [PubMed]
50. Ibaraki, Y.; Murakami, J. Distribution of chlorophyll fluorescence parameter Fv/Fm within individual plants under various stress conditions. In Proceedings of the XXVII International Horticultural Congress-IHC2006: International Symposium on Advances in Environmental Control, Automation, Madrid, Spain, 21–23 June 2006; Volume 761, pp. 255–260.
51. Fernández-Marín, B.; Gago, J.; Clemente-Moreno, M.J.; Flexas, J.; Gulías, J.; García-Plazaola, J.I. Plant pigment cycles in the high-Arctic Spitsbergen. *Polar Biol.* **2019**, *42*, 675–684. [CrossRef]

52. Magagnoli, S.; Masetti, A.; Depalo, L.; Sommaggio, D.; Campanelli, G.; Leteo, F.; Lövei, G.L.; Burgio, G. Cover crop termination techniques affect ground predation within an organic vegetable rotation system: A test with artificial caterpillars. *Biol. Control* **2018**, *117*, 109–114.

53. Stout, M.J.; Zehnder, G.W.; Baur, M.E. Potential for the use of elicitors of plant resistance in arthropod management programs. *Arch. Insect Biochem. Physiol. Publ. Collab. Entomol. Soc. Am.* **2002**, *51*, 222–235.

54. Pappas, M.L.; Steppuhn, A.; Geuss, D.; Topalidou, N.; Zografou, A.; Sabelis, M.W.; Broufas, G.D. Beyond predation: The zoophytophagous predator *Macrolophus pygmaeus* induces tomato resistance against spider mites. *PLoS ONE* **2015**, *10*, e0127251.

55. Perdikis, D.C.; Lykouressis, D.P. *Macrolophus pygmaeus* (Hemiptera: Miridae) population parameters and biological characteristics when feeding on eggplant and tomato without prey. *J. Econ. Entomol.* **2004**, *97*, 1291–1298. [CrossRef]

56. Rasdi, M.Z.; Fauziah, I.; Mohamad, W.W.; Rahman, S.A.S.; Salmah, C.M.; Kamaruzaman, J. Biology of *Macrolophus caliginosus* (Heteroptera: Miridae) Predator of *Trialeurodes vaporariorum* (Homoptera: Aleyrodidae). *Int. J. Biol.* **2009**, *1*, 63.

57. Sylla, S.; Brévault, T.; Diarra, K.; Bearez, P.; Desneux, N. Life-history traits of *Macrolophus pygmaeus* with different prey foods. *PLoS ONE* **2016**, *11*, e0166610. [CrossRef]

58. Usha Rani, P.; Jyothsna, Y. Biochemical and enzymatic changes in rice plants as a mechanism of defense. *Acta Physiol. Plant.* **2010**, *32*, 695–701.

59. War, A.R.; Paulraj, M.G.; War, M.Y.; Ignacimuthu, S. Jasmonic acid-mediated-induced resistance in groundnut (*Arachis hypogaea* L.) against *Helicoverpa armigera* (Hubner) (Lepidoptera: Noctuidae). *J. Plant Growth Regul.* **2011**, *30*, 512–523. [CrossRef]

60. War, A.R.; Paulraj, M.G.; War, M.Y.; Ignacimuthu, S. Herbivore-and elicitor-induced resistance in groundnut to Asian armyworm, *Spodoptera litura* (Fab.) (Lepidoptera: Noctuidae). *Plant Signal. Behav.* **2011**, *6*, 1769–1777. [CrossRef] [PubMed]

61. Silva, D.B.; Jiménez, A.; Urbaneja, A.; Pérez-Hedo, M.; Bento, J.M. Changes in plant responses induced by an arthropod influence the colonization behavior of a subsequent herbivore. *Pest. Manag. Sci.* **2021**, *77*, 4168–4180. [CrossRef] [PubMed]

62. Dias, C.R.; Cardoso, A.C.; Kant, M.R.; Mencalha, J.; Bernardo, A.M.G.; da Silveira, M.C.A.C.; Sarmento, R.A.; Venzon, M.; Pallini, A.; Janssen, A. Plant defences and spider-mite web affect host plant choice and performance of the whitefly *Bemisia tabaci*. *J. Pest Sci.* **2022**, 1–10. [CrossRef]

63. Zarate, S.I.; Kempema, L.A.; Walling, L.L. Silverleaf whitefly induces salicylic acid defenses and suppresses effectual jasmonic acid defenses. *Plant Physiol.* **2007**, *143*, 866–875. [CrossRef]

64. Schulze, A.; Zimmer, M.; Mielke, S.; Stellmach, H.; Melnyk, C.W.; Hause, B.; Gasperini, D. Wound-induced shoot-to-root relocation of JA-Ile precursors coordinates Arabidopsis growth. *Mol. Plant* **2019**, *12*, 1383–1394. [CrossRef] [PubMed]

65. Stenzel, I.; Hause, B.; Maucher, H.; Pitzschke, A.; Miersch, O.; Ziegler, J.; Ryan, C.A.; Wasternack, C. Allene oxide cyclase dependence of the wound response and vascular bundle-specific generation of jasmonates in tomato–amplification in wound signalling. *Plant J.* **2003**, *33*, 577–589.

66. Wasternack, C.; Hause, B. Jasmonates: Biosynthesis, perception, signal transduction and action in plant stress response, growth and development. An update to the 2007 review in Annals of Botany. *Ann. Bot.* **2013**, *111*, 1021–1058.

67. Chauvin, A.; Caldelari, D.; Wolfender, J.L.; Farmer, E.E. Four 13-lipoxygenases contribute to rapid jasmonate synthesis in wounded *Arabidopsis thaliana* leaves: A role for lipoxygenase 6 in responses to long-distance wound signals. *New Phytol.* **2013**, *197*, 566–575.

68. Zhang, N.X.; Messelink, G.J.; Alba, J.M.; Schuurink, R.; Kant, M.R.; Janssen, A. Phytophagy of omnivorous predator Macrolophus pygmaeus affects performance of herbivores through induced plant defences. *Oecologia* **2018**, *186*, 101–113. [CrossRef] [PubMed]

69. Hori, K.; Schaefer, C.W.; Panizzi, A.R. Possible causes of disease symptoms resulting from the feeding of phytophagous Heteroptera. In *Heteroptera of Economic Importance*; CRC Press: Boca Raton, FL, USA, 2000; pp. 11–35.

70. Castañé, C.; Arnó, J.; Gabarra, R.; Alomar, O. Plant damage to vegetable crops by zoophytophagous mirid predators. *Biol. Control* **2011**, *59*, 22–29. [CrossRef]

insects

Article

Effect of Leaf Trichomes in Different Species of Cucurbitaceae on Attachment Ability of the Melon Ladybird Beetle *Chnootriba elaterii*

Valerio Saitta [1], Manuela Rebora [2,*], Silvana Piersanti [2], Elena Gorb [3], Stanislav Gorb [3,†] and Gianandrea Salerno [1,†]

[1] Dipartimento di Scienze Agrarie, Alimentari e Ambientali, University of Perugia, Borgo XX Giugno 74, 06121 Perugia, Italy
[2] Dipartimento di Chimica, Biologia e Biotecnologie, University of Perugia, Via Elce di Sotto 8, 06121 Perugia, Italy
[3] Department of Functional Morphology and Biomechanics, Zoological Institute, Kiel University, Am Botanischen Garten 9, 24098 Kiel, Germany
* Correspondence: manuela.rebora@unipg.it; Tel.: +39-075-5855722
† These authors contributed equally to this work.

Citation: Saitta, V.; Rebora, M.; Piersanti, S.; Gorb, E.; Gorb, S.; Salerno, G. Effect of Leaf Trichomes in Different Species of Cucurbitaceae on Attachment Ability of the Melon Ladybird Beetle *Chnootriba elaterii*. *Insects* **2022**, *13*, 1123. https://doi.org/10.3390/insects13121123

Academic Editors: Ezio Peri and Pasquale Trematerra

Received: 18 October 2022
Accepted: 2 December 2022
Published: 5 December 2022

Publisher's Note: MDPI stays neutral with regard to jurisdictional claims in published maps and institutional affiliations.

Simple Summary: The investigation of insect attachment ability in relation to plant mechanical barriers can shed light on the different steps driving host plant selection in phytophagous insects and help to better understand the complex antagonistic coevolution between insects and plants. In this context, we investigated the attachment ability of the oligophagous melon ladybird beetle *Chnootriba elaterii* at different developmental stages (adult, larva, eggs) to leaves of several Cucurbitaceae species (watermelon, melon, cucumber, zucchini, pumpkin, squirting cucumber, calabash and loofah) characterized by the presence of glandular and non-glandular trichomes with different characteristics (density, length). We used different techniques (scanning electron microscopy, traction force experiments and centrifugal force tests) to characterize the plant leaf surface and insect attachment devices and measure insect attachment ability to the leaf of the different plant species. Data regarding morphological leaf traits of Cucurbitaceae associated with their resistance against phytophagous insects, combined with data regarding the chemical cues involved in the host plant selection, can help to develop environmentally friendly pest control methods.

Abstract: This study investigates the attachment ability of the oligophagous melon ladybird beetle *Chnootriba elaterii* to leaves of several Cucurbitaceae species. Using cryo-SEM, we described adult and larva tarsal attachment devices and leaf surface structures (glandular and non-glandular trichomes) in *Citrullus lanatus, Cucumis melo, Cucumis sativus, Cucurbita moschata, Cucurbita pepo, Ecballium elaterium, Lagenaria siceraria* and *Luffa aegyptiaca*. Using traction force experiments and centrifugal force tests, we measured the friction force exerted by females and larvae on plant leaves. We observed that Cucurbitaceae glandular trichomes do not affect insect attachment ability at both developmental stages, suggesting some adaptation of *C. elaterii* to its host plants, while non-glandular trichomes, when they are dense, short and flexible, heavily reduce the attachment ability of both insect stages. When trichomes are dense but stiff, only the larval force is reduced, probably because the larva has a single claw, in contrast to the adult having paired bifid dentate claws. The data on the mechanical interaction of *C. elaterii* at different developmental stages with different Cucurbitaceae species, combined with data on the chemical cues involved in the host plant selection, can help to unravel the complex factors driving the coevolution between an oligophagous insect and its host plant species.

Keywords: friction; adult; larva; pulvilli; claws; plant trichomes; biomechanics

1. Introduction

Small insect herbivores face the problems of attachment to plant leaves at every stage of their development. On the other hand, in the long antagonistic evolution between phytophagous insects and plants, the latter have developed different chemical and mechanical defenses to protect themselves from consumption. In response to the high insect attachment ability guaranteed by pads (hairy or smooth) and claws typically located in the tarsus and pretarsus [1], many plants have developed anti-adhesive surfaces (see review in [2]). Plant anti-adhesive features can be represented by cell shape [3], wet surfaces [4], cuticular folds [5], epicuticular wax projections (e.g., [6], see review in [7]) or their combinations. These characteristics inducing surface slipperiness are particularly developed in insect-trapping plants [4,8–10]. Trichomes are hair-like protuberances extending from the epidermis, which characterize the surface of many plant species. They can be very different in shape (straight, hooked, branched or unbranched), size, orientation, structure and secretory function [11,12]. Leaf pubescence can play an important role against herbivorous insects [11,13–17]. For instance, the ability of the hooked trichomes of *Phaseolus* plants (Fabaceae) to entrap and kill arthropod species belonging to different insect orders is well known (e.g., [18]; see review in [19]. Another example is the glabrous morph of *Arabidopsis lyrata* (L.) O'Kane and Al-Shehbaz (Brassicaceae) which is subjected to higher damage by phytophagous insects than the morph with trichomes [20]. It has been observed that many plant species respond to insect feeding damage by producing new leaves with an increased density and/or increased number of trichomes (see review in [21]. The effect of leaf pubescence may be different according to the herbivore species [11,22], the trichome type (glandular, with the function to secrete metabolites for the plant or non-glandular) as well as trichomes length and distribution [23]. In general, long-legged sucking insects such as Hemiptera tend to be less deterred by trichomes, and sometimes a positive effect of trichomes in enhancing insect locomotion has been described. This is the case of the generalist stinkbug *Nezara viridula* (L.) (Hemiptera: Pentatomidae), which showed a stronger attachment force on the non-glandular stellate trichomes of *Solanum melongena* L. (Solanaceae) than on smooth glass [24], probably due to the use of claws interlocking to the trichomes and by this improving attachment during pulling. A very good performance on hairy plant surfaces was demonstrated also by the omnivorous mirid bug *Dicyphus errans* (Wolff) (Hemiptera: Miridae) [12] that typically lives on pubescent plants and shows morphological and behavioural adaptations to hairy plants resulting in higher predation, fecundity, and attachment ability, if compared to plants without trichomes. On the opposite, trichomes can have a detrimental effect on species showing a high degree of ventral contact with leaf surfaces such as lepidopteran and sawfly larvae [11,13,22,25,26].

Among Coleoptera, in Coccinellidae, the interaction with plant leaf surface is high in all the developmental stages, both in phytophagous and in predatory species feeding on herbivorous insects. Such plant surface characteristics as wettability and roughness can have an important role in the oviposition site selection in some coccinellids, for example, in *Propylea quatuordecimpunctata* (L.) (Coleoptera: Coccinellidae) [27].

The melon ladybird beetle *Chnootriba elaterii* (Rossi, 1794) (Coleoptera: Coccinellidae) is an oligophagous multivoltine species, widespread in Eurasia, representing a serious pest of Cucurbitaceae in Southern Europe as well as in Near East, Middle East and North Africa [28,29]. The larva and adult of this species feed on leaves of pumpkin, sweet gourd, bitter gourd, cucumber, etc.; sometimes, not only leaves but also flowers or even fruits are destroyed, and seedlings of late sowings can be entirely consumed [30]. Cucurbitaceae, with ca. 800 species under 130 genera [31], are among the most economically important plant families. Most of the species are characterized by the presence of trichomes on their leaves [32]. A recent investigation demonstrated that bifid dentate claws of the adult of *C. elaterii* improve insect attachment to plant surfaces covered by trichomes such as that of *Cucurbita moschata* Duchesne ex Poir. (Cucurbitaceae) [33].

In the present study, in order to better understand the mechanical interaction between *C. elaterii* and its host plants, we measured the force exerted by adults and larvae of

C. elaterii on leaves of different Cucurbitaceae species characterised by trichomes having different densities, lengths and diameters. Since eggs are typically laid on the abaxial side of the leaf, we tested the female attachment ability to the abaxial (lower) side of eight plant species.

2. Materials and Methods

2.1. Insects

Adults and larvae of *C. elaterii* were collected in the field around Perugia (Italy) (43.12822, 12.36097) on wild plants of *Ecballium elaterium* (L.) A. Rich (Cucurbitaceae) in September 2021. Insects were kept in a culture room under controlled conditions (25 ± 2 °C, 45 ± 15% RH, photoperiod 14L:10D) inside net cages 30 × 30 × 30 cm (Vermandel, Hulst, The Netherlands) and fed with plants of *Cucumis melo* var. inodorus Ser. (Cucurbitaceae) obtained from seeds. Eggs, larvae, and adults were maintained in separate cages. Only females and mature larvae (fourth stage) were used in the force experiments. Mated females were kept inside Petri dishes on a leaf of each tested plant species in order to have eggs laid on the abaxial side of different plant surfaces.

2.2. Plants

The abaxial (lower) side of the leaf blades of eight plant species from the family Cucurbitaceae was used in this study: watermelon *Citrullus lanatus* (Thunb.) Matsum. and Nakai, melon (cantaloupe) *Cucumis melo* var. cantalupensis Ser., cucumber *Cucumis sativus* L., squash or pumpkin *C. moschata*, winter squash and pumpkin, summer squash *Cucurbita pepo* L., exploding cucumber *Ecballium elaterium*, calabash *Lagenaria siceraria* (Molina) Standl. var. longissima hort., and sponge gourd *Luffa aegyptiaca* Mill. (=*L. cylindrica* M. Roem).

2.3. Scanning Electron Microscopy

Samples of (1) tarsi of *C. elaterii* at the adult and larval stages, (2) the abaxial sides of the tested plant leaves, and (3) eggs of *C. elaterii* laid on the different tested plant species were examined in a scanning electron microscope (SEM) Hitachi S-4800 (Hitachi High-Technologies Corp., Tokyo, Japan) equipped with a Gatan ALTO 2500 cryo-preparation system (Gatan Inc., Abingdon, UK). Plant material for cryo-SEM examination was collected from the living plants. Leaves were cut off, and 1 cm × 1 cm samples were cut out from the middle region of the leaves. Insect tarsi were dissected from insects anesthetized with carbon dioxide.

Samples were attached to a metal holder and frozen in a cryo stage preparation chamber at −140 °C. Frozen samples were sputter coated with gold-palladium (thickness 6 nm) and studied in the frozen condition (−120 °C) in the cryo-SEM at 3 kV accelerating voltage.

Description of trichomes and determination of their types were performed in accordance with [12]. Types of wax structures were identified based on [34]. Morphometrical variables of surface features were measured from digital images using SigmaScan Pro 5 software (SPSS Inc., Chicago, IL, USA). These data are presented in the text as mean ± SD for $n = 10$.

2.4. Light Microscopy

To visualize the real contact area between the insect tarsal attachment devices and the substrate, living females and larvae of *C. elaterii* walking on hydrophilic glass were examined with reflection contrast microscopy (RCM) using an inverted bright-field microscope ZEISS Axio Observer.A1 (Carl Zeiss Microscopy GmbH, Jena, Germany) in combination with a high-speed camera Photron FASTCAM SA1.1 (VKT Video Kommunikation GmbH—Technisches Fernsehen, Pfullingen, Germany) as described earlier [35–37].

2.5. Contact Angle Measurements

The wettability of the plant surfaces used in the experiments was characterized by measuring the contact angles of water (aqua millipore, 1 µL droplet volume) using a high-speed optical contact angle measuring device OCAH 200 (Dataphysics Instruments GmbH, Filderstadt, Germany). Sessile drop method (*C. lanatus*, *C. melo*, *C. sativus* and *C. pepo*) and sessile needle-in drop method (*C. moschata* and *L. siceraria*), as well as the combination of both (*E. elaterium* and *L. aegyptiaca*) followed by a circle or ellipse fitting, were applied. Data are given in the text as mean $\pm$ SD for n = 10 measurements.

2.6. Force Measurements

The friction (traction) force of *C. elaterii* females on the abaxial side of the leaf blades of eight plant species from the family Cucurbitaceae and on hydrophilic glass (water contact angle of 32.49 $\pm$ 4.17° (mean $\pm$ SD)) was measured using a Biopac force tester (Biopac Systems Ltd., Goleta, CA, USA), while a centrifugal force tester was applied to measure the attachment ability of *C. elaterii* larvae to the same surfaces. Prior to the force measurements, adults and larvae were weighed on a micro-balance (Mettler Toledo AG 204 Delta Range, Greifensee, Switzerland). Experimental insects were anesthetized with CO_2 for 60 s, and females were made incapable of flying by gluing their elytra together with a small droplet of melted bee wax. Before starting the experiments, the treated insects were left to recover for 30 min. All the experiments were performed during the daytime at 25 $\pm$ 2 °C temperature and 50 $\pm$ 5% RH.

The Biopac force tester consisted of a force sensor FORT-10 (10 g capacity; World Precision Instruments Inc., Sarasota, FL, USA) connected to a data acquisition unit MP 160 (Biopac Systems Ltd., Goleta, CA, USA). Data were recorded using AcqKnowledge 5.0 software (Biopac Systems Ltd., Goleta, CA, USA). One end of a fishing thread Gel Spun Polyethylene (Berkley Spirit Lake, IA, USA), 0.02 mm in diameter and about 10 cm long, was fixed with a droplet of molten wax to the insect elytra. The insect was attached to the force sensor by means of the thread and was allowed to move on the test substrate in a direction perpendicular to the force sensor (and parallel to the substrate). The force generated by the insect walking on hydrophilic glass and on the abaxial side of the leaves of the different plant species was measured. Insects pulled on the plant leaf in the direction from its proximal to the distal portion. Force–time curves were used to estimate the maximal pulling force produced by tethered running insects. In total, 20 females were tested on each plant species.

The centrifugal force tester [38] consists of a metal drum covered by a substrate disc to be tested. The drum is driven by a computer-controlled motor. Just above the disc, the fiber-optic sensor monitored by the computer is situated. After positioning the insect on the horizontal disc, the centrifuge drum started the rotation at a speed of 50 rev min^{-1} (0.883 rev s^{-1}). The position of the insect on the drum was monitored by using a combination of a focused light beam and a fiber optical sensor. The drum speed continuously increased until the larva lost its hold on the surface under centrifugal force. The rotational speed at contact loss, the last position of *C. elaterii* larva on the drum (radius of rotation), and the larva weight were used to calculate the maximum frictional component of the attachment force. To test the larva on the abaxial side of leaves of different Cucurbitaceae species, a portion of the leaf was fixed using tape to the metal drum of the centrifuge. In order to avoid potential damage of larvae caused by numerous centrifugal experiments, 15 larvae were tested on the hydrophilic glass as a reference surface and on four of the eight plant species, and the other 15 larvae were tested on glass and on four remaining plant species. For each larva, different plant species were tested in random order.

2.7. Statistical Analysis

The normalized friction force (the ratio between the friction force on leaves and the friction force on the glass of each insect individual) of *C. elaterii* females on different leaf surfaces, the normalized friction force obtained in *C. elaterii* larvae, the density of glandular and non-glandular trichomes as well as the length and diameter of non-glandular trichomes in the different plant species were analyzed using one-way analysis of variance (ANOVA), followed by the Tukey unequal N HSD post hoc test for multiple comparisons between means (Statistica 6.0, Statsoft Inc. 2001, Tulsa, OK, USA). Before the parametric analysis, all the data were subjected to Box-Cox transformations in order to reduce data heteroscedasticity [39].

3. Results

3.1. Adult and Larval Attachment Devices

The tarsi in the female of *C. elaterii* are composed of four tarsal segments. The tarsal attachment organs consist of a pair of pretarsal claws and two hairy pads located on the ventral side of the first and second tarsal segments. The pads are covered with numerous "tenent setae" (hairs modified in the distal portion to increase the actual area of attachment to the surface) (Figure 1a). Each claw is bifid with a basal tooth separated from the claw by a deep cleft (Figure 1b,c). This kind of claw is called bifid dentate or bifid appendiculate.

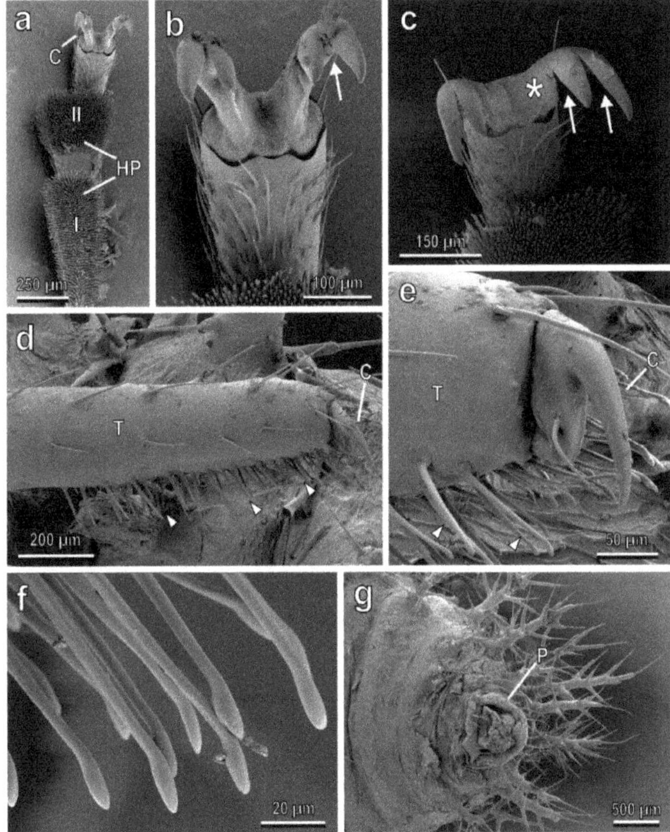

Figure 1. Tarsi of the female (**a–c**) and of the larva (**d–g**) of *Chnootriba elaterii*, cryo-SEM. (**a**) Hairy pads (HP) covered with numerous tenent setae located on the ventral side of the first (I) and second (II) tarsal segments. C, claws. (**b,c**) Details of the bifid claws with a basal tooth (asterisk). Note the deep clefts (arrows). (**d,e**) Single pretarsal claw (C) and tarsal tenent setae (arrowheads) located on the tarsus (T). (**f**) Distal portions of tenent setae. (**g**) Detail of the pygopodium (P) at the end of the abdomen.

The attachment devices in larvae of all four instars are constituted of a single pretarsal claw (Figure 1d,e), tenent setae located on the tarsus (Figure 1d–f) and a pygopodium or "anal organ" at the end of the abdomen (Figure 1g). A widened end plate (spatula) is present in the distal portion of a tenent seta (Figure 1f). Footprints of females walking on hydrophilic glass (Figure 2a,b) reveal that in both the first and second tarsal segments, only distal portions of each pad are in contact with the substrate, whereas larval prints show that both tarsal tenent setae and the pygopodium are responsible for attachment to smooth surfaces (Figure 2c,d).

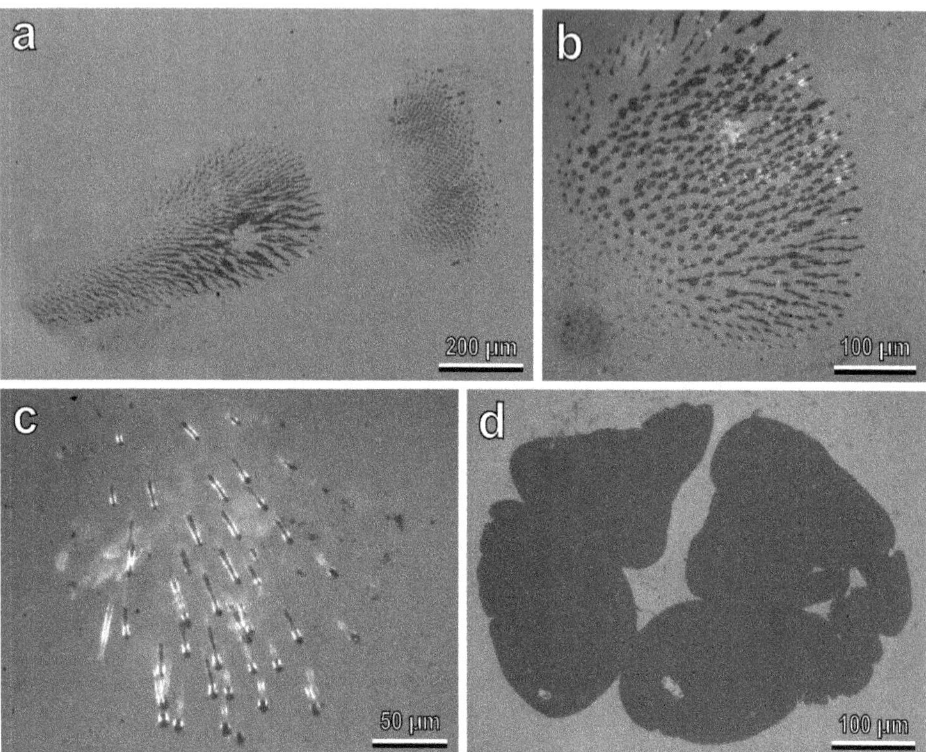

Figure 2. Footprints of females (**a**,**b**) and larvae (**c**,**d**) walking on hydrophilic glass, visualized with reflection contrast microscopy and a high-speed camera. The areas of contact between attachment structures and glass appear dark. (**a**,**b**) Contact area of tenent setae of adhesive pads. (**c**) Contact area of tarsal tenent setae. (**d**) Contact area of pygopodium.

3.2. Characterization of Plant Surfaces

In *C. lanatus*, the abaxial leaf side bears both non-glandular and glandular trichomes, which are regularly distributed over the entire leaf surface (Figure 3a–c). The non-glandular trichomes are multicellular and uniseriate, with big multicellular sockets at the base (Figure 3c,d). They are non-branched, have cone shapes and very sharp tips. These trichomes are bent or curved and accumbent, especially those situated between the leaf veins (Figure 3b,d). On the veins, the non-glandular trichomes are more squarrose being pointed to different directions (Figure 3a,c) and much longer than those between the veins. In the areas between the veins, glandular trichomes are clavate, short-stalked, with the four-celled head region ("characteristic" short-stalked *Cucurbita* type or type I according to [40] (Figure 3e). They are smaller (length: 51.50 ± 10.11 µm) than non-glandular ones and densely cover the surface (abundance: ca. 15 mm^{-2}). The glandular trichomes situated on the veins are capitate, long-stalked, with two- or often also multi(four)-celled heads and a distinct "neck" region. They are much longer (length: 259.66 ± 82.07 µm) than the glandular short-stalked trichomes. With the only difference in the number of head cells (two and more vs. two or three), these trichomes resemble the long-stalked neck-cell type or type II described in

C. pepo var. *styriaca* [40]. Moreover, plentiful stomata are scattered (abundance: up to ca. 500 mm^{-2}) over the leaf surface (Figure 3e).

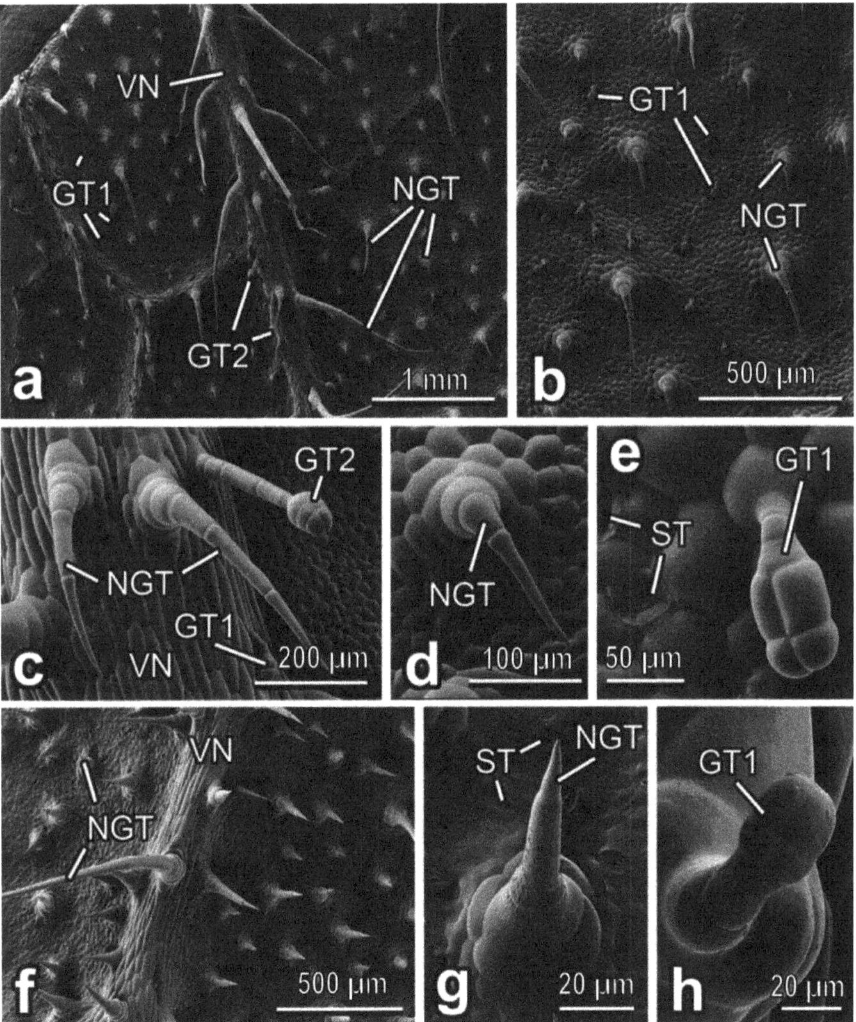

Figure 3. Abaxial leaf side of *Citrullus lanatus* (**a–e**) and *Cucumis melo* (**f–h**), cryo-SEM. (**a,f**) Leaf surface with the prominent vein(s). (**b**) Leaf surface between the prominent veins. (**c**) The vein bearing both non-glandular trichomes and a glandular (long-stalked) one. (**d,g**) Non-glandular trichome from the leaf region between the veins. (**e,h**) Glandular (short-stalked) trichome. Abbreviations: GT1: clavate short-stalked glandular trichome; GT2: capitate long-stalked glandular trichome; NGT: non-glandular trichome; ST: stoma; VN: vein.

The abaxial leaf side in *C. melo* bears numerous, regularly distributed non-glandular trichomes, which are responsible for uniform hairy coverage of the surface (Figure 3f). They are multicellular, uniseriate and sit on rather high multicellular sockets (Figure 3g). These trichomes are non-branched, cone-shaped, with slightly more narrow apical regions and very sharp tips (Figure 3f,g). They are usually straight or gently inclined; on the veins, they are bent and pointed in different directions, being longer here (Figure 3f). The trichome surface, except for the most apical cell, is clearly microstructured with round and ellipsoid knobby protrusions (Figure 3g). Both types of glandular trichomes, such as (1) clavate short stalked, with four-celled heads region (type I [40]) (ca. 70 μm long) (Figure 3h) and (2) capitate long-stalked with multi(four)-celled heads and distinct "neck" region similar to those found in *C. lanatus* (see above) resembling type II [40] (ca. 120 μm long), occur extremely sparsely (almost solitarily). Plenty of stomata (abundance: >300 mm^{-2}) are regularly dispersed over the surface (Figure 3g).

In *C. sativus*, the abaxial side of the leaf blade bears relatively sparse trichome coverage (Figure 4a). Both non-glandular and glandular trichomes of two types are uniformly distributed over the entire leaf surface (Figure 4b,c). Non glandular trichomes are more abundant than glandular ones. These multicellular, uniseriate, conical, non-branched trichomes have rather small multicellular bases and sharp tips (Figure 4b,c). Their surface is microsculptured with plentiful, very prominent rounded/ellipsoid knobby protrusions (Figure 4d). Non-glandular trichomes are straight or slightly bent between the veins but usually bent or curved and pointed in various directions on the veins (Figure 4a–c) Clavate short-stalked glandular trichomes with four-celled heads (type I [40]) are small (length: 50.81 ± 3.55 μm) and tightly accumbent to the surface (Figure 4b,c,e), whereas capitate long-stalked ones with usually four-celled heads and typical "neck" region (similar to those in *C. lanatus* and resembling type II [40]) are much longer (ca. 200–250 μm long), straight or somewhat bent and occur significantly more rare (Figure 4b,c). Stomata (Figure 4c,e) are present in a high number (abundance: >700 mm^{-2}).

The abaxial leaf surface of *C. moschata* is regularly covered by both non-glandular and glandular trichomes situated on both leaf veins and areas between the veins (Figure 4f–h). Whereas plentiful non-glandular trichomes are distributed almost equally on both the veins and between them (Figure 4f–h), scanty glandular trichomes are located mostly in the veins (Figure 4f,h). The non-glandular trichomes are multicellular, uniseriate, with very prominent multicellular sockets, non-branched, cone-shaped, with rather sharp tips and microsculptured surfaces bearing numerous ellipsoid knobby protrusions (Figure 4h,i). These trichomes are either straight or, more often, curved, usually slightly inclined or bent and create relatively uniform and dense leaf pubescence (Figure 4f–h). Glandular clavate, short-stalked trichomes with a four-celled head region (type I [40]) are much smaller (length: 36.27 ± 26.07 μm) than non-glandular ones and occur very sparsely (Figure 4f,h). Numerous (abundance: >250 mm^{-2}) stomata are regularly distributed over the abaxial leaf side (Figure 4j).

The *C. pepo* leaf is noticeably pubescent on its abaxial side, on both the veins and between them (Figure 5a). The uniform hairy coverage is composed of both non-glandular trichomes and glandular ones of two types (Figure 5b,c). Regularly distributed multicellular, uniseriate, non-branched, cone-shaped non-glandular trichomes have multicellular sockets, sharp tips and the surface microsculptured with numerous knobby protrusions of various, but more often elongated shapes (Figure 5d). They are either curved or bent and usually pointed in one preferred direction (Figure 5a–d). Glandular clavate, short-stalked trichomes with four-celled heads (type I [40]) are the smallest among all trichomes (length: 49.88 ± 1.91 μm) (Figure 5e). They are tightly accumbent to the surface and rather sparsely distributed (Figure 5b,c,e). Capitate long-stalked glandular trichomes with typical two-celled heads and slim "necks" (type II [40]) are much longer (length: 191.55 ± 23.48 μm) (Figure 5f) and occur more often than clavate, short-stalked trichomes (Figure 5b,c). Moreover, numerous (abundance: ca. 350 mm^{-2}) stomata are located on the *C. pepo* abaxial leaf surface (Figure 5d,f).

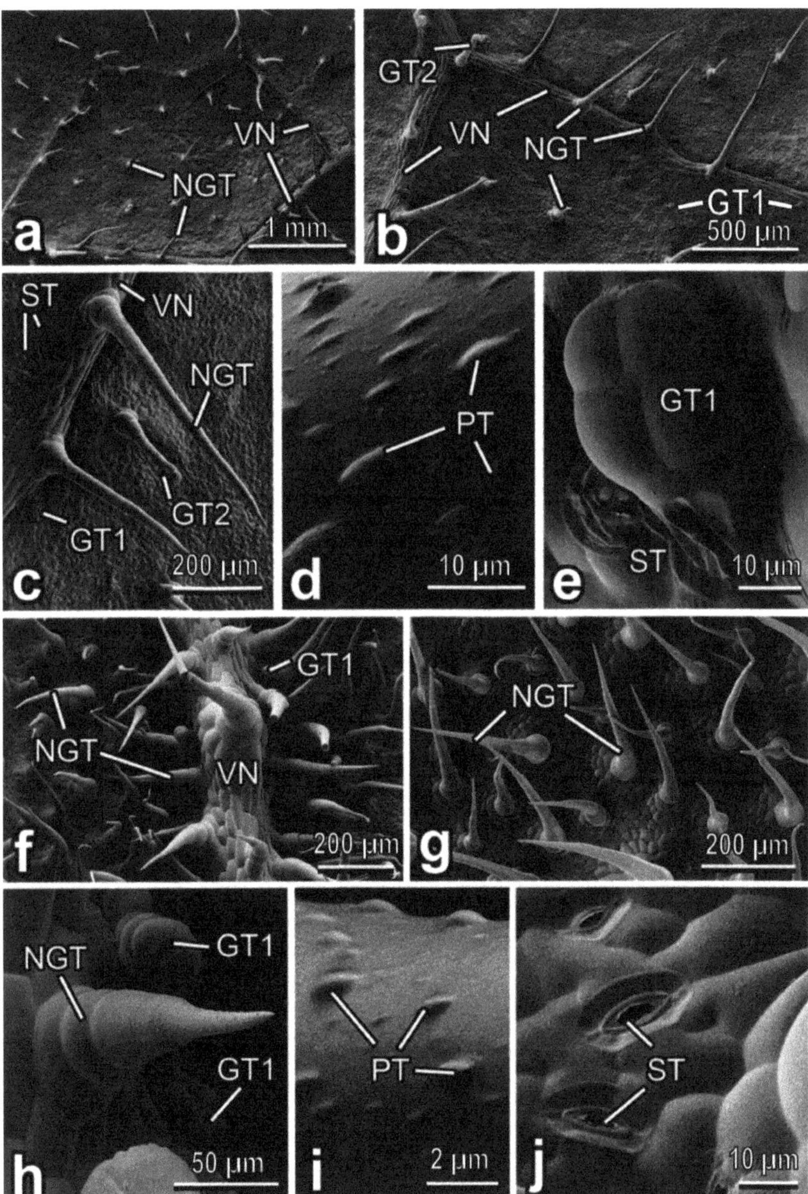

Figure 4. Abaxial leaf side of *Cucumis sativus* (**a–e**) and *Cucurbita moschata* (**f–j**), cryo-SEM. (**a**) General view of the leaf surface. (**b,c,f**) Leaf surface with the vein(s). (**d,i**) Microsculptured surface of the non-glandular trichome. (**e**) Glandular (short-stalked) trichome. (**g**) Leaf surface between the veins. (**h**) The vein bearing both non-glandular and glandular (short-stalked) trichomes. (**j**) Stomata. Abbreviations: GT1: clavate short-stalked glandular trichome; GT2: capitate long-stalked glandular trichome; NGT: non-glandular trichome; PT: protrusion; ST: stoma; VN: vein.

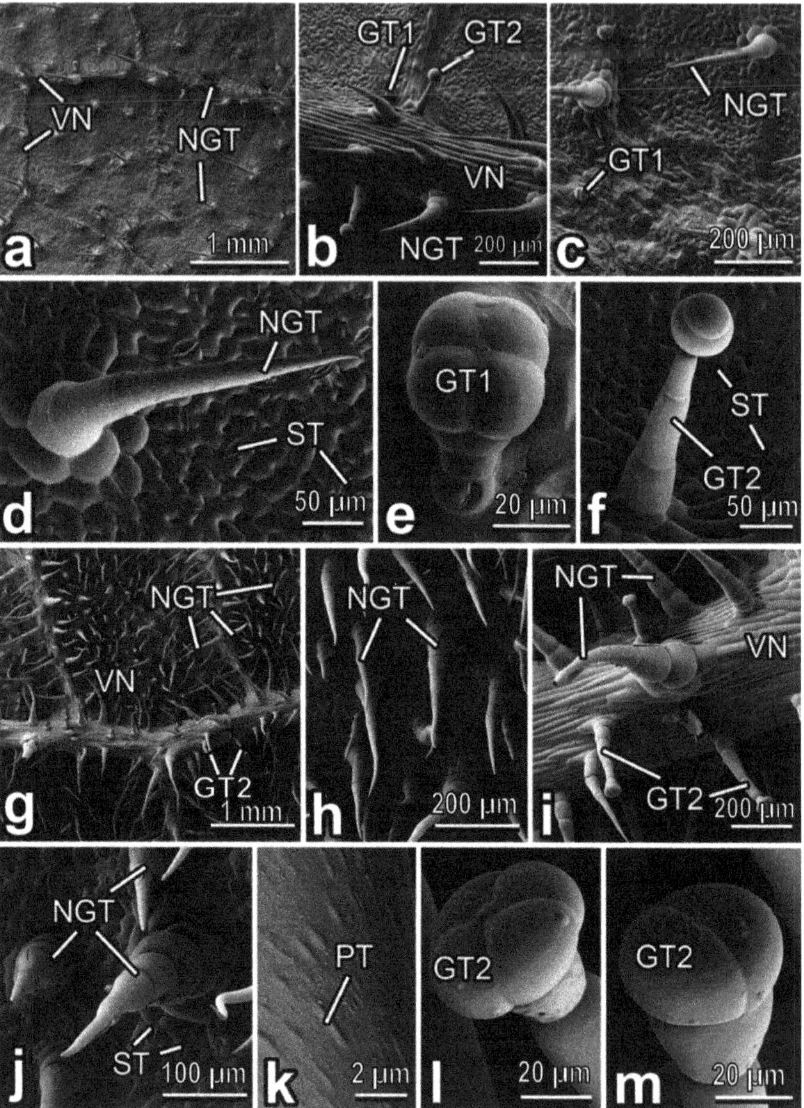

Figure 5. Abaxial leaf side of *Cucurbita pepo* (**a–f**) and *Ecballium elaterium* (**g–m**), cryo-SEM. (**a,g**) General view of the surface. (**b,i**) Leaf surface with the vein. (**c,h**) Leaf surface between the veins. (**d,j**) Non-glandular trichome. (**e**) Glandular (short-stalked) trichome. (**f,l,m**) Glandular (long-stalked) trichome. (**k**) Microsculptured surface of the non-glandular trichome. Abbreviations: GT1: clavate short-stalked glandular trichome; GT2: capitate long-stalked glandular trichome; NGT: non-glandular trichome; PT: protrusion; ST: stoma; VN: vein.

The *E. elaterium* abaxial leaf surface has a wooly appearance because of a trichome coverage created in the main by non-glandular trichomes pointed in different directions (Figure 5g). These densely and uniformly occurring multicellular, uniseriate, non-branched, cone-shaped trichomes have rather low either uni- or multicellular sockets (Figure 5h–j). The cells composing the trichome are much wider in their basal part and become narrower till a very sharp and flexible tip (Figure 5j). Since these cells, except the distal one, have a barrel-like shape, boundaries between the cells are clearly seen. The trichome surface is covered with numerous knobby ellipsoid or elongated protrusions (Figure 5k). On the veins, non-glandular trichomes are large than those located between the veins (Figure 5g). Moreover, glandular trichomes are sparsely located mostly in the veins (Figure 5i). They are capitate, with long multicellular stalks (looking like the central part of non-glandular trichomes in this plant species) (length: 249.83 ± 22.03 μm), distinct "necks", and two- or four-celled heads (Figure 5i,l,m). These trichomes resemble, to some extent, the long-stalked glandular trichomes found in *C. lanatus* (see above) and type II in *C. pepo* var. *styriaca* [40]; however, also typical type II trichomes [40] are also present. Plentiful (abundance: >240 mm^{-2}) stomata are spread over the surface (Figure 5j).

The dense pubescence of the abaxial leaf side in *L. siceraria* consists of numerous non-glandular trichomes distributed rather regularly over the surface, whereas glandular trichomes are located on and near the veins (Figure 6a,b). Non-glandular trichomes are very similar to those found in *E. elaterium*, with usually simple unicellular sockets (Figure 6c). Glandular trichomes of both types, (1) small (length: 51.06 ± 8.61 μm) clavate short-stalked ones with four-celled heads (see the description in *C. lanatus*) and (2) long (length: 118.40 ± 56.20 μm) capitate long-stalked ones with typical both necks and two-celled heads, exactly correspond to types I and II described in *C. pepo* var. *styriaca* [40]. Moreover, stomata in a great number (abundance: ca. 240 mm^{-2}) are present on the surface (Figure 6c).

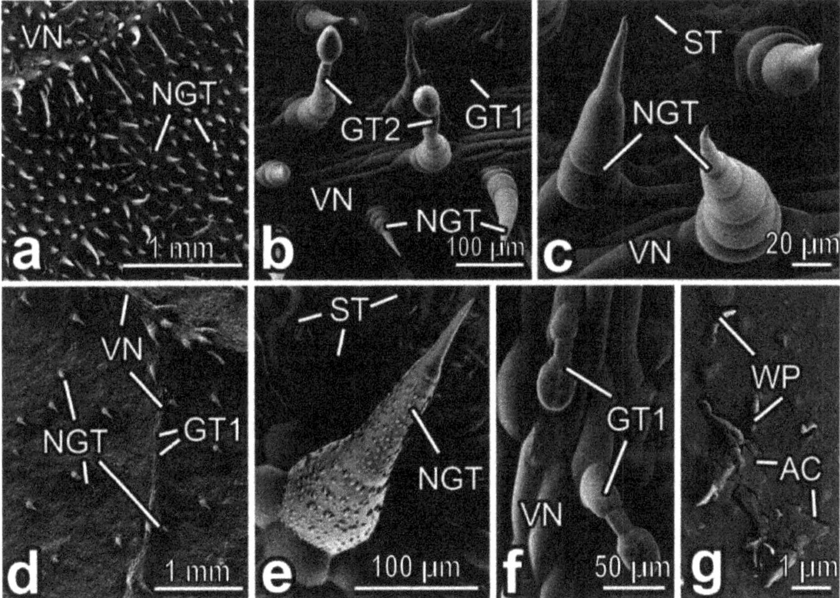

Figure 6. Abaxial leaf side of *Lagenaria siceraria* (**a–c**) and *Luffa aegyptiaca* (**d–g**), cryo-SEM. (**a,d**) General view of the surface. (**b**) Leaf surface with the vein. (**c,e**) Non-glandular trichome(s). (**f**) Glandular (short-stalked) trichomes. (**g**) Epicuticular wax coverage. AC; artificial cracks in the smooth epicuticular wax layer; GT1: clavate short-stalked glandular trichome; GT2: capitate long-stalked glandular trichome; NGT: non-glandular trichome; ST: stoma; VN: vein; WP: epicuticular wax projection.

In *L. aegyptiaca*, the abaxial leaf side bears sparsely but regularly distributed, rather short non-glandular and glandular trichomes (Figure 6d). The non-glandular trichomes are multicellular, uniseriate, non-branched and cone-shaped, with relatively short multicellular sockets and sharp tips (Figure 6e). The microsculpturing (knobby rounded/ellipsoid protrusions) of the trichome surface is very prominent in its basal and central parts but is completely lacking in the distal cell (Figure 6e). Clavate short-stalked glandular trichomes (Figure 6f) are somewhat shorter (length: 79.91 ± 13.09 µm) than non-glandular ones. The heads are mostly four-celled (typical type I [40]) (Figure 6f), while also multicellular heads are present in very few cases. Numerous (abundance: ca. 200 mm^{-2}) stomata are regularly located on the surface between the trichomes (Figure 6e). The cuticle is noticeably covered with a smooth layer of epicuticular wax (detected in SEM due to artificial cracking [40]) and bears solitarily dispersed, microscopic (length: 0.72 ± 0.42 µm), scale-like epicuticular wax projections (Figure 6g).

Data on the density of non-glandular and glandular trichomes and of the length and diameter of the non-glandular trichomes for the studied plant species are presented in Figure 7. In particular, non-glandular trichomes density is significantly higher in both *L. siceraria* and *C. moschata* than in the other species. The latter species show a progressive decrease of non-glandular trichomes density in the following order: *E. elaterium*, *C. melo*, *C. lanatus*, *C. pepo*, *C. sativus*, *L. cylindrica* (Figure 7a) ($F = 383.51$; d.f. $= 7, 23$; $p < 0.0001$). Glandular trichomes density is the highest in *C. lanatus* and the lowest in *E. elaterium*, while it is intermediate in the other plant species (Figure 7b) ($F = 42.84$; d.f. $= 7, 23$; $p < 0.0001$). Non-glandular trichomes are significantly longer in *C. pepo*, shorter in both *C. melo* and *L. siceraria*, and have an intermediate length in *E. elaterium*, *C. moschata*, *C. sativus*, *C. lanatus* and *L. cylindrical* (Figure 7c) ($F = 12.08$; d.f. $= 7, 85$; $p < 0.0001$). Non-glandular trichome diameter is the largest in *L. cylindrica* and the smallest in *L. siceraria*. In the remaining plant species, it is intermediate (Figure 7d) ($F = 13.64$; d.f. $= 7, 85$; $p < 0.0001$).

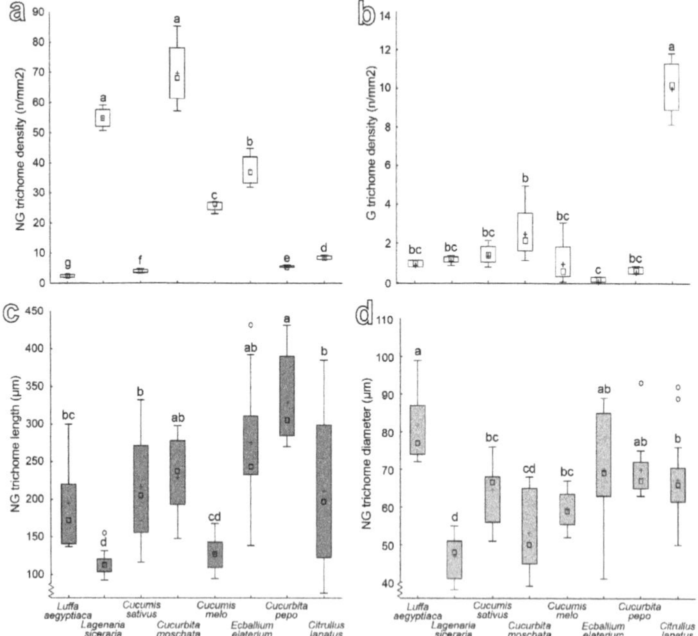

Figure 7. Density of non-glandular (**a**) and glandular trichomes (**b**) and length (**c**) and diameter (**d**) of non-glandular trichomes in the different Cucurbitaceae plant species. Boxplots show the interquartile range and the medians; whiskers indicate 1.5 × interquartile range, "°" is outlier and "+" shows the mean. Boxplots with different letters are significantly different at $p < 0.05$ (Tukey unequal N HSD post hoc test, one-way ANOVA).

All plant surfaces studied were relatively unwettable (contact angle >90°) by water, being either hydrophobic in *C. sativus* (99.5 ± 9.4°), *C. pepo* (99.8 ± 6.0°), *C. lanatus*

(102.6 ± 4.8°), *C. melo* (103.5 ± 4.7°) and *L. aegyptiaca* (119.4 ± 12.0°) or even highly hydrophobic in *E. elaterium* (135.4 ± 9.6°), *L. siceraria* (143.6 ± 7.6°) and *C. moschata* (147.5 ± 5.1°).

3.3. Adult and Larval Attachment Ability to Different Cucurbitaceae Species

The normalized friction force produced by females on leaves of the different plant species is similar on all tested species except for *L. siceraria*, where it is significantly reduced (Figure 8a) (F = 7.84; d.f. = 7, 153; p < 0.0001).

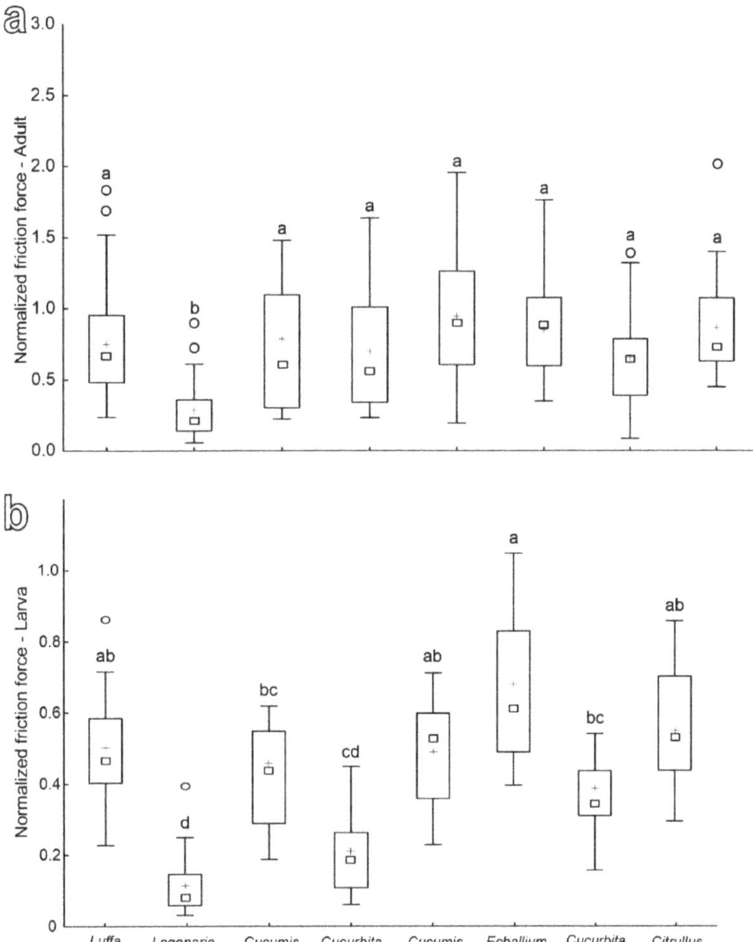

Figure 8. Normalized friction force of females (**a**) and larvae (**b**) of *Chnootriba elaterii* on the abaxial leaf side of different Cucurbitaceae species. Boxplots show the interquartile range and the medians, whiskers indicate 1.5× interquartile range, "°" show the outliers and "+" show the means. Boxplots with different letters are significantly different at p < 0.05 (Tukey unequal N HSD post hoc test, one-way ANOVA).

The values of the normalized friction force registered in larvae on different plant species are the highest on *E. elaterium*, *L. aegyptiaca*, *C. melo* and *C. lanatus* leaves, the lowest on both *L. siceraria* and *C. moschata* and intermediate on the other tested plant species (*C. sativus* and *C. pepo*) (Figure 8b) (F = 12.01; d.f. = 7, 111; p < 0.0001).

The mean friction force of females on hydrophilic glass is 9.77 ± 0.56 mN, while that of larvae reaches 30.59 ± 1.61 mN. As for the mean safety factors on hydrophilic glass, it is 14.73 ± 1.07 in females and rises to 57.48 ± 3.08 in larvae.

3.4. Egg-plant Surface Interaction

The cryo-SEM analysis of the interface between the base of *C. elaterii* eggs and the abaxial side of leaves of different plant species (Figure 9) shows that in the plant species with a dense pubescence such as *L. siceraria* (Figure 9a,b), the glue cannot reach the leaf surface and the trichomes keep the egg away from the leaf surface. In plants with a low trichomes abundance, such as *C. sativus*, the glue readily spreads over the leaf surface (Figure 9c,d).

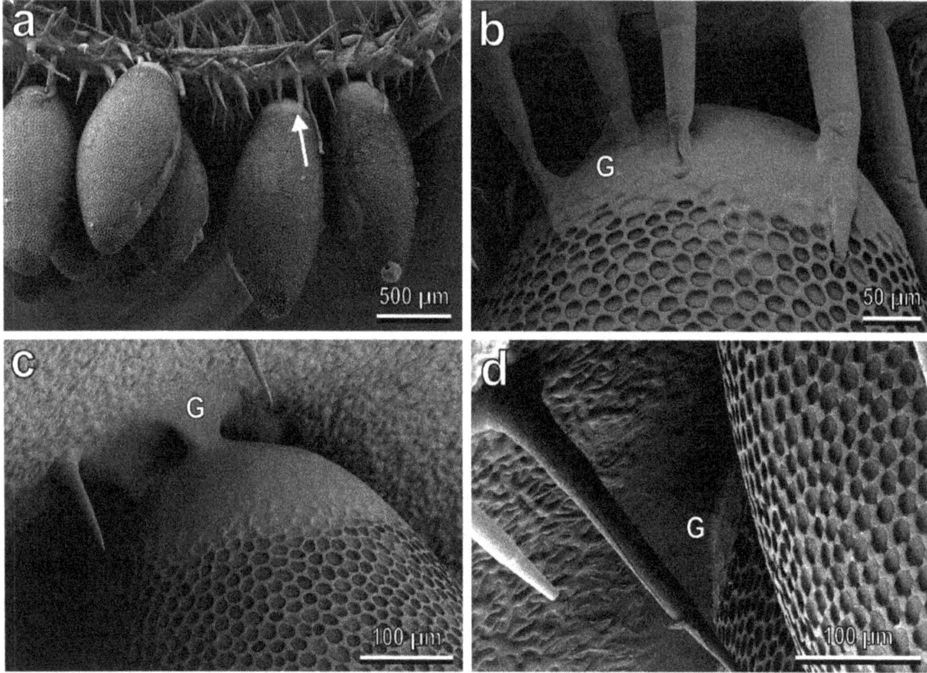

Figure 9. *Chnootriba elaterii* eggs laid on the abaxial side of *Lagenaria siceraria* (**a,b**) and *Cucumis sativus* (**c,d**) leaves, cryo-SEM. (**a,b**) Densely situated trichomes (arrow) do not allow the egg glue (G) to reach the leaf surface. (**c,d**) The egg glue (G) adheres well to the leaf surface, thus replicating the leaf surface microsculpture.

4. Discussion

Leaves of plant species belonging to Cucurbitaceae are covered by glandular and non-glandular trichomes of different shapes and sizes, whose features can be important for taxonomic purposes [41]. The data collected in the present investigation showed that glandular trichomes have much lower density compared with non-glandular trichomes in all tested plant species except *C. lanatus*, where the density of glandular trichomes is considerably higher than in other tested plants and similar to that of non-glandular ones in this species. However, a high number of glandular trichomes in *C. lanatus* seems not to affect the attachment ability of *C. elaterii* at both adult and larval stages: our friction force experiments showed that the insect attachment ability to leaves of this plant was rather good and not significantly lower than to leaves of other species. This result differs from what was observed for predatory coccinellids, such as *Hippodamia convergens* Guérin-Méneville

(Coleoptera: Coccinellidae), in which all instars were affected by the sticky exudates of leaf trichomes in tobacco cultivars [42]. This aspect could be related to (1) different amounts or different chemical compositions of the glandular secretion in these two plant species or (2) special adaptation of *C. elaterii* to the life in the vicinity of glandular trichomes of its host plants. For example, the cuticle of the mirid bug *Pameridea roridulae* Reuter (Heteroptera: Miridae) living on the plant *Roridula gorgonias* Planch. (Roridulaceae), which is rich in adhesive glandular trichomes, bears an epicuticular greasy secretion preventing the formation of contacts between the sticky plant secretion and the solid insect cuticle [43]. Further investigations on *C. elaterii* are needed to clarify this issue.

The attachment forces of the *C. elaterii* adults to different tested plant species were always lower than those recorded on hydrophilic glass. This result can be explained not only by the presence of trichomes, which did not allow the tenent setae of the hairy pads to reach the leaf surface but also by the rather hydrophobic properties of leaves in all tested Cucurbitaceae species. Indeed, for many insect species at both adult and larval stages, a higher attachment ability on hydrophilic smooth surfaces in comparison with smooth hydrophobic ones has been previously observed [27,44–55].

As for the influence of non-glandular trichomes on the mechanical interaction between the ladybird and the plants, female attachment ability was similar on all the tested plants except *L. siceraria*. The attachment to the leaf of this plant was particularly challenging for both the adults and larvae of *C. elaterii*. This can be explained by the combination of a high density of trichomes and their small length and diameter. Moreover, the cells surrounding the bases of the trichome shafts in *L. siceraria* are not enriched by silicon or calcium, differently from other Cucurbitaceae, such as *C. sativus*, *C. melo*, and *C. lanatus*, which show silicon or calcium in this area [56]. For this reason, *L. siceraria* trichomes tend to bend at the base [56]. The above features are responsible for a "velvet like appearance" of the leaf surface in this plant species, which can cause difficulty for an insect when it tries to adhere to this leaf. Our results are in agreement with the data of the previous study testing the friction force of adults in different Coleoptera species, such as *C. elaterii*, *Harmonia axyridis* Pallas (Coccinellidae) and *Chrysolina herbacea* Duftschmid (Chrysomelidae) on a *C. moschata* leaf and its resin replica having higher trichome stiffness [33]. Those experiments revealed that insect attachment ability increased with an increase in trichome stiffness.

Recent investigations on the chemical interaction between *C. elaterii* and the same Cucurbitaceae plant species tested here showed that this insect can perceive the volatile chemicals (VOCs) produced by *L. siceraria*. The electroantennographic study showed that the antennae of *C. elaterii* adults clearly responded to the extract of this plant [57]. At the same time, if females could choose between leaves of *L. siceraria* and other Cucurbitaceae, such as *C. melo*, *C. sativus*, *C. pepo*, *C. moschata* and *C. lanatus*, they never chose *L. siceraria* [58]. Such behavior could be caused by plant chemical defenses (we cannot exclude the presence of gustatory cues on the leaf reducing adult attraction) but also by plant mechanical defenses represented by short, dense and flexible trichomes. Indeed, when adults choose a place for oviposition, plant mechanical features can be relevant. In this regard, it is important to note that a good attachment to the leaf surface is a fundamental prerequisite for oviposition behavior since Coccinellidae species typically lay their eggs upside down on the abaxial leaf surface. In agreement with this, in the whitefly predator ladybird *Serangium japonicum* Chapin (Coleoptera: Coccinellidae), a positive correlation between the female oviposition preference and the attachment force of adults on the plant leaf has been recently demonstrated [59].

A firm attachment of an egg to a plant surface is fundamental for larval survival in a phytophagous species. Leaf trichomes can reduce coccinellid egg adhesion compared to smooth leaf surfaces, as demonstrated for the eggs of *H. axyridis* and *P. quatuordecimpunctata* laid on the stellate trichomes of *S. melongena* [27]. Our cryo-SEM observations revealed that the dense trichome coverage on the leaf of *L. siceraria* reduced the contact of the egg glue with the epidermal leaf surface, thus highly affecting the egg attachment ability to the leaf compared with plants having a rather low density of trichomes, like *C. sativus*, where the egg glue can reach the leaf surface. These results highlight the importance of plant mechanical barriers, which can have a strong impact on shaping the functional composition of the herbivore assemblage [60].

In our experiments, females of *C. elaterii* performed well on all the tested species (except *L. siceraria*), even on *C. moschata* showing a high density of trichomes that are longer, wider and stiffer than those in *L. siceraria*. Such an ability to attach to hairy plant surfaces can be related to the presence of bifid dentate claws with deep clefts, which can interlock with trichomes, as recently demonstrated [33]. In agreement with this, our data demonstrated that the larvae of *C. elaterii* having a single claw cannot adhere to *C. moschata* as well as the adults: larval attachment ability to *C. moschata* was low and not significantly different from that recorded on *L. siceraria*. For the larvae, a dense layer of trichomes (long or short, stiff or flexible) is challenging in any case. This observation confirms that the trichome effect can differ according to the insect developing stage, as already shown in lepidopterous defoliators, where the pubescence of soybean functions as a resistance mechanism to the larval stage, but enhances adult oviposition when compared to plants without pubescence [61].

In conclusion, the data presented here deepen the knowledge of the mechanical interaction of *C. elaterii* at different developmental stages with different Cucurbitaceae species. Such data, combined with the information on the chemical and visual cues involved in the host plant choice, can help to unravel the complex factors driving the coevolution between an oligophagous insect and its host plant species. Moreover, identifying the chemical cues involved in the host plant selection and the morphological and leaf traits of Cucurbitaceae associated with their resistance against phytophagous insects can help to develop environmentally friendly pest control methods.

Author Contributions: Conceptualization, all the authors; methodology, all the authors; formal analysis, G.S. and S.G.; investigation, S.G., E.G., V.S., G.S., S.P. and M.R. writing—original draft preparation, M.R., E.G. and V.S.; writing—review and editing, all the authors; supervision, S.G. and G.S. All authors have read and agreed to the published version of the manuscript.

Funding: This research received no external funding.

Data Availability Statement: The data presented in this study are openly available in Mendeley Data at https://data.mendeley.com/datasets/mnpv3mmftz/1 (accessed on 20 October 2022).

Acknowledgments: We are very grateful to Chiara Brozzi and Valentina Marciano for their help in data collection.

Conflicts of Interest: The authors declare no conflict of interest.

References

1. Beutel, R.; Gorb, S. Ultrastructure of attachment specializations of hexapods (Arthropoda): Evolutionary patterns inferred from a revised ordinal phylogeny. *J. Zool. Syst. Evol. Res.* **2001**, *39*, 117–207. [CrossRef]
2. Gorb, E.; Gorb, S.N. Anti-adhesive surfaces in plants and their biomimetic potential. In *Materials Design Inspired by Nature: Function through Inner Architecture*; Royal Society of Chemistry: Cambridge, UK, 2013; pp. 282–309.
3. Prüm, B.; Florian Bohn, H.; Seidel, R.; Rubach, S.; Speck, T. Plant surfaces with cuticular folds and their replicas: Influence of microstructuring and surface chemistry on the attachment of a leaf beetle. *Acta Biomater.* **2013**, *9*, 6360–6368. [CrossRef] [PubMed]
4. Bauer, U.; Bohn, H.F.; Federle, W. Harmless nectar source or deadly trap: *Nepenthes pitchers* are activated by rain, condensation and nectar. *Proc. R. Soc. B Biol. Sci.* **2008**, *275*, 259–265. [CrossRef]
5. Prüm, B.; Seidel, R.; Bohn, H. Plant surfaces with cuticular folds are slippery for beetles. *J. R. Soc. Interface* **2012**, *9*, 127–135. [CrossRef]
6. Stork, N.E. Role of waxblooms in preventing attachment to brassicas by the mustard beetle, *Phaedon cochleariae. Entomol. Exp. Appl.* **1980**, *28*, 100–107. [CrossRef]
7. Gorb, E.; Gorb, S.N. Anti-adhesive effects of plant wax coverage on insect attachment. *J. Exp. Bot.* **2017**, *68*, 5323–5337. [CrossRef]
8. Daltorio, K.; Horchler, A.; Gorb, S.; Ritzmann, R.; Quinn, R. A Small Wall-Walking Robot with Compliant, Adhesive Feet. In Proceedings of the 2005 IEEE/RSJ International Conference on Intelligent Robots and Systems, Edmonton, AB, Canada, 2–6 August 2005; pp. 3648–3653.

9. Oelschlägel, B.; Gorb, S.; Wanke, S.; Neinhuis, C.; Oelschlä, B. Structure and biomechanics of trapping flower trichomes and their role in the pollination biology of *Aristolochia plants* (Aristolochiaceae). *New Phytol.* **2009**, *184*, 988–1002. [CrossRef]
10. Poppinga, S.; Koch, K.; Bohn, H.; Barthlott, W. Comparative and functional morphology of hierarchically structured anti-adhesive surfaces in carnivorous plants and kettle trap flowers. *Funct. Plant Biol.* **2010**, *37*, 952–961. [CrossRef]
11. Southwood, S. Plant surfaces and insects-an overview. *Insects Plant Surf.* **1986**, *6*, 1–22.
12. Voigt, D.; Gorb, E.; Gorb, S. Plant surface–bug interactions: *Dicyphus errans* stalking along trichomes. *Arthropod. Plant. Interact.* **2007**, *1*, 221–243. [CrossRef]
13. Levin, D.A. The Role of Trichomes in Plant Defense. *Q. Rev. Biol.* **1973**, *48*, 3–15. [CrossRef]
14. Karban, R.; Baldwin, I.T. *Induced Responses to Herbivory*; University of Chicago Press: Chicago, IL, USA, 1997.
15. Baur, R.; Binder, S.; Benz, G. Nonglandular leaf trichomes as short-term inducible defense of the grey alder, *Alnus incana* (L.), against the chrysomelid beetle, *Agelastica alni* L. *Oecologia* **1991**, *87*, 219–226. [CrossRef]
16. Fordyce, J.A.; Agrawal, A.A. The role of plant trichomes and caterpillar group size on growth and defence of the pipevine swallowtail *Battus philenor*. *J. Anim. Ecol.* **2001**, *70*, 997–1005. [CrossRef]
17. Traw, B.; Dawson, T. Differential induction of trichomes by three herbivores of black mustard. *Oecologia* **2002**, *131*, 526–532. [CrossRef]
18. Johnson, B. The injurious effects of the hooked epidermal hairs of French beans (*Phaseolus vulgaris* L.) on *Aphis craccivora* Koch. *Bull. Entomol. Res.* **1953**, *44*, 779–788. [CrossRef]
19. Rebora, M.; Salerno, G.; Piersanti, S.; Gorb, E.; Gorb, S.N. Entrapment of Bradysia paupera (Diptera: Sciaridae) by *Phaseolus vulgaris* (Fabaceae) plant leaf. *Arthropod. Plant. Interact.* **2020**, *14*, 499–509. [CrossRef]
20. Løe, G.; Toräng, P.; Gaudeul, M.; Ågren, J. Trichome production and spatiotemporal variation in herbivory in the perennial herb *Arabidopsis lyrata*. *Oikos* **2007**, *116*, 134–142. [CrossRef]
21. Dalin, P.; Ågren, J.; Björkman, C.; Huttunen, P.; Kärkkäinen, K. Leaf trichome formation and plant resistance to herbivory. In *Induced Plant Resistance to Herbivory*; Springer: Dordrecht, The Netherlands, 2008; pp. 89–105.
22. Hare, J.; Elle, E. Variable impact of diverse insect herbivores on dimorphic *Datura wrightii*. *Ecology* **2002**, *83*, 2711–2720. [CrossRef]
23. Andres, M.R.; Connor, E.F. The community-wide and guild-specific effects of pubescence on the folivorous insects of manzanitas Arctostaphylos spp. *Ecol. Entomol.* **2003**, *28*, 383–396. [CrossRef]
24. Salerno, G.; Rebora, M.; Gorb, E.; Gorb, S. Attachment ability of the polyphagous bug *Nezara viridula* (Heteroptera: Pentatomidae) to different host plant surfaces. *Sci. Rep.* **2018**, *8*, 1–14. [CrossRef]
25. Haddad, N.; Hicks, W. Host Pubescence and the Behavior and Performance of the Butterfly *Papilio Troilus* (Lepidoptera: Papilionidae). *Environ. Entomol.* **2000**, *29*, 299–303. [CrossRef]
26. Lill, J.; Marquis, R.; Forkner, R.; Le Corff, J.; Holmberg, N.; Barber, N. Leaf pubescence affects distribution and abundance of generalist slug caterpillars (Lepidoptera: Limacodidae). *Environ. Entomol.* **2006**, *35*, 797–806. [CrossRef]
27. Salerno, G.; Rebora, M.; Piersanti, S.; Buescher, T.H.; Gorb, E.; Gorb, S.N. Oviposition site selection and attachment ability of *Propylea quatuordecimpunctata* and *Harmonia axyridis* from the egg to the adult stage. *Physiol. Entomol.* **2022**, *47*, 20–37. [CrossRef]
28. Liotta, G. Contributo alla conoscenza della biologia dell' *Epilachna chrysomelina* F. Sicilia (Col. Coccinellidae). *Boll. Inst. Entomol. Agr. Osser. Fitopathol. Palermo* **1964**, *5*, 235–262.
29. Akandeh, M.; Shishehbor, P. Life history traits of melon ladybeetle, *Epilachna chrysomelina* (Col.: Coccinellidae), on four host plant species. *J. Entomol. Soc. Iran* **2011**, *31*, 17–27.
30. Al-Digail, S.A.; Assagaf, A.I.; Mahyoub, J.A. Effect of Temperature and Humidity on the Population Abundance of Spotted Oriental Cucumber Beetle *Epilachna chrysomelina* (F.) (Coccinellidae: Coleoptera) In Al–Qunfudah Western Saudi Arabia. *Curr. World Environ.* **2012**, *7*, 7–12. [CrossRef]
31. Jeffrey, C. A new system of Cucurbitaceae. *Bot. Zhurn.* **2005**, *90*, 332–335.
32. Ali, M.; Al-Hemaid, F. Taxonomic significance of trichomes micromorphology in cucurbits. *Saudi J. Biol. Sci.* **2011**, *18*, 87–92. [CrossRef]
33. Salerno, G.; Rebora, M.; Piersanti, S.; Saitta, V.; Gorb, E.V.; Gorb, S.N. Coleoptera claws and trichome interlocking. *J. Comp. Physiol. A* **2022**, *208*, 1–14. [CrossRef]
34. Barthlott, W.; Neinhuis, C.; Cutler, D.; Ditsch, F.; Meusel, I.; Theisen, I.; Wilhelmi, H. Classification and terminology of plant epicuticular waxes. *Bot. J. Linn. Soc.* **1998**, *126*, 237–260. [CrossRef]
35. Ploem, J.S. Reflection-contrast microscopy as a tool for investigation of the attachment of living cells to a glass surface. In *Mononuclear Phagocytes in Immunity, Infection and Pathology*; Blackwell: Oxford, UK, 1975; pp. 405–421.
36. Federle, W.; Riehle, M.; Curtis, A. An integrative study of insect adhesion: Mechanics and wet adhesion of pretarsal pads in ants. *Integr. Comp. Biol.* **2002**, *42*, 1100–1106. [CrossRef] [PubMed]
37. Heepe, L.; Kovalev, A.; Gorb, S. Direct observation of microcavitation in underwater adhesion of mushroom-shaped adhesive microstructure. *Beilstein J. Nanotechnol.* **2014**, *5*, 903–909. [CrossRef] [PubMed]
38. Gorb, S.; Gorb, E.; Kastner, V. Scale effects on the attachment pads and friction forces in syrphid flies (Diptera, Syrphidae). *J. Exp. Biol.* **2001**, *204*, 1421–1431. [CrossRef] [PubMed]
39. Sokal, R.R.; Rohlf, F.J. *Biometry: The Principles and Practice of Statistics in Biological Research*; W. H. Freeman: New York, NY, USA, 1998.
40. Kolb, D.; Müller, A. Light, Conventional and Environmental Scanning Electron Microscopy of the Trichomes of *Cucurbita pepo* subsp. pepo var. styriaca and Histochemistry of Glandular secretory products. *Ann. Bot.* **2004**, *94*, 515–526. [CrossRef] [PubMed]

41. Inamdar, J.; Gangadhara, M.; Shenoy, K.N. Structure, ontogeny, organographic distribution, and taxonomic significance of trichomes and stomata in the Cucurbitaceae. In *Biology and utilization of the Cucurbitaceae*; Bates, D.M., Robinson, R.W., Joffrey, C., Eds.; Cornell University Press Ithaca. New York, NY, USA, 2019; Volume 17, pp. 1–209.

42. Belcher, D.; Thurston, R. Inhibition of movement of larvae of the convergent lady beetle by leaf trichomes of tobacco. *Environ. Entomol.* **1982**, *11*, 91–94. [CrossRef]

43. Voigt, D.; Gorb, S. An insect trap as habitat: Cohesion-failure mechanism prevents adhesion of *Pameridea roridulae* bugs to the sticky surface of the plant *Roridula gorgonias*. *J. Exp. Biol.* **2008**, *211*, 2647–2657. [CrossRef] [PubMed]

44. Gorb, E.; Gorb, S. Effects of surface topography and chemistry of *Rumex obtusifolius* leaves on the attachment of the beetle *Gastrophysa viridula*. *Entomol. Exp. Appl.* **2009**, *130*, 222–228. [CrossRef]

45. Salerno, G.; Rebora, M.; Piersanti, S.; Gorb, E.; Gorb, S. Mechanical ecology of fruit-insect interaction in the adult Mediterranean fruit fly *Ceratitis capitata* (Diptera: Tephritidae). *Zoology* **2020**, *139*, 125748. [CrossRef]

46. Salerno, G.; Rebora, M.; Piersanti, S.; Matsumura, Y.; Gorb, E.V.; Gorb, S.N. Variation of attachment ability of *Nezara viridula* (Hemiptera: Pentatomidae) during nymphal development and adult aging. *J. Insect Physiol.* **2020**, *127*, 104–117. [CrossRef]

47. Rebora, M.; Salerno, G.; Piersanti, S.; Gorb, E.; Gorb, S. Role of Fruit Epicuticular Waxes in Preventing Bactrocera oleae (Diptera: Tephritidae) Attachment in Different Cultivars of *Olea europaea*. *Insects* **2020**, *11*, 189. [CrossRef]

48. Gorb, E.; Gorb, S.N. Attachment ability of females and males of the ladybird beetle *Cryptolaemus montrouzieri* to different artificial surfaces. *J. Insect Physiol.* **2020**, *121*, 104011. [CrossRef] [PubMed]

49. Lüken, D.; Voigt, D.; Gorb, S.; Zebitz, C.P.W. Die Tarsenmorphologie und die Haftfähigkeit des Schwarzen Batatenkäfers *Cylas puncticollis* (Boheman) auf glatten Oberflächen mit unterschiedlichen physiko-chemischen Eigenschaften. *Mitt. Dtsch. Ges. Allg. Angew. Entomol.* **2009**, *17*, 109–113.

50. Gorb, E.V.; Hosoda, N.; Miksch, C.; Gorb, S.N. Slippery pores: Anti-adhesive effect of nanoporous substrates on the beetle attachment system. *J. R. Soc. Interface* **2010**, *7*, 1571–1579. [CrossRef] [PubMed]

51. Hosoda, N.; Gorb, S.N. Underwater locomotion in a terrestrial beetle: Combination of surface de-wetting and capillary forces. *Proc. R. Soc. B Biol. Sci.* **2012**, *279*, 4236–4242. [CrossRef] [PubMed]

52. Voigt, D.; Gorb, S. Attachment ability of sawfly larvae to smooth surfaces. *Arthropod Struct. Dev.* **2012**, *41*, 145–153. [CrossRef]

53. Friedemann, K.; Kunert, G.; Gorb, E.; Gorb, S.; Beutel, R. Attachment forces of pea aphids (*Acyrthosiphon pisum*) on different legume species. *Ecol. Entomol.* **2015**, *40*, 732–740. [CrossRef]

54. Zurek, D.; Gorb, S.; Voigt, D. Locomotion and attachment of leaf beetle larvae *Gastrophysa viridula* (Coleoptera, Chrysomelidae). *Interface Focus* **2015**, *5*, 20140055. [CrossRef]

55. Salerno, G.; Rebora, M.; Gorb, E.; Kovalev, A.; Gorb, S. Attachment ability of the southern green stink bug *Nezara viridula* (Heteroptera: Pentatomidae). *J. Comp. Physiol. A Neuroethol. Sens. Neural Behav. Physiol.* **2017**, *203*, 601–611. [CrossRef]

56. Abe, J. Silicon deposition in leaf trichomes of Cucurbitaceae horticultural plants: A short report. *Am. J. Plant Sci.* **2019**, *10*, 486–490. [CrossRef]

57. Piersanti, S.; Saitta, V.; Rebora, M.; Salerno, G. Olfaction in phytophagous ladybird beetles: Antennal sensilla and sensitivity to volatiles from host plants in *Chnootriba elaterii* (Coleoptera Coccinellidae). *Arthropod. Plant. Interact.* **2022**, *16*, 617–630. [CrossRef]

58. Piersanti, S.; Saitta, V.; Rebora, M.; Salerno, G. Plant selection and development in the phytophagous ladybird *Chnootriba elaterii*. *Physiol. Entomol.* **2022**. submitted.

59. Yao, F.-L.; Lin, S.; Wang, L.-X.; Mei, W.-J.; Monticelli, L.S.; Zheng, Y.; Desneux, N.; He, Y.-X.; Weng, Q.-Y. Oviposition preference and adult performance of the whitefly predator *Serangium japonicum* (Coleoptera: Coccinellidae): Effect of leaf microstructure associated with ladybeetle attachment ability. *Pest Manag. Sci.* **2021**, *77*, 113–125. [CrossRef] [PubMed]

60. Peeters, P. Correlations between leaf structural traits and the densities of herbivorous insect guilds. *Biol. J. Linn. Soc.* **2002**, *77*, 43–65. [CrossRef]

61. Lambert, L.; Beach, R.M.; Kilen, T.C.; Todd, J.W. Soybean pubescence and its influence on larval development and oviposition preference of lepidopterous insects. *Crop Sci.* **1992**, *32*, 463–466. [CrossRef]

Article

Attachment Performance of Stick Insects (Phasmatodea) on Plant Leaves with Different Surface Characteristics

Judith Burack, Stanislav N. Gorb and Thies H. Büscher *

Department of Functional Morphology and Biomechanics, Zoological Institute, Kiel University, 24118 Kiel, Germany
* Correspondence: tbuescher@zoologie.uni-kiel.de

Simple Summary: Herbivorous insects and plants greatly affected each other's evolution due to their close interactions. This resulted in the development of a variety of adaptations on both sides. Through the need for protection against herbivorous insects, surfaces with lower attachment ability evolved in many plants. As a response, the attachment systems of insects have developed numerous specializations. Stick insects (Phasmatodea) have an attachment system, consisting of paired claws, arolium (attachment pad between the claws) and euplantulae (paired attachment pads on the tarsomeres), which is well adapted to different natural surfaces. We used measurements of pull-off and traction force in two species (*Medauroidea extradentata* and *Sungaya inexpectata*) representing the most common microstructures used for attachment within stick insects (nubby and smooth) to quantify the attachment ability of Phasmatodea on natural surfaces. Plant leaves with different surface properties (smooth, trichome-covered, hydrophilic and covered with crystalline waxes) were selected as substrates. Wax-crystal-covered fine-roughness substrates revealed the lowest, whereas strongly structured substrates showed the highest attachment performance among the stick insects studied. Removing the claws of the insects resulted in lower attachment ability on structured substrates. Furthermore, claw removal revealed that the attachment performance of the pads is less reduced by contaminating wax crystals in the species with nubby attachment structures. Long-lasting effects of the leaves on the attachment ability were briefly investigated, but not confirmed.

Abstract: Herbivorous insects and plants exemplify a longstanding antagonistic coevolution, resulting in the development of a variety of adaptations on both sides. Some plant surfaces evolved features that negatively influence the performance of the attachment systems of insects, which adapted accordingly as a response. Stick insects (Phasmatodea) have a well-adapted attachment system with paired claws, pretarsal arolium and tarsal euplantulae. We measured the attachment ability of *Medauroidea extradentata* with smooth surface on the euplantulae and *Sungaya inexpectata* with nubby microstructures of the euplantulae on different plant substrates, and their pull-off and traction forces were determined. These species represent the two most common euplantulae microstructures, which are also the main difference between their respective attachment systems. The measurements were performed on selected plant leaves with different properties (smooth, trichome-covered, hydrophilic and covered with crystalline waxes) representing different types among the high diversity of plant surfaces. Wax-crystal-covered substrates with fine roughness revealed the lowest, whereas strongly structured substrates showed the highest attachment ability of the Phasmatodea species studied. Removal of the claws caused lower attachment due to loss of mechanical interlocking. Interestingly, the two species showed significant differences without claws on wax-crystal-covered leaves, where the individuals with nubby euplantulae revealed stronger attachment. Long-lasting effects of the leaves on the attachment ability were briefly investigated, but not confirmed.

Keywords: adhesion; leaf surface; tarsus morphology; trichomes; surface free energy; mechanoecology; ecomorphology

Citation: Burack, J.; Gorb, S.N.; Büscher, T.H. Attachment Performance of Stick Insects (Phasmatodea) on Plant Leaves with Different Surface Characteristics. *Insects* **2022**, *13*, 952. https://doi.org/10.3390/insects13100952

Academic Editor: Ezio Peri

Received: 3 October 2022
Accepted: 16 October 2022
Published: 19 October 2022

Publisher's Note: MDPI stays neutral with regard to jurisdictional claims in published maps and institutional affiliations.

1. Introduction

Insects and plants have interacted for a long period of time, leading to coevolution that resulted in diverse interactions between them [1–4]. On the plant side, the attraction of pollinators or vectors for seed dispersal plays a role, but also mechanisms for protection against herbivorous insects have evolved [3–5]. The plants' defensive strategies against damage due to herbivory or oviposition vary from chemical defense to structural surface modifications associated with mechanical protection [6,7]. While some insects developed specialized attachment systems, plants responded accordingly with attachment-diminishing strategies [1,7]. Evolutionary novel acquisitions do not necessarily fulfill only one single purpose, but can be a result of different selective pressures. The following characteristics have been found to serve as anti-attachment surface modifications. Some plant surfaces make use of modified cell shape, cell orientation or wet coverage as it can be found in the pitcher rim of some *Nepenthes* species (Nepenthaceae) [4,8]. Trichomes of some plants are also known to reduce the attachment ability of insects to the plant surface by decreasing the actual contact area for the feet. Glandular trichomes are capable of either chemically poisoning the insect or mechanically impeding its movement ability by various secretions [9–11]. However, there are also observations of trichomes increasing the attachment ability of insects by providing an additional 'foothold', e.g., [4,12–14]. Another plant surface characteristic that is also reported to affect insect attachment is cuticular folds [15]. Similar to trichomes, they can either reduce the contact area or increase the potential interlocking sites for claws depending on insect species [13]. A very effective way to decrease attachment ability of insects on plant surfaces are epicuticular wax crystals [4,15,16]. These specialized surface coatings are found, for example, on the leaves of *Eucalyptus* species (Myrtaceae) and presumably cause loss of adhesion through increased roughness and impairment of the attachment system by contamination, wax-dissolving or fluid-adsorption [5,17]. As shown for some insects, wax crystals can contaminate the attachment system at first contact, but they can be subsequently removed during the next footsteps [4,17].

Previous studies on the attachment of insect species on different plant surfaces showed that the surface of plants affects the attachment of insects. For example, the bug *Nezara viridula* (Linnaeus, 1758) (Heteroptera: Pentatomidae) showed a significantly lower attachment ability on leaves covered with cuticular waxes compared to leaves with a trichome coverage [6]. Salerno et al. [18] showed a stronger attachment of two ladybird species onto hydrophilic surfaces compared to hydrophobic ones [18]. Voigt et al. [13] investigated the attachment ability of *Dicyphus errans* (Wolff, 1804) (Heteroptera: Miridae) on the surfaces of six different plants. This bug showed reduced attachment to plant surfaces with wax crystals and strong attachment on plant surfaces with either nonglandular or glandular trichomes. Consequently, most literature sources conclude that the plant-surface type has a significant influence on the insect attachment ability [13,19,20].

The focal insect group in this study is Phasmatodea, which are also known as stick and leaf insects. It is a lineage of insects with more than 3400 known species and worldwide distribution [21,22]. Stick and leaf insects are herbivorous and many camouflage themselves as parts of plants, such as bark, twigs, leaves or moss [23]. Phasmids inhabit most habitats in various tropical and subtropical ecosystems and are distributed from the ground to the canopy within forests [24]. Owing to close coevolution with plants and due to the different conditions of substrates in the respective environments, different adaptations to attachment evolved within Phasmatodea. The ground pattern of phasmid attachment structures always include two types of attachment pads, the pretarsal arolium and the tarsal euplantulae, together with paired claws. [25]. The two pads work in complementary directions: the arolium provides adhesion to the substrate and the euplantulae generate friction due to shear forces when pressed onto the surface. The combination of these pads and the adjustability of the amount of involved euplantulae results in a highly adaptive attachment system [25–30]. While the arolium of Phasmatodea is usually smooth, the euplantulae show a high diversity of microscopic surface structures [25], most likely to adapt best to the corresponding substrate conditions prevailing in different habitats. The most common microstructures are smooth and nubby, whereas other euplantula microstructures are characterized by different pattering and aspect ratios resulting in several potential functions, such as frictional anisotropy, randomization of pattern directionality and coping with water or particular contaminations [25,28,29]. The adhesion and traction of the nubby and smooth euplantulae have already been experimentally tested on artificial surfaces with different roughness. These studies showed that smooth euplantular microstructures provide better attachment on smooth surfaces and nubby euplantular microstructures perform comparatively better on microrough surfaces [26,27,31]. It has been shown that the claws are important for interlocking with rough surfaces and work complementarily with the attachment pads [26,32]. The experimental studies on stick insect attachment so far comprise exclusively artificial substrates to elucidate the basic functionality of the attachment system of these insects. Plants, however, offer various influences on the attachment performance on the one hand, and are the most important substrate for these herbivores.

The aim of this study is to investigate the attachment ability of Phasmatodea on plant surfaces with different characteristics. Specifically, we try to answer, whether there is a difference in attachment ability on natural substrates between two species with different attachment microstructures: *Medauroidea extradentata* (Brunner von Wattenwyl, 1907) (Phasmatidae) with smooth and *Sungaya inexpectata* (Zompro, 1996) (Heteropterygidae) with nubby euplantula microstructure. We used attachment force measurements to compare (I) if the attachment ability is different on the leaves of the four selected plant species, (II) if there are differences between the performances of the two species and (III) whether the performance was influenced by removing the claws. Furthermore, we explored a potential long-lasting contamination of the attachment system by plant surfaces.

2. Materials and Methods

2.1. Species

We used the same two species as used in Büscher and Gorb [26], because of their difference in euplantular microstructure [26]. *Medauroidea extradentata* has a smooth arolium and smooth euplantulae and *Sungaya inexpectata* has nubby euplantulae and a smaller smooth arolium (Figure 1) [25,26]. *Sungaya inexpectata* has an accessory euplantula on the fifth tarsomere (Figure 1A), which is found in some phasmid species [26]. The animals were taken from the laboratory cultures of the Department of Functional Morphology and Biomechanics (Kiel University, Kiel, Germany). The age of the individuals was not further considered, as they were selected by adequate size oriented to the used leaves (mean body mass: *M. extradentata* 144 $\pm$ 69 mg; *S. inexpectata* 332 $\pm$ 174 mg). During the time of the experiment, the animals were kept in adequate boxes: 10–20 animals and fed with hazelnut and blackberry leaves ad libitum. Each individual was checked for intactness of legs and tarsus before the measurements and was replaced in case of damage.

2.2. Experimental Substrates

The leaves of *Epipremnum aureum* (Linden et André) (Alismatales: Araceae), *Tibouchina urvilleana* (Cogniaux) (Myrtales: Melastomataceae), *Hoffmannia ghiesbreghtii* (Lemaire) (Gentianales: Rubiaceae) and *Eucalyptus globulus* (Labillardière) (Myrtales: Myrtaceae) were used as substrates for the attachment of the tested animals. They were chosen because of their surface characteristics and for their suitable leaf size. *Epipremnum aureum* (Figure 2A,B) has leaves with a smooth glossy to sometimes dull lamina, the upper side of the blade [33]. The leaves of *T. urvilleana* (Figure 2C,D) are abundantly covered with nonglandular trichomes [34]. *Hoffmannia ghiesbreghtii* (Figure 2E,F) has leaves that are glabrous on the adaxial side [35] and also very hydrophilic (Figure 2F). *Eucalyptus globulus* leaves (Figure 2G,H) have a smooth surface covered with wax in different amounts depending especially on the age of the leaves and different components [36,37]. A glass plate represented the control substrate. The leaves were collected on the day of use from the Botanical Garden Kiel (Christian-Albrechts-University Kiel, Kiel, Germany). It was always assured that the leaves were intact and in a fresh state.

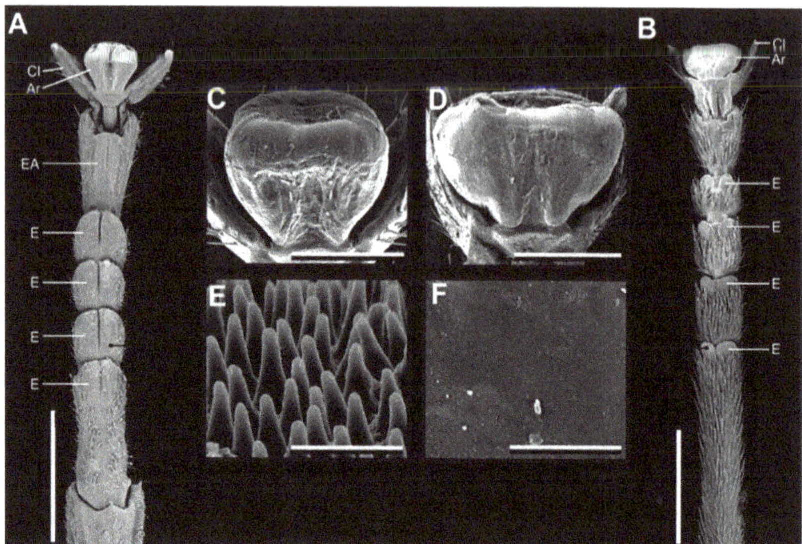

Figure 1. Tarsal morphology. Scanning electron micrographs of the feet of *Sungaya inexpectata* (**A,C,E**) and *Medauroidea extradentata* (**B,D,F**). (**A,B**) Ventral overviews. (**C,D**) Arolia. (**E,F**) microstructure of the euplantulae. Ar, arolium; EA, accessory euplantula (5th euplantula); E, euplantula; Cl, claw. Scale bars: 1 mm (**A,B**), 300 μm (**C,D**), 3 μm (**E,F**). Figure from Büscher and Gorb 2019 [26] reproduced with permission.

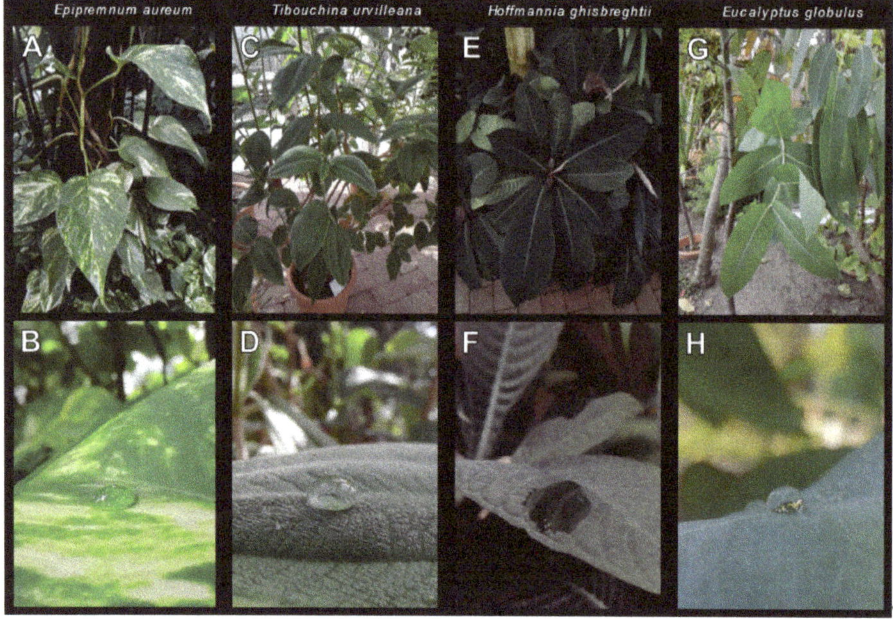

Figure 2. Plant habitus and a closeup of adaxial leaf surface with a waterdrop. (**A,B**) *Epipremnum aureum.* (**C,D**) *Tibouchina urvilleana* (**E,F**) *Hoffmannia ghiesbreghtii.* (**G,H**) *Eucalyptus globulus.* The contact angle of the droplet indicates the hydrophilicity/hydrophobicity of the leaf surface.

2.3. Force Measurement

The force measurement setup (Figure 3) followed the one used by Büscher and Gorb [26] and Büscher et al. [38] for similar measurements. A BIOPAC Model MP100 and a BIOPAC TCI-102 system (BIOPAC Systems, Inc., Goleta, CA, USA) connected to a force transducer (25 g capacity; FORT25, World Precision Instruments Inc., Sarasota, FL, USA) were used for the force measurements. The plant leaves were fixed with double-sided adhesive tape on a glass board adaxial side up and held in a fresh state with a wetted paper towel secured with parafilm during the experiment. Throughout the whole experiment, the temperature ranged between 21 and 24 °C and the ambient humidity between 17 and 52%.

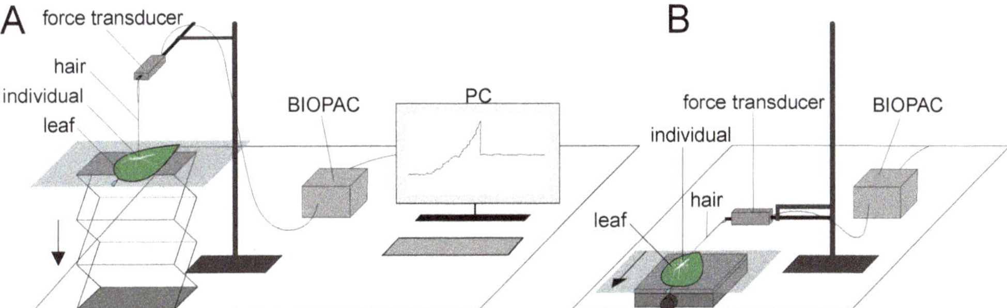

Figure 3. Experimental setup for the measurement of the pull-off force (**A**) and traction force (**B**). For simplification, the PC design is omitted in (**B**). The arrows indicate the direction in which the testing substrates were moved.

The experimental procedure was identical for all species and substrates. First, an insect was randomly chosen, weighed and anesthetized with CO_2. While sedated, a human hair with a loop on one side was attached using a drop of melt wax onto the metanotum. The loop of the hair was then secured onto the force transducer. After full recovery of the stick insect, the force measurement was started. For pull-off force measurements, the force transducer was aligned perpendicular to the substrate and the substrate was constantly lowered with an approximate speed of 0.5–1.0 cm/s through a laboratory scissor jack until the insect completely lost contact to the surface. The traction force was measured by aligning the force transducer horizontal to the substrate. The phasmid was held in position facing the stem of the leaf, while the surface was constantly pulled away from the transducer so that the animal was pulled along its body axis (approx. 200–300 μm/s) until the individual completely stretched the legs and showed loss of grip. The force–time curves were recorded with AcqKnowledge 3.7.0 software (BIOPAC Systems Inc., Goleta, CA, USA). The value of the highest peak of the curves represented the maximum pull-off or traction force ($F_{Ind,Sub}$). For both measurements, the measuring was repeated three times per individual and the mean value was calculated to reduce intraindividual variability. Every phasmid was given sufficient resting time in between the measurements. To generate a value independent of the body mass (m_{Ind}) of the individuals, the safety factor for pull-off and traction force was calculated with the following equation, by normalizing the force by the insect's body mass (1):

$$\text{Safety factor}_{Ind,Sub} = F_{Ind,Sub} \times (m_{Ind} \times g)^{-1} \tag{1}$$

These measurements with no further specifications were done with 15 individuals ($n = 45$, $N = 15$) of each species.

2.3.1. Claw Manipulations

The above measurements were also repeated with 10 individuals of both species with amputated claws for the minimization of potential interlocking between the substrate and the claws. The claws were cut off just above the claw base with fine scissors while the animal was sedated with CO_2 and attention was paid to not injure the arolium or other parts of the attachment system.

2.3.2. Contamination Effects

To find possible long-lasting effects of the plant leaves on the attachment ability of each species, another set of force measurements was performed. In consideration of potential effects due to the wax crystals of *Eu. globulus*, the lasting measurement on those leaves was done with 15 individuals of *M. extradentata* and 15 individuals of *S. inexpectata*. The other substrates (glass, *Ep. aureum*, *T. urvilleana*, *H. ghiesbreghtii*) were tested with 5 individuals of each species. All individuals were prepared as previously described, but were tested with a different order of substrates: first, the force was measured once on glass as a reference, then once on the respective plant substrate (*Ep. aureum*, *T. urvilleana*, *H. ghiesbreghtii* or *Eu. globulus*) and subsequently measured twice on glass. These two last measurements were pooled in the further data processing. This was conducted only once with each individual for pull-off and traction force on each substrate. The glass was wiped with a paper towel after each individual. Based on the obtained values, the safety factors (e.g., SF_{glass1}, $SF_{Ep.aureum}$, SF_{glass2}, SF_{glass3}) were similarly calculated in the other experiments.

An additional control measurement with only glass as substrate was performed for both species.

2.4. Claw Metrics

The size of the claws was measured by dissecting them in five individuals of *M. extradentata* and *S. inexpectata*. The claws were air dried, mounted to aluminum stubs and coated with 10 nm thickness gold–palladium. Claws were studied by using a scanning electron microscope (SEM; Hitachi TM3000, Hitachi High-technologies Corp., Tokyo, Japan) at 15 kV acceleration voltage. The inner claw curvature (r_{in}) and the claw tip sharpness (d_{tip}) were measured with ImageJ [39], following Büscher and Gorb [26].

2.5. Surface Characterization

To characterize the surface of the substrates, the roughness and the height of leaf veins were measured with the optical surface scanner macroscope (VR-3000 Series, Keyence, Osaka, Japan). For this purpose, seven to ten areas on the midvein (*Ep. aureum*, *Eu. globulus*: 10, *T. urvilleana*: 7, *H. ghiesbreghtii*: 8) and ten areas on secondary veins on the adaxial side of one leaf per species were selected (Figure S1A). We measured arithmetic mean roughness (R_a), maximum profile peak height (R_p), maximum profile valley depth (R_v) and maximum roughness (R_z). For *Ep. aureum* and *Eu. globulus*, a high magnification ($40\times$) was used, whereas for *T. urvilleana* and *H. ghiesbreghtii* a lower magnification ($12\times$) was used due to the more pronounced surface geometry. The adaxial side of the leaves was examined for further characterization of the surface structures by using the SEM Hitachi S-4800 (Hitachi High-Technologies Corp., Tokyo, Japan) equipped with a Gatan ALTO 2500 cryopreparation system (Gatan Inc., Cambridge, UK). See Gorb and Gorb [40] for details on the preparation method. Whole mounts of the leaves were sputter-coated with 10 nm gold–palladium while frozen and examined in the cryostage of the microscope at 3 kV acceleration voltage and a temperature of $-120\ ^\circ$C.

2.6. Statistical Analysis

For the statistical analysis, SigmaPlot 12.0 (Systat Software Inc., San José, CA, USA) was used. All results were tested for normal distribution (Shapiro–Wilk test) and, if passed, tested with an Equal Variance Test (Levene). The means of the measurements of the two species on the plant leaves were compared using Kruskal–Wallis one-way analysis of variance (ANOVA) on ranks and Tukey's post hoc test with a significance level of 0.05. The results obtained on the claw-amputated individuals were analyzed using

Kruskal–Wallis ANOVA followed by all pairwise multiple comparison procedures (Dunn's method) within the species or Tukey's post hoc test between the two species. The data of the contamination measurement were tested with one-way ANOVA and all pairwise multiple comparison procedures (Holm–Šídák method) or Kruskal–Wallis ANOVA and Dunn's method accordingly to the distribution.

3. Results

3.1. Force Measurements

The attachment performance of *M. extradentata* and *S. inexpectata* (Figure 4) revealed higher traction force than pull-off force for both species on all substrates.

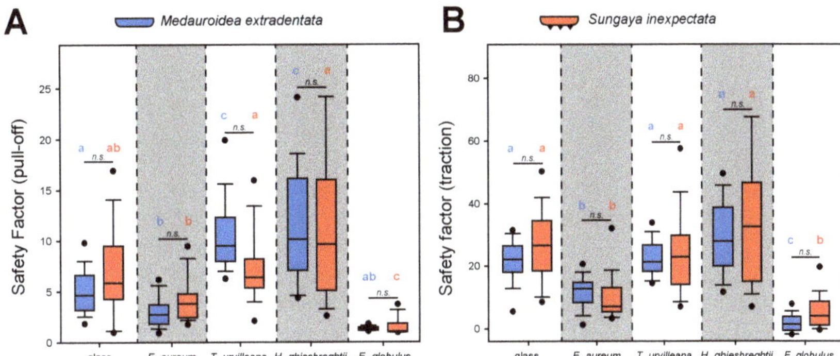

Figure 4. Force measurements on *M. extradentata* (blue) and *S. inexpectata* (red). (**A**,**B**) Safety factors: (**A**) pull-off force, (**B**) traction force. Statistically equal groups within each species are marked with the same letter and between the species with *n.s.* in comparison on the same substrate. Kruskal–Wallis one-way ANOVA on ranks, ($p \leq 0.001$; pull-off: H = 287.832, d.f. = 9, $N = 15$; traction: H = 269.977, d.f. = 9, $N = 45$) and Tukey's test ($p < 0.05$). The boxes indicate the 25th and 75th percentiles, whiskers define the 10th and 90th percentiles and the line within the boxes represents the median. Outliers are shown as individual points.

The pull-off measurement (Figure 4A) showed the highest mean for both phasmid species on *H. ghiesbreghtii* with 11.814 ± 4.840 (mean ± s.d.) for *M. extradentata* and 13.287 ± 13.057 for *S. inexpectata*. The second highest and statistically not different pull-off forces were measured on *T. urvilleana* for both species. For the pull-off measurement on *Ep. aureum* and *Eu. globulus*, mean values of 3.064 ± 1.208 and 1.357 ± 0.220 occurred for *M. extradentata* and mean values of 4.212 ± 1.658 and 1.523 ± 0.827 for *S. inexpectata*. The pull-off force measurement on glass with *M. extradentata* revealed a mean statistically different from all other substrates besides the measurement on *Ep. aureum* with a value of 5.014 ± 1.774. The control glass with *S. inexpectata* revealed a mean of 7.945 ± 4.716, only statistically different to the measurement of this insect species on *Eu. globulus*. The pull-off measurements of *M. extradentata* were not statistically different between *T. urvilleana* and *H. ghiesbreghtii*, and between *Ep. aureum* and glass. All other combinations differ significantly from each other. The pull-off force of *S. inexpectata* between the measurements on *Eu. globulus* and each substrate and between *Ep. aureum* and each substrate, except glass, were statistically different. All other combinations showed no significant differences for *S. inexpectata*. (H = 287.832, d.f. = 9, $N = 15$, $p \leq 0.001$; Tukey's test, $p < 0.05$).

In the traction force measurements (Figure 4B), the highest traction force of *M. extradentata* was reached on the leaves of *H. ghiesbreghtii* with a mean of 29.300 ± 10.951 and for *S. inexpectata* on the same substrate with a mean of 34.514 ± 19.799. On *E. globulus*, the lowest traction force was obtained with a mean of 1.936 ± 2.475 and 5.736 ± 5.451 of *M. extradentata* and *S. inexpectata*. The leaves of *Ep. aureum* and *T. urvilleana* and the glass control revealed a mean of 11.731 ± 4.661, 22.410 ± 5.207 and 19.952 ± 7.520 for the traction force of *M. extradentata*. For *M. extradentata*, there were no significant differences among glass, *T. urvilleana* and *H. ghiesbreghtii*. This is similar with *S. inexpectata*, with means of 10.067 ± 5.039 on *Ep. aureum*, 24.631 ± 12.885 on *T. urvilleana* and 26.828 ± 11.164 on glass. Solely the measurements on *Ep. aureum* were not significantly different compared to the traction force on *Eu. globulus*. (H = 269.977, d.f. = 9, N = 15, $p \leq 0.001$; Tukey's test, $p < 0.05$).

The comparison of the pull-off and traction force between the two insect species (Figure 4) revealed that the values of *S. inexpectata* are slightly higher than those of *M. extradentata*, except by *T. urvilleana* and *H. ghiesbreghtii*, as substrates for the pull-off force measurement and *Ep. aureum* as the substrate for the traction force measurement. The statistical analysis showed no difference between *M. extradentata* and *S. inexpectata* on any tested substrate for both types of force measurements (pull-off: H = 287.832, d.f. = 9, N = 45, $p \leq 0.001$; Tukey's test, $p < 0.05$; traction: H = 269.977, d.f. = 9, N = 15, $p \leq 0.001$; Tukey's test, $p < 0.05$).

3.1.1. Force Measurement after Claw Manipulations

The individuals without claws of *M. extradentata* showed overall lower attachment force values for the pull-off force measurements, if compared with individuals of the same species with intact claws (Figure 5A). The highest mean pull-off force, however, was similar to the measurements with claws on the leaves of *H. ghiesbreghtii* (3.584 ± 1.791) followed by the mean measured on *T. urvilleana* (2.999 ± 0.903), on glass (2.356 ± 0.821), on *Ep. aureum* (1.530 ± 0.432) and the lowest mean pull-off force on *Eu. globulus* (0.910 ± 0.097). The statistical analysis revealed a significant difference between the individuals with and without claws on the glass control, *T. urvilleana* and *H. ghiesbreghtii* for the pull-off of *M. extradentata* (Kruskal–Wallis one-way ANOVA on ranks, H = 283.535, d.f. = 9, $N_{without\_claws}$ = 10, N_{with_claws} = 15, $p \leq 0.001$; Dunn's method, $p < 0.05$).

In the case of the pull-off force measurement of *S. inexpectata*, the individuals with claws also showed higher values except on the leaves of *Eu. globulus* (Figure 5B) and the same ranking order as *M. extradentata*. The only statistical differences were found between the individuals with and without claws on glass and *T. urvilleana* (H = 179.328, d.f. = 9, $N_{without\_claws}$ = 10, N_{with_claws} = 15, $p \leq 0.001$; Dunn's method, $p < 0.05$).

The traction forces of *M. extradentata* were also higher for individuals with intact claws (Figure 5C). However, different from the measurement with intact claws, the highest mean for individuals without claws occurred on glass (11.800 ± 8.524) and not on *H. ghiesbreghtii* (7.034 ± 5.374). The lowest traction was found on *Eu. globulus* with a mean of 0.359 ± 0.823. The values for *M. extradentata* with and without claws were statistically significantly different on glass, *T. urvilleana* and *H. ghiesbreghtii* (H = 276.410, d.f. = 9, $N_{without\_claws}$ = 10, N_{with_claws} = 15, $p \leq 0.001$; Dunn's method, $p < 0.05$).

Sungaya inexpectata without claws had mostly lower traction force, if compared with individuals of the species with intact claws (Figure 5D). However, the mean on *Ep. aureum* (12.352 ± 7.596) and *Eu. globulus* (6.874 ± 3.266) was higher for *S. inexpectata* individuals without claws. There was a significant difference between the traction force of *S. inexpectata* with and without claws on *H. ghiesbreghtii*, *T. urvilleana* and glass, but not between *Ep. aureum* and *Eu. globulus* (H =171.182, d.f. = 9, $N_{without\_claws}$ = 10, N_{with_claws} = 15, $p \leq 0.001$; Dunn's method, $p < 0.05$)

The comparison of individuals without claws between the two species revealed higher values of pull-off and traction force for *S. inexpectata* (Figure 5E,F). Nevertheless, the only significant difference between *M. extradentata* and *S. inexpectata* was found on *Eu. globulus* for both types of force measurements (traction and pull-off). Between all other substrates, no significant differences were found ($H_{pull-off}$ = 126.899, $H_{traction}$ = 100.812, d.f. = 9, N = 10, $p \leq 0.001$; Tukey's test, $p < 0.05$).

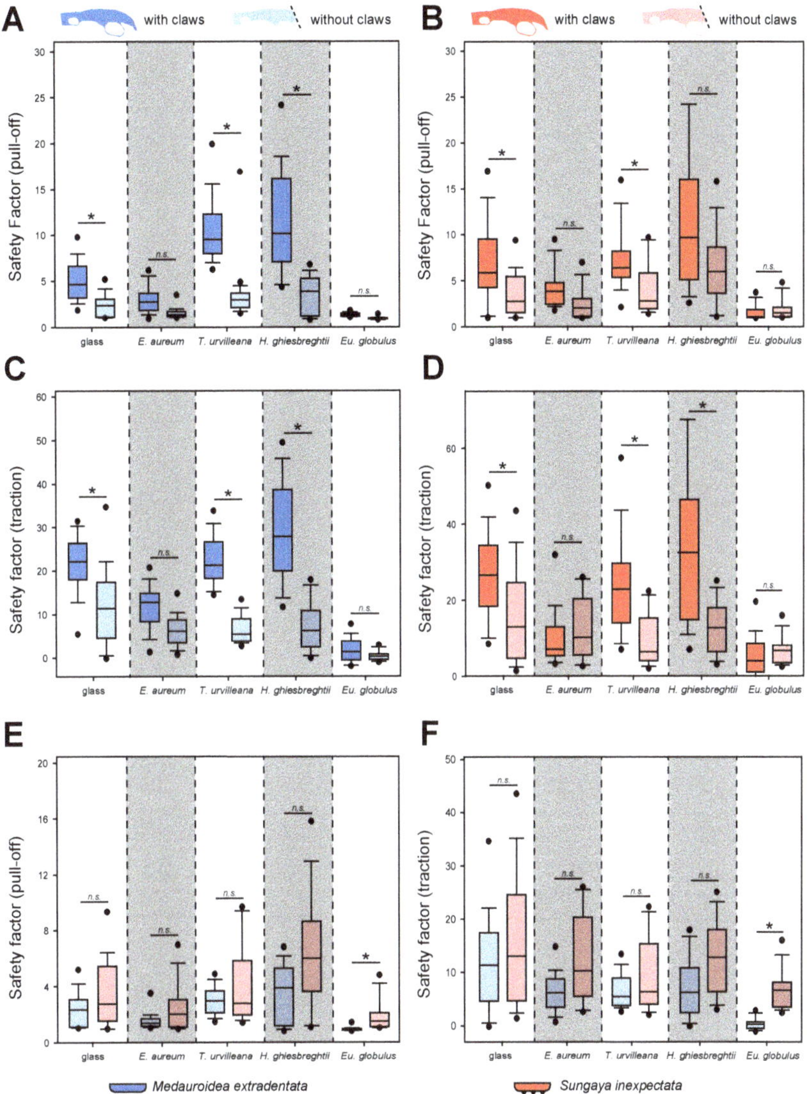

Figure 5. Safety factors of *M. extradentata* and *S. inexpectata* with claws and without claws on different substrates. (**A**) Pull-off force measurement of *M. extradentata* with and without claws. (**B**) Pull-off force measurement of *S. inexpectata* with and without claws. (**C**) Traction force measurement of *M. extradentata* with and without claws. (**D**) Traction force measurement of *S. inexpectata* with and without claws. (**E**) Pull-off force measurement of both species without claws. (**F**) Traction force measurement of both species without claws. Groups which do not differ significantly from each other are marked *n.s.* and groups which differ with a significance of $p > 0.05$ are marked * for each substrate. Kruskal–Wallis one-way ANOVA on ranks ((**A**): H = 287.832, (**B**): H = 276.410, (**C**): H = 179.328, (**D**): H =171.182, (**E**): H = 126.899, (**F**): H = 100.812; $p \leq 0.001$, d.f. = 9, $N_{without\_claws}$ = 10, N_{with_claws} = 15) and Dunn's method (**A–D**; $p < 0.05$) or Turkey's test (**E**,**F**; $p < 0.05$). The boxes indicate 25th and 75th percentiles, whiskers define the 10th and 90th percentiles and the line within the boxes represents the median. Outliers are shown as individual points.

3.1.2. Contamination Effects

The measurements testing the lasting of effects toward the attachment system of *M. extradentata* (Figure 6) showed no statistical difference between the values on glass before and after the substrate for any of the plant species studied, except for the pull-off measurement with *Ep. aureum*. The pull-off force on *Ep. aureum* and glass 2 were statistically not different, the first measurement on glass was significantly different from both of them (one-way ANOVA, F = 8.453, d.f. = 2, $N_{glass1,Ep.aureum}$ = 5, N_{glass2} = 10, p = 0.003; Holm–Šídák method, $p < 0.05$). For the traction measurement with *M. extradentata* on *Ep. aureum* and the pull-off measurements on *T. urvilleana* and on *H. ghiesbreghtii*, all the means were not significantly different from each other (one-way ANOVA, F < 2.76, d.f. = 2, $N_{glass1,plants}$ = 5, N_{glass2} = 10, $p < 0.37$). The traction force measured on the plant leaves of *T. urvilleana* and *H. ghiesbreghtii* were significantly different from the traction force on glass, but glass 1 and glass 2 did not differ significantly from each other (*T. urvilleana*: one-way ANOVA, F = 7.722, d.f. = 2, $N_{glass1,T.urvillena}$ = 5, N_{glass2} = 10, p = 0.004; Holm–Šídák method, $p < 0.05$; *H. ghiesbreghtii*: Kruskal–Wallis one-way ANOVA on ranks, H = 1.054, d.f. = 2, $N_{glass1,H.ghiesbreghtii}$ = 5, N_{glass2} = 10, p = 0.590). Likewise, the pull-off and traction force at the measurement with *Eu. globulus* on glass 1 and glass 2 were not significantly different from each other; the values on *Eu. globulus* were significantly different to the values on glass (Kruskal–Wallis one-way ANOVA on ranks, H < 21.68, d.f. = 2, $N_{glass1,Eu.globulus}$ = 15, N_{glass2} = 30, $p \leq 0.001$; Dunn's method, $p < 0.05$).

The measurement with *S. inexpectata* testing for contamination effects to the attachment system (Figure 7) revealed for neither of the substrates a statistical difference between the values on glass before and after the leaf. In the pull-off measurement with *S. inexpectata* on *Ep. aureum*, on *T. urvilleana* and each measurement on *H. ghiesbreghtii*, the mean values were not statistically different from each other (Kruskal–Wallis one-way ANOVA on ranks, H < 4.82, d.f. = 2, $N_{glass1,plants}$ = 5, N_{glass2} = 10, $p \leq 0.91$). The traction measurement with *Ep. aureum* in-between glass, the values on glass 1 and glass 2 and the values on glass 1 and *Ep. aureum* were statistically not different from each other, while those on *Ep. aureum* and glass 2 were significantly different (one-way ANOVA, F = 4.013, d.f. = 2, $N_{glass1,Ep.aureum}$ = 5, N_{glass2} = 10, p = 0.037; Holm–Šídák method, $p < 0.05$). The traction force on glass 1 and glass 2 in the measurements on *Eu. globulus* and *T. urvilleana* were not significantly different from each other. The value on the plant leaves were statistically different from the measurements on glass (*Eu. globulus*: Kruskal–Wallis one-way ANOVA on ranks, H < 25.81, d.f. = 2, $N_{glass1,Eu.globulus}$ = 15, N_{glass2} = 30, $p \leq 0.001$; Dunn's method, $p < 0.05$; *T. urvilleana*: one-way ANOVA, F = 5.927, d.f. = 2, $N_{glass1,T.urvillena}$ = 5, N_{glass2} = 10, p = 0.011; Holm–Šídák method, $p < 0.05$).

The additional control measurement with only glass as substrate revealed no significant differences between the attachment ability on the three glasses for both species (Figure S2).

3.2. Leaf Surface Characteristics

The surfaces of the adaxial sides of the leaves used as substrates were visualized using cryoscanning electron microscopy (Figure 8). The surface of *Ep. aureum* was the smoothest and revealed almost no surface protrusions. However, it contains a rather thick but smooth wax layer (Figure 8A,B). *Tibouchina urvilleana* leaves are more structured and the surface is densely covered with elongated trichomes with a length of 432.6 ± 151.3 µm (mean ± SD, n = 10) with pointed tips (Figure 8C,D). The adaxial leaf surface of *H. ghiesbreghtii* has an undulating profile and convex hemispherically shaped cells. Occasionally trichomes are found at the leaf edge (Figure 8E,F). Additionally, this surface was partially fouled with microorganisms. The adaxial surface of the leaves of *Eu. globulus* is covered with wax crystals. These wax protrusions are tubular and elongated, with a length of 3.1 ± 1.4 µm (Figure 8E,F).

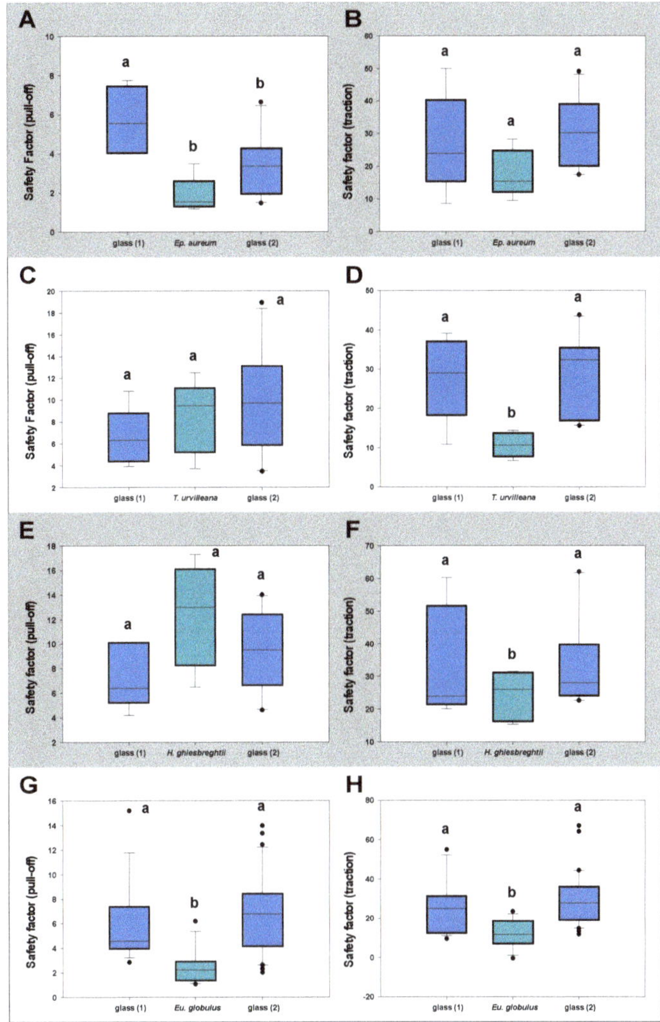

Figure 6. Contamination effect of the leaf substrates tested with *M. extradentata* shown through safety factor. (**A**) *Ep. aureum*, pull-off force. (**B**) *Ep. aureum*, traction force. (**C**) *T. urvilleana*, pull-off force. (**D**) *T. urvilleana* traction force. (**E**) *H. ghiesbreghtii*, pull-off force. (**F**) *H. ghiesbreghtii*, traction force. (**G**) *Eu. globulus*, pull-off force. (**H**) *Eu. globulus*, traction force. Statistically similar groups within a graph are marked with the same lowercase letter, tested through (**A**): one-way ANOVA (F = 8.453, d.f. = 2, $N_{glass1,Ep.aureum}$ = 5, N_{glass2} = 10, p = 0.003) Holm–Šídák method ($p < 0.05$); (**B**): one-way ANOVA (F = 2.215, d.f. = 2, $N_{glass1,Ep.aureum}$ = 5, N_{glass2} = 10, p = 0.140); (**C**): one-way ANOVA (F = 1.062, d.f. = 2, $N_{glass1,T.urvilleana}$ = 5, N_{glass2} = 10, p = 0.368); (**D**): one-way ANOVA (F = 7.722, d.f. = 2, $N_{glass1,T.urvillena}$ = 5, N_{glass2} = 10, p = 0.004), Holm–Šídák method ($p < 0.05$); (**E**): one-way ANOVA (F = 2.761, d.f. = 2, $N_{glass1,H.ghiesbreghtii}$ = 5, N_{glass2} = 10, p = 0.092); (**F**): Kruskal–Wallis one-way ANOVA on ranks (H = 1.054, d.f. = 2, $N_{glass1,H.ghiesbreghtii}$ = 5, N_{glass2} = 10, p = 0.590); (**G**): Kruskal–Wallis one-way ANOVA on ranks (H = 21.674, d.f. = 2, $N_{glass1,Eu.globulus}$ = 15, N_{glass2} = 30, $p \leq 0.001$), Dunn's method ($p < 0.05$); (**H**): Kruskal–Wallis one-way ANOVA on ranks (H = 21.001, d.f. = 2, $N_{glass1,Eu.globulus}$ = 15, N_{glass2} = 30, $p \leq 0.001$), Dunn's method ($p < 0.05$). The boxes indicate 25th and 75th percentiles, whiskers define the 10th and 90th percentiles and the line within the boxes represents the median. Outliers are shown as individual points.

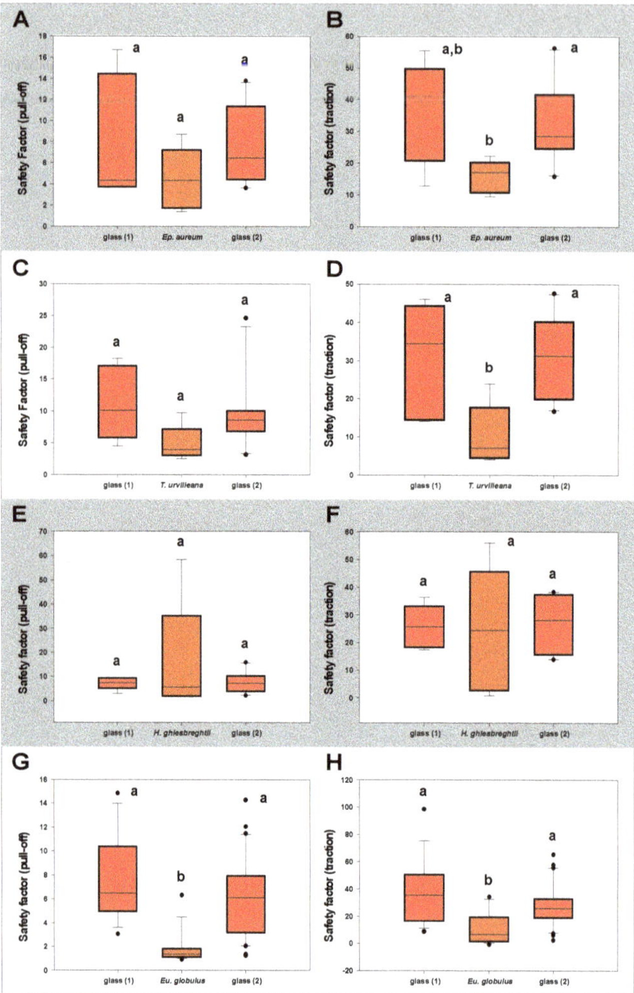

Figure 7. Contamination effect of the substrates tested with *S. inexpectata* shown through safety factor. (**A**) *Ep. aureum*, pull-off force. (**B**) *Ep. aureum*, traction force. (**C**) *T. urvilleana*, pull-off force. (**D**) *T. urvilleana*, traction force. (**E**) *H. ghiesbreghtii*, pull-off force. (**F**) *H. ghiesbreghtii*, traction force. (**G**) *Eu. globulus*, pull-off force. (**H**) *Eu. globulus,* traction force. Statistically similar groups within a graph are marked with the same lowercase letter. (**A**): Kruskal–Wallis one-way ANOVA on ranks (H = 1.054, d.f. = 2, $N_{glass1,Ep.aureum}$ = 5, N_{glass2} = 10, p = 0.331); (**B**): one-way ANOVA (F = 4.013, d.f. = 2, $N_{glass1,Ep.aureum}$ = 5, N_{glass2} = 10, p = 0.037; Holm–Šídák method, $p < 0.05$); (**C**): Kruskal–Wallis one-way ANOVA on ranks (H = 4.817, d.f. = 2, $N_{glass1,T.urvillena}$ = 5, N_{glass2} = 10, p = 0.09); (**D**): one-way ANOVA (F = 5.927, d.f. = 2, $N_{glass1,T.urvillena}$ = 5, N_{glass2} = 10, p = 0.011); Holm–Šídák method ($p < 0.05$); (**E**): one-way ANOVA (F = 5.927, d.f. = 2, $N_{glass1,T.urvillena}$ = 5, N_{glass2} = 10, p = 0.011); Holm–Šídák method ($p < 0.05$); (**F**): Kruskal–Wallis one-way ANOVA on ranks (H = 0.189, d.f. = 2, $N_{glass1,H.ghiesbreghtii}$ = 5, N_{glass2} = 10, p = 0.910); (**G**): Kruskal–Wallis one-way ANOVA on ranks (H = 25.805, d.f. = 2, $N_{glass1,Eu.globulus}$ = 15, N_{glass2} = 30, $p \leq 0.001$), Dunn's method ($p < 0.05$); (**H**): Kruskal–Wallis one-way ANOVA on ranks (H = 17.517, d.f. = 2, $N_{glass1,Eu.globulus}$ = 15, N_{glass2} = 30, $p \leq 0.001$; Dunn's method ($p < 0.05$). The boxes indicate 25th and 75th percentiles, whiskers define the 10th and 90th percentiles and the line within the boxes represents the median. Outliers are shown as individual points.

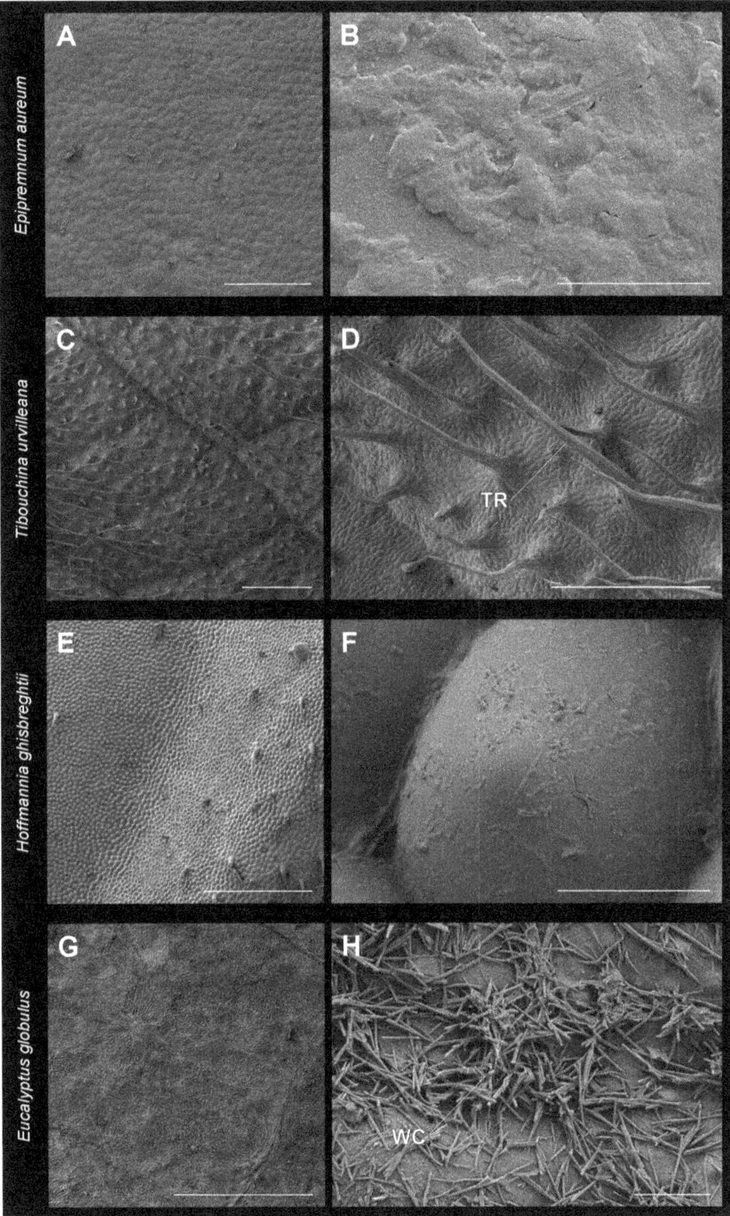

Figure 8. Cryoscanning electron microscopy images of the adaxial leaf surface. (**A**,**B**) *Epipremnum aureum*, (**C**,**D**) *Tibouchina urvilleana*, (**E**,**F**) *Hoffmannia ghiesbreghtii*, (**G**,**H**) *Eucalyptus globulus*. Scale bars: (**A**,**D**) = 400 μm; (**B**,**H**) = 5 μm; (**C**,**E**,**G**) = 1 mm, (**F**) = 20 μm. TR: trichome; WC: wax crystals.

The surface texture of the substrates was characterized by measurements of the roughness and the vein height on different areas on the adaxial side of the leaves (Figures S1 and S3). Figure 9A shows the arithmetical mean roughness R_a and Figure 9B shows the heights of the middle and secondary vein of each leave combined. The leaf of *H. ghiesbreghtii* shows the highest arithmetical mean roughness (514.5 ± 146.7 μm) and mean vein height (1470.5 ± 997.4 µm). With less than a tenth of the mean R$_a$ of *H. ghiesbreghtii*, *Eu. globulus* has the lowest arithmetical mean roughness with 44.1 ± 22.1 µm and with 109.0 ± 80.4 µm, which is also the lowest mean value of vein heights. The leaves of *T. urvilleana* and *Ep. aureum* show intermediate roughness compared to the previous two species, whereas *T. urvilleana* has a higher mean R$_a$ (106.2 ± 49.3 µm) and mean vein height (408.6 ± 184.2 µm) than *Ep. aureum* with a mean vein height of 309.5 ± 196.6 µm and mean R$_a$ of 90.3 ± 53.4 µm.

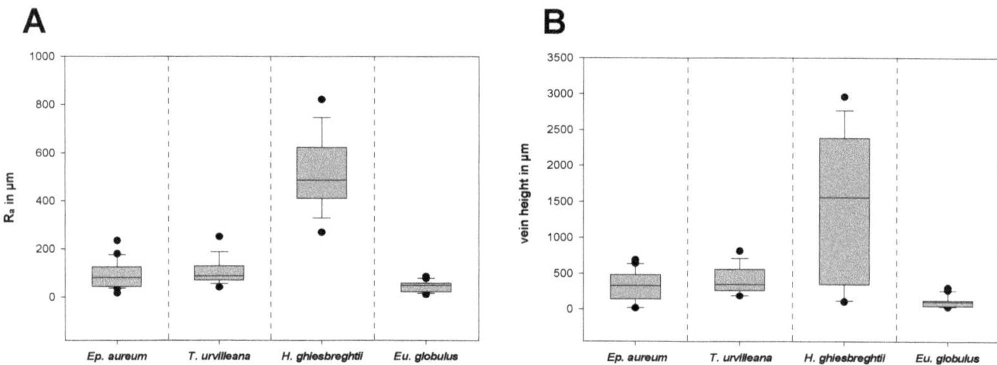

Figure 9. Characterization of the surface texture of the leaves. (**A**) Arithmetical mean roughness R$_a$ and (**B**) vein height. *n* = 17 (*T. urvilleana*), 18 (*H. ghiesbreghtii*), 20 (*Ep. aureum, Eu. globulus*) for both (**A,B**). The boxes indicate 25th and 75th percentiles, whiskers define the 10th and 90th percentiles and the line within the boxes represents the median. Outliers are shown as individual points.

4. Discussion

4.1. Attachment Performance on the Different Plant Leaves

We investigated the influence of different plant leaf characteristics on the attachment performance of stick insects. In general, the performance varied across the leaves used as substrates. The highest pull-off and traction forces were reached on the microstructured, hydrophilic leaves of *H. ghiesbreghtii* followed by the values on the trichome-covered leaves of *T. urvilleana*. Both substrates are relatively rough and are either covered with trichomes (*T. urvilleana*, Figure 8C,D) or present an undulating surface profile due to the leaf venation (*H. ghiesbreghtii*, Figure 8E,F). The elevations of the leaves of *H. ghiesbreghtii* seemed to generate a better holding surface for phasmids. Phasmids already showed an increased pull-off force on curved artificial substrates, if compared to flat surfaces [38]. Similar effects have been observed in other insects [41–45] and frogs [46,47]. In contrast to all other leaves on which the phasmids engaged the arolium only during the pull-off measurement, and similar to the individuals measured on flat artificial substrates in Büscher and Gorb [26], some individuals brought their euplantulae into contact during the pull-off measurement on *H. ghiesbreghtii*. The waviness of the surface additionally has an influence on the peeling angle of the tarsus during detachment and consequently on the resulting attachment force [38,48]. Furthermore, the leaves of *H. ghiesbreghtii* seem to be less stiff and might enable penetration of stiff claws. The claws of *S. inexpectata* were observed to cause small scratches during the traction measurement. The hydrophilic properties of the leaves enhanced adhesion of Phasmatodea species studied [49]. Phasmatodea have adhesive secretions [50–53] that could be positively affected by the hydrophilicity of the substrate [49].

Trichomes on plant surfaces have already been shown to generate an additional "foothold", thereby increasing the insect attachment, e.g., [12,13,16,17,54]. The trichomes of *T. urvilleana* provide good support for mechanical interlocking of the claws because they are long enough to represent such a "foothold" for the tested individuals of both species. The attachment ability on the relatively smooth leaves of *Ep. aureum* (Figure 8A,B) of both species was lower than on *H. ghiesbreghtii* and *T. urvilleana*. The pull-off and traction force on *Ep. aureum* were generally lower than on glass, but the difference was found significant only for the traction measurement with individuals with intact claws. The slightly lower attachment performance can be due to the different surface free energy and presumably lower stiffness of *Ep. aureum* leaves compared to glass. Studies showed higher attachment force on smooth substrates generally, explained through a high real contact area compared to rough surfaces, e.g., [13,17,54–59], since the claws cannot interlock on smooth substrates [26,32,41]. In the case of Phasmatodea, the force produced by successful mechanical interlocking of the claws surpasses the attachment by the attachment pads, at least in the case of pronounced trichomes of *T. urvilleana*.

The lowest traction and pull-off force, and therefore the lowest attachment of phasmids tested herein were measured on the leaves of *Eu. globulus*. Those leaves are covered with wax crystals, but smooth underneath the wax (Figure 8G,H). They showed by far the most significant differences to the other substrates. The leaves were also the stiffest ones in compression (pers. observation) and did not show any sites for claw interlocking. The insect attachment on surfaces with epicuticular waxes was analyzed in previous studies with most studies showing a low attachment performance, e.g., [6,16,17,48,49,60,61]. This can be due to the increased roughness, impairment of the attachment system with contamination, wax-dissolving or fluid absorption [5,17,62]. During the measurements of this study, the Phasmatodea showed the most effort to get off the leaves of *Eu. globulus*, if compared to the other substrates. This may also be a behavioral response to the reduced attachment ability on the leaves. We found no indication for an effect of ambient humidity during the measurements, although such an effect is reported for other insects [50,63] did not find an effect of ambient humidity on stick insect attachment as well. However, very high ambient humidity could have an effect, but was not reached during our experiments. The effect of ambient humidity could be experimentally tested in a systematic way as it has been done for presence of water films and wettability of the substrate for phasmids [49,62].

4.2. Impact of the Claws

Especially on leaves with strongly expressed surface structure, claws of insects are able to interlock and provide strong attachment. The interlocking of the claws is dependent on their structure, e.g., tip diameter, density and curvature [26,32,54,64]. In this study, the statistical difference between the attachment ability of individuals with intact claws and individuals without claws for both species was measured on glass and *T. urvilleana* for traction and pull-off force. On *H. ghiesbreghtii*, every measurement except pull-off with *S. inexpectata* was significantly different as well. Consequently, the effect of the claws has a greater impact on structured surfaces such as trichome-covered (*T. urvilleana*, Figure 8C,D) or leaves with strongly expressed microstructure (*H. ghiesbreghtii*, Figure 8E,F).

Trichomes occur on the leaves of various plant lineages and serve different functions [65]. For the attachment ability of insects, however, the trichomes can potentially generate a "foothold" that the claws can grab onto, e.g., [12,13,16,17,54]. Nevertheless, the role of trichomes for insect attachment is also a question of the dimension of the plant feature in relation to the insect attachment system or even the insect itself. In general, trichomes are reported to hinder locomotion and negatively affect the oviposition, acceptance of the plant as food or attachment to the surface [66–70], but these effects can be different for differently sized insects. Insects which are small enough can simply step between trichomes [14], while a suitable size difference between trichome and tarsal claws can also lead to specialized clamping devices on the claws that use the trichomes for interlocking [71]. Clamping fibrillar surface features can strongly increase the attachment, as shown for parasitic insects, e.g., [72]. Stick insects are some of the largest insects and are several times larger than the insects reported to be repelled by plant puberance and also the tarsi of stick insects are usually much larger than the trichomes. Some aschiphasmatine stick insects possess pectinate claws as well, e.g., [30,73], but the two species investigated herein have simple pointed claws (Figure 1). Removing these revealed the noted influences, i.e., on trichome-bearing as well as corrugated surfaces and on glass. In our two focal stick insect species, the inner radius of claw curvature (Figure S4) is larger in both species than the thickness of the trichomes (Figure 8). Accordingly, the claws can grasp around single trichomes. The mechanical interlocking of claws at higher roughness or structured surfaces was already shown in other studies, e.g., [26,60,74]. For example, the removal of claws of the beetle *Chrysolina polita* (Chrysomelidae, Coleoptera) resulted in lower pull-off forces on cloth but not on glass [54]. However, our measurements presented in this study show that the attachment performance of individuals without claws was also reduced on glass for both species. This could be due to the mechanical support of the claws to the arolium. The claws provide stability and proper placement of the arolium; therefore, in their absence, the attachment ability of the arolium can be reduced [75].

4.3. Comparison of Attachment Performance between the Two Species

The main difference between attachment systems of the two species is that *M. extradentata* has smooth euplantulae and *S. inexpectata* has nubby microstructures on the euplantulae [25]. It has been shown in previous studies that a smooth euplantula provides stronger adhesion to smooth surfaces compared to a euplantula with a nubby microstructure [26,31]. Büscher and Gorb [26] also showed that the traction force was higher for *S. inexpectata* at different defined roughnesses (0.3, 1, 12, 425 µm) and *M. extradentata* performed better on smooth and coarse substrates in the pull-off experiment [26]. In contrast to the previous measurements on standardized surfaces, the performance of the two species on natural leaves did not differ statistically on most substrates. Actual plant surfaces have a wide spectrum of different structures and properties. For example, the hydrophilicity of *H. ghiesbreghtii* leaves can affect the attachment ability of both phasmids independent of their adaptations toward different roughness. Solely the wax-crystal-covered leaves of *Eu. globulus* revealed a significant difference between individuals of *M. extradentata* and *S. inexpectata* without claws (Figure 5E,F). The attachment force of *S. inexpectata* with nubby euplantulae was significantly higher than that for the species with smooth euplantulae. Consequently, the nubby attachment pads seem to be less affected by the contamination of the wax crystals. Nubby or hairy attachment structures have already been assumed to be less influenced by contamination than smooth attachment structures [74]. When the function of the claws is excluded, the attachment pads are the main organ involved in attachment of Phasmatodea [25,26]. Species with nubby euplantula microstructures within Phasmatodea are more frequently found in species that are somewhat ground-associated, which are either known to dwell on the ground, feed on dropped leaves or bury their eggs into the soil [25,28,30]. This preference could be accompanied by potential contaminations from the soil, at least in species using their tarsi to dig holes for egg deposition (e.g., some Heteropterygidae [76]). Further experiments testing the influence of contaminations on both types of euplantula microstructures would be helpful to investigate the role of attachment microstructure for contaminations.

4.4. Lasting of Contaminating Effects

The coverage of plant leaves, such as waxes [25] and excretions of glandular trichomes, e.g., [65,77–79] can potentially reduce the attachment ability of insects instantaneously, but often also cause persistent contaminations that last longer and remain on the feet of the insects [80]. None of the substrates used in this study seem to have a long-lasting effect on the attachment of Phasmatodea. For *Eu. globulus*, a possible persistent decrease in attachment due to contaminating effect of wax crystals on the surface of the leaves [4] could not be observed. However, this observation does not exclude the presence of some contamination. The effect of wax-crystal contamination could already be overcome before or during the first contact during the force measurement. The wax crystals of *Eu. globulus* are large in comparison to the attachment system of the phasmid studied and therefore might enable a quick removal of them. It is quite likely that the initial attachment forces might have been reduced, but with the repeating contact during walking, the attachment is regenerated due to shedding off the wax crystals. The only noteworthy difference due to possible lasting effects of the substrate was observed in the pull-off force measurement with *M. extradentata* on leaves of *Ep. aureum* in between two measurements on the glass as a control. The reason for the significant difference between the control measurements on glass with *Ep. aureum* in between could be a result of other plant leaf secretions.

5. Conclusions

The attachment of Phasmatodea is influenced by different features on the surface of leaves.

(I)   The highest attachment ability is observed on the microstructured, hydrophilic leaves of *H. ghiesbreghtii* and the trichome-covered leaves of *T. urvilleana*. Strong surface corrugations and high substrate waviness are beneficial for the function of both claws and their combination with attachment pads. Epicuticular wax crystals on the surface of *Eu. globulus* leaves caused the lowest attachment ability for the Phasmatodea species studied.

(II) The claws of the insects did have the strongest impact on the attachment on leaves with trichomes or strong surface corrugations. There was no significant difference between the two tested species of stick insects (*M. extradentata* and *S. inexpectata*) with intact claws despite different size of the claws.

(III) Removing the claws showed better performance of attachment pads with nubby microstructures in face of wax-crystal-covered leaves that potentially contaminate the attachment pads.

(IV) The long-lasting effects of the leaf surfaces on attachment were not evidenced. In summary, the attachment of Phasmatodea was affected by the different substrates, with rough surface and trichome coverage being beneficial for attachment. The wax crystals on the plant surface provided the highest potential for protection of the plant through decreasing the attachment of insects.

Supplementary Materials: The following supporting information can be downloaded at: https://www.mdpi.com/article/10.3390/insects13100952/s1, Figure S1: Surface characterization supplementary results; Figure S2: Contamination effect glass measurement; Figure S3: Surface characterization; Figure S4: Claw metrics.

Author Contributions: Conceptualization, S.N.G. and T.H.B.; methodology, J.B., S.N.G. and T.H.B.; validation, J.B., S.N.G. and T.H.B.; formal analysis, J.B. and T.H.B.; investigation, J.B.; resources, S.N.G. and T.H.B.; data curation, T.H.B. and J.B.; writing—original draft preparation, J.B.; writing—review and editing, T.H.B. and S.N.G.; visualization, J.B. and T.H.B.; supervision, T.H.B.; project administration, T.H.B.; funding acquisition, S.N.G. All authors have read and agreed to the published version of the manuscript.

Funding: This research was partially funded by the German Science Foundation, DFG grant GO 995/34-1.

Data Availability Statement: Raw data of the claw metrics, contamination measurements and leaf surface characterization are provided as supplementary files.

Acknowledgments: We thank Esther Appel, Alexander Kovalev and Benedikt Josten (Functional Morphology and Biomechanics, Kiel University) for technical assistance. Susanne Petersen (Botanical Garden, Kiel University) and the staff of the Botanical Garden of Kiel University are thanked for cultivation of the plants used in the experiments.

Conflicts of Interest: The authors declare no conflict of interest.

References

1. Southwood, T.R.E. The insect/plant relationship—An evolutionary perspective. *Symp. R. Entomol. Soc. Lond.* **1973**, *6*, 3–30.
2. Scott, A.C.; Paterson, S. Techniques for the study of plant/arthropod interactions in the fossil record. *Geobios* **1984**, *17*, 449–457. [CrossRef]

3. Chartier, M.; Gibernau, M.; Renner, S.S. The evolution of pollinator-plant interaction types in the *Araceae*. *Evolution* **2013**, *68*, 1533–1543. [CrossRef]
4. Gorb, E.V.; Gorb, S.N. Anti-adhesive surfaces in plants and their biomimetic potential. *RSC Smart Mater.* **2013**, *4*, 282–309. [CrossRef]
5. Ohmart, C.P.; Edwards, P.B. Insect herbivory on *Eucalyptus*. *Annu. Rev. Entomol.* **1991**, *36*, 637–657. [CrossRef]
6. Gorb, E.V.; Gorb, S.N. Attachment ability of the beetle *Chrysolina fastuosaon* various plant surfaces. *Entomol. Exp. Appl.* **2002**, *105*, 13–28. [CrossRef]
7. Rebora, M.; Salerno, G.; Piersanti, S.; Gorb, E.V.; Gorb, S.N. Role of Fruit Epicuticular Waxes in Preventing *Bactrocera oleae* (Diptera: Tephritidae) Attachment in Different Cultivars of *Olea europaea*. *Insects* **2020**, *11*, 189. [CrossRef]
8. Gaume, L.; Gorb, S.N.; Rowe, N. Function of epidermal surfaces in the trapping efficiency of *Nepenthes alata* pitchers. *New Phytol.* **2002**, *156*, 479–489. [CrossRef]
9. Gorb, E.V.; Kastner, V.; Peressadko, A.; Arzt, E.; Gaume, L.; Rowe, N.; Gorb, S.N. Structure and properties of the glandular surface in the digestive zone of the pitcher in the carnivorous plant *Nepenthes ventrata* and its role in insect trapping and retention. *J. Exp. Biol.* **2004**, *207*, 2947–2963. [CrossRef]
10. Levin, D.A. The role of trichomes in plant defense. *Q. Rev. Biol.* **1973**, *48*, 3–15. [CrossRef]
11. Stipanovic, R.D. Function and chemistry of plant trichomes and glands in insect resistance: Protective chemicals in plant epidermal glands and appendages. *Am. Chem. Soc.* **1983**, *208*, 69–100. [CrossRef]
12. Johnson, H.B. Plant pubescence: An ecological perspective. *Bot. Rev.* **1975**, *41*, 233–258. [CrossRef]
13. Lee, Y.I.; Kogan, M.; Larsen, J.R. Attachment of the potato leafhopper to soybean plant surfaces as affected by morphology of the pretarsus. *Entomol. Exp. Appl.* **1986**, *42*, 101–107. [CrossRef]
14. Voigt, D.; Gorb, E.V.; Gorb, S.N. Plant surface-bug interactions: *Dicyphus errans* stalking along trichomes. *Arthropod-Plant Interact.* **2007**, *1*, 221–243. [CrossRef]
15. Koch, K. Design of hierarchically sculptured biological surfaces with anti-adhesive properties. In Proceedings of the Beilstein Bozen Symposium on Functional Nanoscience, Bozen, Italy, 17–21 May 2010; pp. 167–178.
16. Chang, G.C.; Neufeld, J.; Eigenbrode, S.D. Leaf surface wax and plant morphology of peas influence insect density. *Entomol. Exp. Appl.* **2006**, *119*, 197–205. [CrossRef]
17. Salerno, G.; Rebora, M.; Gorb, E.V.; Gorb, S.N. Attachment ability of the polyphagous bug *Nezara viridula* (Heteroptera: Pentatomidae) to different host plant surfaces. *Sci. Rep.* **2018**, *8*, 10975. [CrossRef]
18. Salerno, G.; Rebora, M.; Piersanti, S.; Büscher, T.H.; Gorb, E.V.; Gorb, S.N. Oviposition site selection and attachment ability of *Propylea quatuordecimpunctata* and *Harmonia axyridis* from the egg to the adult stage. *Physiol. Entomol.* **2022**, *47*, 20–37. [CrossRef]
19. Jones, L.C.; Rafter, M.A.; Walter, G.H. Host interaction mechanisms in herbivorous insects—Life cycles, host specialization and speciation. *Biol. J. Linn. Soc.* **2022**, *137*, 1–14. [CrossRef]
20. Salerno, G.; Rebora, M.; Piersanti, S.; Gorb, E.V.; Gorb, S.N. Mechanical ecology of fruit-insect interaction in the adult Mediterranean fruit fly *Ceratitis capitata* (Diptera: Tephritidae). *Zoology* **2020**, *139*, 125748. [CrossRef] [PubMed]
21. Brock, P.D.; Büscher, T.H.; Baker, E. Phasmida species file online: Phasmida species file version 5.0/5.0. In *Species 2000 and ITIS Catalogue of Life*; Roskov, Y., Kunze, T., Paglinawan, L., Orrell, T., Nicolson, D., Culham, A., Bailly, N., Kirk, P., Bourgoin, T., Baillargeon, G., et al., Eds.; Naturalis: Leiden, The Netherlands, 2017. Available online: www.catalogueoflife.org/col (accessed on 15 September 2022).
22. Brock, P.D.; Büscher, T.H. *Phasmids of the World*; NAP Editions: Verrières-le-Buisson, France, 2022.
23. Bedford, G.O. Biology and ecology of the Phasmatodea. *Annu. Rev. Entomol.* **1978**, *23*, 125–149. [CrossRef]
24. Gottardo, M.; Vallotto, D.; Beutel, R.G. Giant stick insects reveal unique ontogenetic changes in biological attachment devices. *Arthropod Struct. Dev.* **2015**, *44*, 195–199. [CrossRef] [PubMed]
25. Büscher, T.H.; Buckley, T.R.; Grohmann, C.; Gorb, S.N.; Bradler, S. The Evolution of tarsal adhesive microstructures in stick and leaf insects (Phasmatodea). *Front. Ecol. Evol.* **2018**, *6*, 69. [CrossRef]
26. Büscher, T.H.; Gorb, S.N. Complementary effect of attachment devices in stick insects (Phasmatodea). *J. Exp. Biol.* **2019**, *222*, jeb209833. [CrossRef] [PubMed]
27. Labonte, D.; Federle, W. Functionally different pads on the same foot allow control of attachment: Stick insects have load-sensitive "heel" pads for friction and shear-sensitive "toe" pads for adhesion. *PLoS ONE* **2014**, *8*, 81943. [CrossRef] [PubMed]
28. Büscher, T.H.; Kryuchkov, M.; Katanaev, V.L.; Gorb, S.N. Versatility of Turing patterns potentiates rapid evolution in tarsal attachment microstructures of stick and leaf insects (Phasmatodea). *J. R. Soc. Interface* **2018**, *15*, 20180281. [CrossRef] [PubMed]
29. Büscher, T.H.; Gorb, S.N. A review: Physical constraints lead to parallel evolution of micro- and nanostructures of animal adhesive pads. *Beilstein J. Nanotechnol.* **2021**, *12*, 725–743. [CrossRef]
30. Büscher, T.H.; Grohmann, C.; Bradler, S.; Gorb, S.N. Tarsal attachment pads in Phasmatodea (Hexapoda: Insecta). *Zoologica* **2019**, *164*, 1–94.
31. Bußhardt, P.; Wolf, H.; Gorb, S.N. Adhesive and frictional properties of tarsal attachment pads in two species of stick insects (Phasmatodea) with smooth and nubby euplantulae. *Zoology* **2012**, *115*, 135–141. [CrossRef] [PubMed]
32. Song, Y.; Dai, Z.; Ji, A.; Gorb, S.N. The synergy between the insect inspired claws and adhesive pads increases the attachment ability on various rough surfaces. *Sci. Rep.* **2016**, *6*, 26219. [CrossRef]

33. Boyce, P.C. The genus *Epipremnum* Schott (Araceae-Monsteroideae-Monstereae) in West and Central Malesia. *Blumea* **1998**, *43*, 183–213.
34. Gardner, R. *Tibouchina* (Melastomataceae) the glory bushes. *Bot. Soc. J.* **1992**, *47*, 45–47.
35. Standley, P.C.; Williams, L.O.; Nash, D.L. *Flora of Guatemala*; Field Museum of Natural History: Chicago, IL, USA, 1974; Volume 24, p. 106.
36. Baumert, K. Experimentelle Untersuchungen über Lichtschutz an grünen Blättern. *Beitr. Biol. Pflanz.* **1907**, *9*, 83–162.
37. Steinbauer, M.J.; Davies, N.W.; Gaertner, C.; Derridj, S. Epicuticular waxes and plant primary metabolites on the surfaces of juvenile *Eucalyptus globulus* and *E. nitens* (Myrtaceae) leaves. *Aust. J. Bot.* **2009**, *57*, 474–485. [CrossRef]
38. Bücher, T.H.; Becker, M.; Gorb, S.N. Attachment performance of stick insects (Phasmatodea) on convex substrates. *J. Exp. Biol.* **2020**, *223*, jeb226514. [CrossRef] [PubMed]
39. Schneider, C.A.; Rasband, W.S.; Eliceiri, K.W. NIH Image to ImageJ: 25 years of image analysis. *Nat. Methods* **2012**, *9*, 671–675. [CrossRef]
40. Gorb, E.V.; Gorb, S.N. Functional surfaces in the pitcher of the carnivorous plant *nepenthes alata*: A cryo-sem approach. In *Functional Surfaces in Biology: Adhesion Related Phenomena*; Gorb, S.N., Ed.; Springer: Dordrecht, The Netherlands, 2009; Volume 2, pp. 205–238.
41. Gorb, S.N. *Attachment Devices of Insect Cuticle*; Kluwer Academic Publishers: Dordrecht, The Netherlands, 2001.
42. Arzt, E.; Gorb, S.N.; Spolenak, R. From micro to nano contacts in biological attachment devices. *Proc. Natl. Acad. Sci. USA* **2003**, *100*, 10603–10606. [CrossRef]
43. Bußhardt, P.; Kunze, D.; Gorb, S.N. Interlocking-based attachment during locomotion in the beetle *Pachnoda marginata* (Coleoptera, Scarabaeidae). *Sci. Rep.* **2014**, *4*, 6998. [CrossRef]
44. Voigt, D.; Takanashi, T.; Tsuchihara, K.; Yazaki, K.; Kuroda, K.; Tsubaki, R.; Hosoda, N. Strongest grip on the rod: Tarsal morphology and attachment of Japanese pine sawyer beetles. *Zool. Lett.* **2017**, *3*, 16. [CrossRef] [PubMed]
45. Voigt, D.; Goodwyn, P.; Sudo, M.; Fujisaki, K.; Varenberg, M. Gripping ease in southern green stink bugs *Nezara viridula* L. (Heteroptera: Pentatomidae): Coping with geometry, orientation and surface wettability of substrate. *Entomol. Sci.* **2019**, *22*, 105–118. [CrossRef]
46. Bijma, N.N.; Gorb, S.N.; Kleinteich, T. Landing on branches in the frog *Trachycephalus resinifictrix* (Anura: Hylidae). *J. Comp. Physiol. A* **2016**, *202*, 267–276. [CrossRef]
47. Hill, I.D.C.; Dong, B.; Barnes, W.J.P.; Ji, A.; Endlein, T. The biomechanics of tree frogs climbing curved surfaces: A gripping problem. *J. Exp. Biol.* **2018**, *221*, jeb.168179. [CrossRef] [PubMed]
48. Gu, Z.; Li, S.; Zhang, F.; Wang, S. Understanding surface adhesion in nature: A peeling model. *Adv. Sci.* **2016**, *3*, 1500327. [CrossRef] [PubMed]
49. Thomas, J.; Gorb, S.N.; Büscher, T.H. Influence of surface free energy of the substrate and flooded water on the attachment performance of stick insects (Phasmatodea) with different adhesive surface microstructures. under review.
50. Drechsler, P.; Federle, W. Biomechanics of smooth adhesive pads in insects: Influence of tarsal secretion on attachment performance. *J. Comp. Physio. A* **2006**, *192*, 1213–1222. [CrossRef] [PubMed]
51. Dirks, J.-H.; Federle, W. Fluid-based adhesion in insects—Principles and challenges. *Soft Matter* **2011**, *7*, 11047–11053. [CrossRef]
52. Dirks, J.-H.; Clemente, C.J.; Federle, W. Insect tricks: Two-phasic foot pad secretion prevents slipping. *J. R. Soc. Interface* **2010**, *7*, 587–593. [CrossRef]
53. Kaimaki, D.-M.; Adrew, C.N.S.; Attipoe, A.E.L.; Labonte, D. The physical properties of the stick insect pad secretion are independent of body size. *J. R. Soc. Interface* **2022**, *19*, 20220212. [CrossRef]
54. Stork, N.E. Experimental analysis of adhesion of *Chrysolina polita* (Chrysomelidae, Coleoptera) on a variety of surfaces. *J. Exp. Biol.* **1980**, *88*, 91–107. [CrossRef]
55. Eigenbrode, S.D. Plant surface waxes and insect behaviour. In *Plant Cuticles—An Integral Functional Approach*; Kerstiens, G., Ed.; BIOS: Oxford, UK, 1996; pp. 201–222.
56. Way, M.J.; Murdie, G. An example of varietal resistance of Brussel sprouts. *Ann. Appl. Biol.* **1965**, *56*, 326–328. [CrossRef]
57. Perez-Goodwyn, P.; Peressadko, A.; Schwarz, H.; Kastner, V.; Gorb, S.N. Material structure, stiffness, and adhesion: Why attachment pads of the grasshopper (*Tettigonia viridissima*) adhere more strongly than those of the locust (*Locusta migratoria*) (Insecta: Orthoptera). *J. Comp. Physiol. A* **2006**, *192*, 1233–1243. [CrossRef]
58. Peisker, H.; Michels, J.; Gorb, S.N. Evidence for a material gradient in the adhesive tarsal setae of the ladybird beetle *Coccinella septempunctata*. *Nat. Commun.* **2013**, *4*, 1661. [CrossRef]
59. Bennemann, M.; Backhaus, S.; Scholz, I.; Park, D.; Mayer, J.; Baumgartner, W. Determination of the Young's modulus of the epicuticle of the smooth adhesive organs of *Carausius morosus* using tensile testing. *J. Exp. Biol.* **2014**, *217*, 3677–3687. [CrossRef] [PubMed]
60. Edwards, P.B. Do waxes of juvenile Eucalyptus leaves provide protection from grazing insects? *Aust. J. Ecol.* **1982**, *7*, 347–352. [CrossRef]
61. Federle, W.; Maschwitz, U.; Fiala, B.; Riederer, M.; Hölldobler, B. Slippery ant-plants and skilful climbers: Selection and protection of specific ant partners by epicuticular wax blooms in Macaranga (Euphorbiaceae). *Oecologia* **1997**, *112*, 217–224. [CrossRef] [PubMed]

62. Labonte, D.; Robinson, A.; Bauer, U.; Federle, W. Disentangling the role of surface topography and intrinsic wettability in the prey capture mechanism of Nepenthes pitcher plants. *Acta Biomater.* **2021**, *119*, 225–233. [CrossRef] [PubMed]

63. Heepe, L.; Wolff, J.O.; Gorb, S.N. Influence of ambient humidity on the attachment ability of ladybird beetles (*Coccinella septempunctata*). *Beilstein J. Nanotechnol.* **2016**, *7*, 1322–1329. [CrossRef]

64. Pattrick, J.G.; Labonte, D.; Federle, W. Scaling of claw sharpness: Mechanical constraints reduce attachment performance in larger insects. *J. Exp. Biol.* **2018**, *221*, jeb188391. [CrossRef]

65. Uphof, J.C.T.; Hummel, K. Plant hairs. *Encycl. Plant Anat.* **1962**, *5*, 1–292.

66. Ringlund, K.; Everson, E.H. Leaf Pubescence in Common Wheat, *Triticum aestivum* L., and Resistance to the Cereal Leaf Beetle, *Oulema melanopus* (L.). *Crop Sci.* **1968**, *8*, 705–710. [CrossRef]

67. Gibson, R.W. Glandular hairs providing resistance to aphids in certain wild potato species. *Ann. Appl. Biol.* **1971**, *68*, 113–119. [CrossRef]

68. Hoxie, R.P.; Wellso, S.G.; Webster, J.A. Cereal Leaf Beetle Response to Wheat Trichome Length and Density. *Environ. Entomol.* **1975**, *4*, 365–370. [CrossRef]

69. Zvereva, E.L.; Kozlov, M.V.; Niemelä, P. Effects of leaf pubescence in *Salix borealis* on host-plant choice and feeding behaviour of the leaf beetle, *Melasoma lapponica*. *Entomol. Exp. Appl.* **1998**, *89*, 297–303. [CrossRef]

70. Ranger, C.M.; Hower, A.A. Glandular trichomes on perennial alfalfa affect host-selection behavior of *Empoasca fabae*. *Entomol. Exp. Appl.* **2002**, *105*, 71–81. [CrossRef]

71. Salerno, G.; Rebora, M.; Piersanti, S.; Saitta, V.; Gorb, E.V.; Gorb, S.N. Coleoptera claws and trichome interlocking. *J. Comp. Physiol. A* **2022**, 1–14. [CrossRef] [PubMed]

72. Büscher, T.H.; Petersen, D.S.; Bijma, N.N.; Bäumler, F.; Pirk, C.W.W.; Büsse, S.; Heepe, L.; Gorb, S.N. The exceptional attachment ability of the ectoparasitic bee louse *Braula coeca* (Diptera, Braulidae) on the honeybee. *Physiol. Entomol.* **2021**, *47*, 83–95. [CrossRef]

73. Bradler, S. *Die Phylogenie der Stab-und Gespenstschrecken (Insecta: Phasmatodea)*; Universitätsverl: Göttingen, Germany, 2009; Volume 2, pp. 1–139.

74. Bullock, J.M.; Federle, W. The effect of surface roughness on claw and adhesive hair performance in the dock beetle *Gastrophysa viridula*. *Insect Sci.* **2011**, *18*, 298–304. [CrossRef]

75. Winand, J.; Gorb, S.N.; Büscher, T.H. Gripping performance in the stick insect *Sungaya inexpectata* in dependence on the pretarsal architecture. *J. Comp. Physiol. A* **2022**, 1–11. [CrossRef]

76. Bank, S.; Buckley, T.R.; Büscher, T.H.; Bresseel, J.; Constant, J.; de Haan, M.; Dittmar, D.; Dräger, H.; Kahar, R.S.; Kang, A.; et al. Reconstructing the nonadaptive radiation of an ancient lineage of ground-dwelling stick insects (Phasmatodea: Heteropterygidae). *Syst. Entomol.* **2021**, *46*, 487–507. [CrossRef]

77. Gregory, P.; Tingey, W.M.; Ave, D.A.; Bouthyette, P.Y. Potato Glandular Trichomes: A Physicochemical Defense Mechanism Against Insects. In *Natural Resistance of Plants to Pests*; ACS: Washington, DC, USA, 1986; pp. 160–167.

78. Lapointe, S.L.; Tingey, W.M. Glandular Trichomes of *Solanum neocardenasii* Confer Resistance to Green Peach Aphid (Homoptera: Aphididae). *J. Econ. Entomol.* **1986**, *79*, 1264–1268. [CrossRef]

79. Avé, D.A.; Gregory, P.; Tingey, W.M. Aphid repellent sesquiterpenes in glandular trichomes of *Solanum berthaultii* and *S. tuberosum*. *Entomol. Exp. Appl.* **1987**, *44*, 131–138. [CrossRef]

80. Broadway, R.M.; Duffey, S.S.; Pearce, G.; Ryan, C.A. Plant Proteinase inhibitors: A defense against herbivorous insects? *Entomol. Exp. Appl.* **1986**, *41*, 33–38. [CrossRef]

insects

Article

Floral Volatile Organic Compounds and a List of Pollinators of *Fallopia baldschuanica* (Polygonaceae)

Anna Jakubska-Busse [1,*], **Mariusz Dziadas** [2], **Iwona Gruss** [3] and **Michał J. Kobyłka** [2]

1 Department of Botany, Faculty of Biological Sciences, University of Wroclaw, 50-328 Wroclaw, Poland
2 Faculty of Chemistry, University of Wroclaw, 50-353 Wrocław, Poland
3 Department of Plant Protection, Wroclaw University of Environmental and Life Sciences, 50-363 Wroclaw, Poland
* Correspondence: anna.jakubska-busse@uwr.edu.pl; Tel.: +48-71-375-40-81

Simple Summary: *Fallopia baldschuanica* (Polygonaceae) is an Asian plant growing wild in parts of Europe and North and Central America as an introduced taxon. Although *F. baldschuanica* is considered a potentially invasive alien plant species, little is known about its pollination biology in climatic conditions in Europe. In this study, we identified the volatile organic compounds emitted from *F. baldschuanica* flowers, from which some are important insect attractants. We also described the pollinator populations of this plant. We confirm that the chemical composition of floral aroma in *F. baldschuanica* attracts a large group of potential pollinators, which in addition to the intensive growth of the plant is a feature enabling the species to rapidly expand.

Abstract: *Fallopia baldschuanica* (Polygonaceae) is an Asian plant growing wild in parts of Europe and North and Central America as an introduced taxon, in many countries it is considered a potentially invasive species. This article presents the list of 18 volatile organic compounds (VOCs) emitted by the flowers of *F. baldchuanica* and identified by headspace gas chromatography/mass spectrometry (HS-GC/MS) analyzes, and a list of flower-visiting and pollinating insects that have been observed in the city center of Wrocław (SW Poland). β-ocimene, heptanal, nonanal, α-pinene, 3-thujene, and limonene, were detected as the floral scent's most important aroma compounds. *F. baldschuanica* also produces the aphid alarm pheromones, i.e., β-farnesene and limonene, that repels aphids. Additionally, the pollinators of *F. baldschuanica* were indicated, based on two years of observations in five sites in the urban area. It was found, that the pollinators of this plant with the highest species stability are: Diptera from families Syrphidae (*Chrysotoxum bicinctum, Eristalis pertinax, Eupeodes corollae, Episyrphus balteatus, Eristalis tenax, Syrphus ribesii, Eristalis intricaria*), Muscidae (*Musca domestica*), Sarcophagidae (*Sarcophaga* spp.), Calliphoridae (*Lucilia sericata, Lucilia caesar*), Hymenoptera from families Vespidae (*Vespula vulgaris*), and Apidae (*Apis* sp., *Bombus* sp.). The key role of VOCs in adaptation to plant expansion is discussed.

Keywords: *Fallopia aubertii*; Bukhara fleeceflower; floral scent; HS-GC/MS; VOCs

Citation: Jakubska-Busse, A.; Dziadas, M.; Gruss, I.; Kobyłka, M.J. Floral Volatile Organic Compounds and a List of Pollinators of *Fallopia baldschuanica* (Polygonaceae). *Insects* **2022**, *13*, 904. https://doi.org/10.3390/insects13100904

Academic Editors: Gianandrea Salerno, Manuela Rebora, Stanislav N. Gorb and Lukasz L. Stelinski

Received: 25 July 2022
Accepted: 3 October 2022
Published: 5 October 2022

1. Introduction

Fallopia baldschuanica (Regel) Holub (syn. *Bilderdykia baldschuanica* (Regel) D.A.Webb, *Fagopyrum baldschuanicum* (Regel) Gross, *Fagopyrum baldschuanicum* H. Gross, *Polygonum baldschuanicum* Regel, *Reynoutria baldschuanica* (Regel) Moldenke) [1] also known as "Bukhara fleeceflower", "Russian Vine", "Fleece Flower", "Fleece Vine", "China Fleece Vine", "Silver Fleece Vine", "Silverlace Vine", or "Mile—a minute Plant" is a woody deciduous climber that belongs to the Knotweed family [2,3]. The taxonomic status of this plant is unclear. Currently, *F. baldschuanica* is treated as a synonym of *Fallopia aubertii* [4–6], although until recently these species had been separated as a different taxon [7,8]. The plant is native to Asia, mainly distributed in China, Russia, Kazakhstan, Afghanistan, Tajikistan, Pakistan, and Iran [7,9,10]. This species was probably introduced into Europe from Baldshuan

Khanate in Turkestan [11]. In the wild, plants grow at altitudes from 500 (900) m to 3200 m a.s.l. [7,12]. *F. baldschuanica* is also used in traditional and folk medicine to treat fever, pneumonia, and gout [13].

Fallopia baldschuanica is grown as an ornamental plant, that is often used by architects in Europe due to its extremely vigorous growth habit, as a cover for ugly structures, unsightly fences, and other garden structures. It often occurs as discarded material on waste ground, but until recently it was thought to be rarely well naturalized [14,15]. It can be found growing wild in parts of Europe and North and Central America [9]. It was first recorded outside cultivation in 1942 in the sea dunes in Duinbergen (Knokke, Belgium) [16]. This alien species was also classified as "thugs" by the Royal Horticultural Society (RHS) [17]. This term refers to garden plants that are easily available to buy, and that have the potential to become a nuisance [17]. Plants grow at a tremendous rate, and can put on over 4 m in one year; they may smother any other plants in their way [10].

Today, in many regions of Europe, *F. baldschuanica* is treated as an invasive plant, rapidly spreading beyond its intended borders [18]. *F. baldscuanica*, like other species of knotweeds, can reproduce sexually through seed production and/or clonally propagated by rhizomes and stem cuttings [19]. This species has been known to form a hybrid with *Fallopia japonica*, the hybrid was named *F. × conollyana* [19]. A characteristic feature of this species is the white, fragrant, long flowering (from July until the first frost) flowers, which provide nectar and pollen to various pollinating and visiting groups of insects.

Although *F. baldschuanica* is considered a potentially invasive plant species, little is known about its pollination biology in climatic conditions in Europe. Recently, scientists have been very interested in issues related to the hybridization of *Fallopia* species, important driving forces of invasive processes [19,20]. It should be remembered that pollinators play a key role in this process. Additionally, the occurrence of alien plant species can negatively affect the number of pollinators visiting native species. It was found that in urban areas pollinators choose more frequently the invasive plants, in comparison to similar pollinator communities in natural areas. Therefore, native plants in urban areas are less visited by insects and their diversity may decline in the future, which is a major aspect of its negative impact on the environment [21]. Additionally, pollinators could be more specialized in urban, than in natural areas [22].

In this study, we investigated: (i) what volatile organic compounds (VOCs) are emitted by *F. baldschuanica*; (ii) which groups of insects visit and pollinate plants of *F. baldschuanica*, in an urban area of Wrocław, Poland; as well as, based on data reported in the literature, (iii) what role the identified VOCs can play in the biology of the observed flower-visiting insects.

2. Materials and Methods

2.1. Plant Material

Fully open flowers of *F. baldschuanica* at the same developmental stage were collected from individuals of the five populations (sites 1–5) located in the center of Wrocław city, between 7 and 24 October 2020, and were used for the chemical analyzes. The study sites were located in the Old Town of Wrocław (Figure 1). Plants grew on the fence (sites 1 and 4) and in the vicinity of the dumpsters (sites 2, 3, 5). The location of sampling sites with GPS (Global Positioning System) coordinates: (1) 51°06′40.0″ N 17°02′13.1″ E; (2) 51°06′43.3″ N 17°02′15.3″ E; (3) 51°06′27.5″ N 17°02′00.1″ E; (4) 51°06′38.8″ N 17°02′10.4″ E; (5) 51°06′43.6″ N 17°01′58.1″ E. The largest distance between the sites was 600 m.

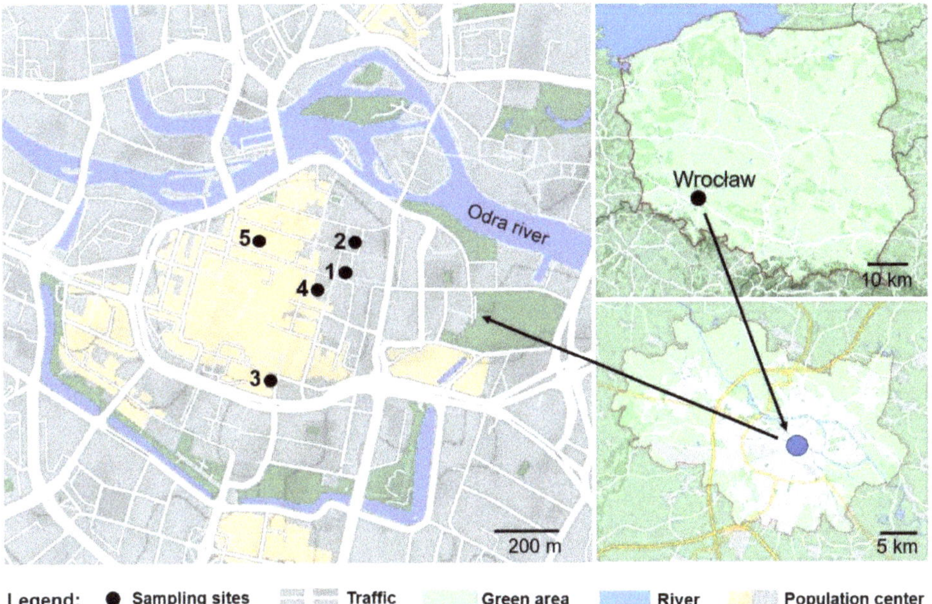

Figure 1. Location map of investigation area in the center of Wrocław: sampling sites: (1) 5 Kotlarska St., (2) 2–4 Jodłowa St., (3) 78–80 Szewska St., (4) Łaciarska St. (backyard), (5) 21 Kotlarska St.

2.2. HS-GC/S Analysis of Volatiles Fractions

The analysis of volatiles from the sample was carried out using GC-MS QP 2010 Ultra system (Shimadzu, Kyoto, Japan) equipped with headspace autosampler HS-20 (Shimadzu Corporation, Kyoto, Japan). A fresh sample of flowers (2 g) was weighed directly into a clean headspace vial (20 mL) with 10 mL pure water Merck Millipore (Merck Millipore, Warsaw, Poland) containing 180 ug 2-octanol (Sigma-Aldrich, Poznan, Poland) as internal standard, and closed using a screw cap with butyl septa (Sigma-Aldrich, Poznan, Poland). The sample was analyzed in triplicate. Table 1 shows the standard deviation (SD) of three replicates.

Program of head space autosampler: oven temperature 80°C, sample line 150°C, transfer line 150°C, equilibration time 10 min, pressurizing time 0.5 min (60 kPa), load time 0.5 min, injection time 1.00 min, needle flush time 2.00 min, shaking level 2. Chromatography analysis was carried out using ZB-5 ms capillary column (30 m × 0.25 diam., 0.25 film, Phenomenex, Torrance, CA, USA) with 1 mL/min flow of helium 6.0 purity (Linde Gas, Kraków, Poland) with split 1:2. Oven program: 40°C–0.00 min, 4°C/min to 140°C hold 0.00 min, 15 °C/min to 320°C hold 0.00 min. Single quadrupole mass detector operates in 38.00 to 488 scan range with 20,000 scan speed. The temperature of the ion source was 220°C, the interface was 260°C, and the solvent cut time was 1 min. The LabSolution ver 4.20 (Shimadzu, Kyoto, Japan) was used as software for data processing with NIST libraries 14 and 17 as databases. The tentative identification of compounds was based on a comparison with the mass spectral library and is presented in Table 1 and Supplementary Material S1.

The odor characteristics of the chemical compounds that are components of the scent of the analyzed plant, which were identified during the chromatographic study, were based on information available online [23].

Table 1. List of volatile organic compounds (VOCs), its chemical structure, and odor characteristic identified in *Fallopia baldschuanica*. Abbreviation: RT—retention time/minutes; SD—standard deviation.

	Chemical Name	RT	μg/2 g Sample	SD	NIST Search [%]	Kovats Index	Structure
1	hexanal	4.31	0.66	0.08	90	824	
	Odor characteristic: green, fruity, acorn, tallowy, fishy, grassy, herbal, leafy						
2	hex-3-en-1-ol	5.67	7.44	0.56	95	857	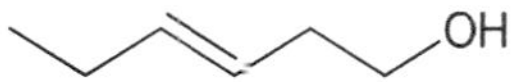
	Odor characteristic: fresh, green, grassy, leafy						
3	nonane	6.75	1.02	0.12	94	900	
	Odor characteristic: fusel-like						
4	heptanal	6.87	1.66	0.19	94	904	
	Odor characteristic: citrus, green, fatty, dry fish, pesticide, solvent, smoky, rancid, fruity						
5	3-thujene (5-isopropyl-2-methylbicyclo[3.1.0]hex-2-ene)	7.55	0.24	0.04	92	928	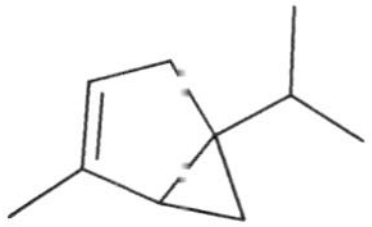
	Odor characteristic: woody, herbal, green						
6	α-pinene (2,6,6-trimethylbicyclo[3.1.1]hept-2-ene)	7.75	2.48	0.19	94	935	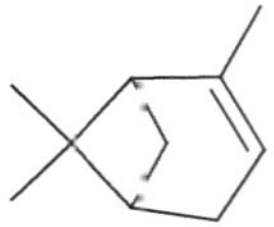
	Odor characteristic: terpeny, fruity, sweet, green, woody, pine, citrus, lime, camphor						

Table 1. *Cont.*

	Chemical Name	RT	µg/2 g Sample	SD	NIST Search [%]	Kovats Index	Structure
7	β-myrcene (7-methyl-3-methylene-1,6-octadiene)	9.65	0.16	0.03	92	991	
	Odor characteristic: metallic, musty, geranium, sweet, fruity, ethereal, soapy, lemon, spicy, woody						
8	decane	10	1.1	0.11	95	1000	
	Odor characteristic: fusel-like, fruity, sweet						
9	hex-3-ene-1-ol acetate	10.21	1.34	0.18	90	1008	
	Odor characteristic: green, freshly cut grass, slightly fruity						
10	limonene (1-methyl-4-(1-methylethenyl)-cyclohexene)	11.01	2.12	0.33	93	1036	
	Odor characteristic: licorice, green, citrus, ethereal, fruity						
11	β-ocimene ((E)-3,7-dimethyl-1,3,6-octatriene)	11.63	4.02	0.76	96	1056	
	Odor characteristic: herbal, mild, citrus, floral, woody, sweet, orange, lemon						
12	γ-terpinene (1-isopropyl-4-methyl-cyclohexa-1,4-diene)	12.05	0.06	0.01	90	1069	
	Odor characteristic: citrus, terpeny, herbal, fruity, sweet						

Table 1. *Cont.*

	Chemical Name	RT	µg/2 g Sample	SD	NIST Search [%]	Kovats Index	Structure
13	nonanal	13.78	0.72	0.1	97	1118	
	Odor characteristic: gravy, green, tallowy, fruity, gas, chlorine, floral (rose, orris), waxy, sweet, melon, soapy, fatty, lavender, citrus fruit						
14	hex-3-ene-1-ol butanoate	15.12	0.52	0.1	97	1153	
	Odor characteristic: sweet green, freshly cut grass, slightly fruity						
15	dodecane	17.18	0.16	0.03	93	1200	
	Odor characteristic: fusel-like						
16	butanoic acid, 2-methyl, 3-hexenyl ester (3-hexen-1-yl 2-methyl butyrate)	18.19	16	2.05	96	1233	
	Odor characteristic: fresh green apple sweet fruity pear						
17	2-butenoic acid, 2-methyl, (3Z)-3-hexen-1-yl	21.45	1.56	0.23	91	1325	
	Odor characteristic: leafy, green, vegetable						
18	β-farnesene ((E)-7,11-dimethyl-3-methylene-1,6,10-dodecatriene)	26.59	0.14	0.02	90	1458	
	Odor characteristic: woody, green						

2.3. Field Observations of Insects Activity

The observations were conducted during the flowering season (more precisely the peak of the plant flowering period) from the beginning of August to the end in October in 2019, and from the middle of July to the middle of October in 2020, in the urban area of Wrocław (SW Poland), in the five mentioned above sites (sites 1–5), located in Wrocław city center (Figure 1). Observations were made over a span of 2–6 h, covering daylight hours (9:00 a.m.–6:00 p.m.). Flower visitors were observed, with a total observation time of >60 h. The pollinators and visitor insects were photographed/documented using a Canon digital camera D50 camera (Canon EOS 50D, Canon Inc., Tokyo, Japan) with a Tamron 90 mm f /2.8 SP Di Macro lens, captured in field conditions by A.J-B. and identified by specialists. Only bumblebee species protected by law in Poland were not caught, they were photographed on flowers of *F. baldschuanica* and identified by entomologists on the basis of macrophotographs.

The insect abundance and related ecological indices between sites were compared using the non-parametric Kruskal-Wallis test by ranks, performed in SAS University Edition. The analyzes were performed separately for 2019 and 2020. The following indices were calculated: the dominance index (d), the Berger-Parker dominance index (D) [24], the Shannon-Weaver index (H′) [25], the Pielou index (J) [26], Margalef's species richness index (S) [27], species stability index C [28], and the Jaccard similarity index (SJ) [29]. The formulas used for calculations of the indices are included in the Supplementary Material S2.

The taxa abundance was correlated with the date, time of observation, and site, using the constrained analysis Canonical Correspondence Analysis (CCA). The analyzes were performed in Canoco 5.0. the significance of the axes was tested using the Monte-Carlo test.

3. Results

3.1. Identification of VOCs

The analysis of volatiles from the sample was performed by headspace analysis from fresh flowers using gas chromatography coupled with mass spectrometry (HS-GC/MS). The obtained chromatogram is presented in Figure 2.

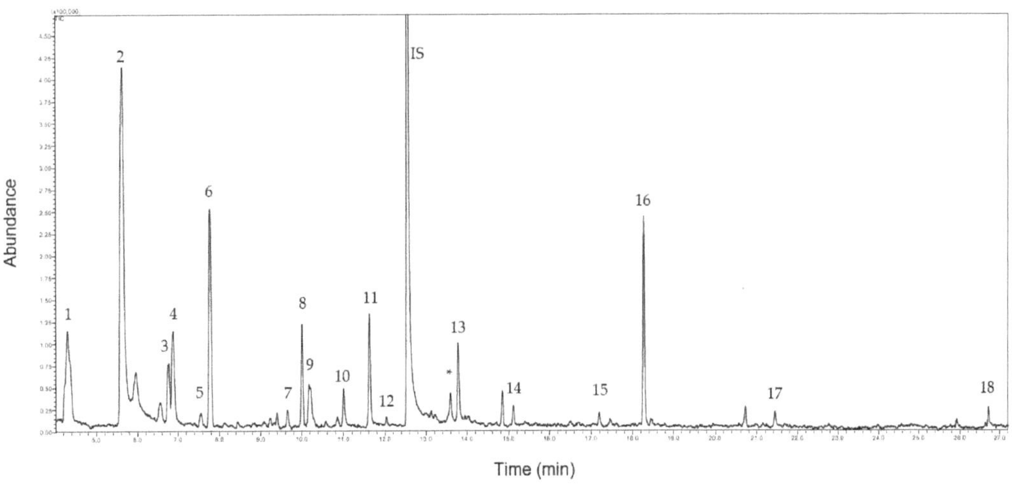

Figure 2. The chromatogram obtained during the analysis of the volatile organic compounds (VOCs) emitted by *Fallopia baldschuanica*. Abbreviation: *—unknown compound; IS—internal standard. The numbers in the chromatogram correspond to the numbers of the identified compounds in Table 1. Unnumbered signals are from the column filling (stationary phase) and/or from air pollution in the environment in which the flowers were packed into vials.

The list of identified volatile organic compounds (VOCs) together with its formulae and the odor characteristic of the identified compounds emitted by *F. baldschuanica* is presented in Table 1.

The analysis of volatiles from the sample reveals that *F. baldschuanica* emits mainly hydrocarbons: saturated (3, 8, and 15) and unsaturated commonly classified as monoterpenes (5, 6, 7, 10, 11, and 12) and sesquiterpene (18). Additionally, three aldehydes (1, 4, and 13), one unsaturated alcohol (2), and esters (9, 14, 16, and 17) having unsaturated chains are being produced.

A detailed analysis of the presented chromatogram reveals that there is another volatile compound present in the scent of *F. baldschuanica*. Its retention time is 13.60 min (abundance 18 ppm with SD—0.05). Unfortunately, we were not able to identify this chemical, thus its structure remains unknown.

3.2. True Pollinators and Visitors Insects

The analyzes were performed separately for 2019 and 2020. First, we aimed to find out if the pollinator's pool changed during the research. Secondly, each season is characterized by different weather and vegetation conditions, and therefore it was more appreciated to analyze each season separately. The pollinator community was the same in both years of the study, accounting for 24 species. However, the specific population indices differ between the seasons and are further described for the season 2019 and 2020. Some of the pollinators observed during the two-year study period are presented in Figure 3.

2019: The number of plant visitors, as well as the number of species, differed significantly between sites (Table 2). The largest abundances and species numbers were found in sites 1 and 4. Analyzing the species diversity indices, the Margalef's index and Shannon-Weaver index show better pollinator diversity in site 4 in comparison to measures in four other sites. The Pielou's index, representing the species' evenness, shows the lowest evenness for site 1, despite the large pollinator abundances. In total, 24 pollinator species were observed. At each site, four eudominants occurred, with the dominance index of >0.1. The Berger-Parker Dominance, which expresses the importance of the most abundant species, was the highest in site 4. There were several species which stability index (C) was > 100, taking into account all sites. Among those species observed in 2019 are Diptera from the families Syrphidae (*Chrysotoxum bicinctum*, *Eupeodes corollae*, *Episyrphus balteatus*, *Eristalis tenax*, *Syrphus ribesii*), Muscidae (*Musca domestica*), Calliphoridae (*Lucilia sericata*, *Lucilia caesar*), and Hymenoptera from the families Vespidae (*Vespula vulgaris*) and Apidae (*Apis* sp., *Bombus* sp.) (Table S1, Figure 3).

Table 2. The species and community responses of pollinators observed on *Fallopia baldschuanica* flowers in 2019.

Site	1		2		3		4		5		Total	C
	n	d	n	d	n	d	N	d	N	d		
Episyrphus balteatus	150.80	**0.13** *	29.78	**0.15**	7.00	0.04	202.40	**0.15**	87.89	**0.14**	219.62	**170.83**
Sarcophaga spp.	129.00	**0.11**	19.85	**0.17**	49.25	**0.34**	180.70	**0.13**	145.89	**0.25**	207.02	**191.67**
Musca domestica	149.70	**0.13**	26.47	**0.32**	37.75	**0.28**	164.10	**0.12**	121.44	**0.19**	195.06	**200.00**
Chrysotoxum bicinctum	97.40	**0.11**	0.00	0.00	0.00	0.00	110.00	0.08	52.75	0.07	172.14	**116.67**
Apis sp.	116.20	0.09	4.38	0.03	9.00	0.07	138.10	**0.10**	6.22	0.01	127.14	**170.83**
Syrphus ribesii	93.10	0.07	4.33	0.04	4.00	0.00	118.80	0.09	27.86	0.03	124.05	**154.17**
Lucilia caesar	95.33	0.07	13.93	**0.22**	23.67	**0.15**	116.00	0.08	59.56	**0.10**	122.61	**187.50**

Table 2. *Cont.*

Site	1		2		3		4		5		Total	C
	n	**d**	**n**	**d**	**n**	**d**	**N**	**d**	**N**	**d**		
Lucilia sericata	95.11	0.06	16.00	0.03	0.00	0.00	53.00	0.04	24.50	0.03	109.03	**116.67**
Eupeodes corollae	97.75	0.05	2.00	0.03	66.00	0.08	56.60	0.04	13.00	0.01	101.79	**116.67**
Eristalis tenax	76.10	0.05	6.33	0.01	0.00	0.00	58.30	0.04	13.80	0.01	98.76	**116.67**
Eristalis pertinax	72.90	0.04	5.00	0.01	0.00	0.00	38.78	0.02	17.75	0.01	89.15	**104.17**
Eristalis intricaria	53.63	0.03	0.00	0.00	0.00	0.00	25.11	0.02	84.67	0.03	86.57	83.33
Sphaerophoria scripta	31.83	0.01	0.00	0.00	0.00	0.00	33.33	0.02	68.13	0.09	86.33	95.83
Stomoxys calcitrans	43.67	0.01	1.50	0.01	0.00	0.00	28.80	0.02	20.25	0.01	55.13	91.67
Cerceris rybyensis	36.80	0.01	0.00	0.00	0.00	0.00	9.00	0.00	1.00	0.00	45.11	33.33
Cantharis pellucida	39.33	0.01	0.00	0.00	0.00	0.00	11.22	0.01	40.00	0.01	37.00	54.17
Bombus sp.	24.20	0.02	2.25	0.01	8.50	0.03	28.20	0.02	6.56	0.01	33.83	**145.83**
Bombus terrestris	8.50	0.00	0.00	0.00	1.00	0.00	4.80	0.00	14.00	0.00	12.17	45.83
Vespula vulgaris	5.00	0.00	0.00	0.00	7.00	0.01	3.70	0.00	1.86	0.00	7.29	**112.50**
Gasteruption assectator	5.75	0.00	0.00	0.00	0.00	0.00	2.00	0.00	0.00	0.00	6.60	37.50
Aglais io	3.00	0.00	0.00	0.00	0.00	0.00	1.00	0.00	0.00	0.00	3.00	20.83
Gaurotes virginea	2.00	0.00	0.00	0.00	0.00	0.00	0.00	0.00	0.00	0.00	2.00	4.17
Polistes dominula	1.00	0.00	0.00	0.00	0.00	0.00	1.00	0.00	1.00	0.00	1.50	12.50
Gasteruption spp.	0.00	0.00	0.00	0.00	0.00	0.00	1.00	0.00	0.00	0.00	1.33	8.33

Community indices

Site	1		2		3		4		5		Chi-Square	p
Total	1288.50	a	86.07	b	142.75	b	1362.10	a	639.22	Ab	33.02	0.0001
Species number	16.50	a **	5.80	b	6.00	b	18.40	a	12.67	Ab	37.88	0.0001
Berger-Parker dominance (D)	2.36	b	2.49	b	2.43	b	6.33	a	3.95	B	33.09	0.0001
Margalef (S)	1.50	ab	1.25	b	1.01	b	2.41	a	1.81	B	29.57	0.0001
Shannon − Weaver (H')	1.23	b	1.40	b	1.41	b	2.43	a	1.97	Ab	33.63	0.0001
Pielou (J)	0.44	b	0.82	a	0.80	a	0.84	a	0.78	A	17.20	0.0018

The n-mean abundance of the species in a particular site; N—mean number of all taxa in a particular site; d-dominance index; Total—an abundance of the species in all stands; C-species stability index; Chi-square, p—results of Kruskal-Wallis test; * The bold values indicate the dominance D > 0.1 and the species stability C > 100; ** Different lowercase letters in rows indicate significant differences between treatments, Kruskal-Wallis test, $p \leq 0.05$.

The CCA biplot shows the species abundance in relation to the date and time of observation, as well as the site (Figure S1 and Table S2). The total variance explained by the variables was 25.24 %, while mostly the CCA1 (variance explained = 16.28%) corresponds to the taxa abundance. It was found that taxa abundance decreases from the beginning (August) to the end (October) of the season. The community composition of pollinators of site 5 was the most universal, site 3 the most unique (Figure S1, Table 3).

Table 3. The species similarity index between sites in 2019.

Site	1	2	3	4	5
1	x	0.52	0.50	0.92	0.86
2		x	0.62	0.52	0.88
3			x	0.41	0.50
4				x	0.92
5					x

Figure 3. Pollinators and visitors of *Fallopia baldschuanica* (Polygonaceae) were observed in 2019–2020 in the center of Wrocław, SW Poland. (**A**) *Syrphus ribesii*, (**B**), *Sphaerophoria scripta*, (**C**) *Episyrphus balteatus*, (**D**) *Eristalis tenax*, (**E**) *Eristalis pertinax*, (**F**) *Sarcophaga* sp. (**G**) *Lucilia sericata*, (**H**) *Apis mellifera*, (**I**) *Gasteruption assectator*, (**J**) *Vespula vulgaris*, (**K**) *Polistes dominula*, and (**L**) *Aglais io*.

2020: The total number of plant visitors was significantly higher in sites 3 and 4, while the species number was significantly lower in sites 4 and 5 in comparison to other treatments (Table 4). Additionally, the diversity indices, Margalef, Shannon-Weaver, and Pielou, show better biodiversity responses in sites 3 and 4 than in the remaining populations. Similar to 2019, 24 species were observed. There were 3–4 eudominants (species accounting for more than 0.1 of total species occurrence) observed on each site. The species, with the stability index accounting for more than 100 are: Diptera from the families Syrphidae (*Chrysotoxum bicinctum, Eristalis pertinax, Eupeodes corollae, Episyrphus balteatus, Eristalis tenax, Syrphus ribesii, Eristalis intricaria*), Muscidae (*Musca domestica*), Sarcophagidae (*Sarcophaga* spp.), Calliphoridae (*Lucilia sericata, Lucilia caesar*), and Hymenoptera from families Vespidae (*Vespula vulgaris*) and Apidae (*Apis* sp., *Bombus* spp.) (Table S1, Figure 3). The species with the highest stability index were similar in both years, except for *Sarcophaga* spp., in which stability increased in 2020. Analyzing the CCA biplot, the taxa were uniformly distributed along with the CCA 1 and CCA 2 (Figure S2, Table S3). The total variance explained by variables is 26.18%, while CCA 1 explained 9.93% and CCA 2 explained 7.44% of the variance. The similarity between sites was low. Similar to 2020, most of the species were negatively distributed over time (the taxa abundance decreased during the season going). The time of sampling during the day has a minor impact. In the second year of the observations, the species similarity between sites increased and the similarity index between all sites was more than 0.8 (Table 5).

Table 4. The species and community responses of pollinators observed on *Fallopia baldschuanica* flowers in 2020.

Site	1		2		3		4		5		Total	C
	N	d	n	d	n	d	n	d	n	d		
Musca domestica	170.10	**0.17**	233.40	**0.22**	212.58	**0.17**	119.33	0.07	115.09	**0.15**	171.65	150.72
Episyrphus balteatus	114.92	**0.17**	187.64	0.19	178.50	0.15	180.33	**0.16**	153.25	**0.23**	162.51	**159.46**
Apis spp.	94.75	**0.13**	179.18	**0.20**	193.42	**0.14**	140.75	**0.11**	35.58	0.04	127.88	**200.00**
Sarcophaga spp.	81.38	0.05	88.78	0.06	102.29	0.02	137.67	0.12	128.17	**0.20**	111.58	**135.21**
Syrphus ribesii	30.10	0.03	89.20	0.07	152.10	0.06	171.45	**0.11**	63.89	0.05	103.50	**126.58**
Lucilia caesar	62.08	0.08	137.56	**0.11**	100.55	0.07	98.73	0.09	58.80	0.06	89.87	158.21
Sphaerophoria scripta	5.20	0.00	8.67	0.00	122.88	0.04	167.33	0.08	62.09	0.06	83.33	**100.00**
Lucilia sericata	54.33	0.09	84.90	0.08	69.10	0.04	126.60	0.07	52.57	0.03	78.08	**144.12**
Eupeodes corollae	40.83	0.02	3.88	0.00	150.75	0.06	5.00	0.00	44.40	0.02	61.04	76.71
Chrysotoxum bicinctum	95.70	**0.11**	4.60	0.00	58.42	0.07	70.00	0.07	8.67	0.00	60.64	**116.67**
Eristalis tenax	32.50	0.03	8.90	0.01	82.27	0.04	72.71	0.03	55.92	0.06	49.31	**136.84**
Eristalis pertinax	13.63	0.01	2.83	0.00	90.00	0.04	38.86	0.01	48.78	0.03	43.43	**103.90**
Eristalis intricaria	29.33	0.02	19.71	0.02	63.64	0.03	50.00	0.01	19.11	0.02	36.51	**104.00**
Bombus sp.	18.64	0.02	11.09	0.01	50.67	0.03	60.25	0.05	20.42	0.02	32.81	**193.33**
Stomoxys calcitrans	47.25	0.02	4.00	0.00	25.88	0.01	76.00	0.00	54.33	0.01	31.56	77.14
Polistes dominula	5.43	0.00	8.86	0.01	23.56	0.01	4.00	0.00	12.50	0.00	13.07	84.85
Vespula vulgaris	7.17	0.01	10.27	0.01	9.80	0.00	4.55	0.00	4.40	0.00	7.24	**166.15**
Bombus terrestris	5.00	0.00	4.20	0.00	8.33	0.00	11.13	0.00	6.00	0.00	6.88	**108.20**

Table 4. *Cont.*

Site	1		2		3		4		5		Total	C
	N	d	n	d	n	d	n	d	n	d		
Cantharis pellucida	4.00	0.00	2.40	0.00	12.86	0.00	0.00	0.00	5.00	0.00	6.50	55.00
Cerceris rybyensis	4.50	0.00	7.38	0.01	5.00	0.00	4.00	0.00	2.00	0.00	5.46	77.42
Gaurotes virginea	3.00	0.00	1.25	0.00	5.17	0.00	14.00	0.00	0.00	0.00	3.89	46.91
Gasteruption assectator	2.00	0.00	1.33	0.00	0.00	0.00	2.00	0.00	1.00	0.00	1.57	22.22
Aglais io	1.50	0.00	1.50	0.00	1.00	0.00	2.00	0.00	0.00	0.00	1.40	24.39
Gasteruption spp.	1.25	0.00	1.25	0.00	1.00	0.00	0.00	0.00	0.00	0.00	1.18	34.38
Community indices												
Site	1		2		3		4		5		Chi-Square	*p*
Total	784.58	b	995.64	b	1481.25	a	1255.25	a	777.50	B	33.47	0.0001
Species number	17.25	a	16.73	a	17.00	a	12.75	b	12.92	B	37.88	0.0001
Berger-Parker dominance (d)	2.69	b	3.88	ab	5.26	a	5.58	a	4.42	ab	33.09	0.0001
Margalef (S)	1.66	b	2.28	a	2.26	a	1.70	b	1.83	B	29.57	0.0001
Shannon − Weaver (H')	1.37	b	1.90	ab	2.33	a	2.21	a	2.03	ab	33.63	0.0001
Pielou (J)	1.37	b	1.90	ab	2.33	a	2.21	a	2.03	ab	33.63	0.0001
Total	0.48	b	0.69	ab	0.83	a	0.87	a	0.80	ab	17.20	0.0017

The n-mean abundance of the species in a particular site; N—mean number of all taxa in a particular site; d-dominance index; Total—an abundance of the species in all stands; C-species stability index; Chi-square, *p*—results of Kruskal-Wallis test; The bold values indicate the dominance D > 0.1 and the species stability C > 100; Different lowercase letters in rows indicate significant differences between treatments, Kruskal-Wallis test, $p \le 0.05$.

Table 5. The species similarity index between sites in 2020.

	1	2	3	4	5
1	x	1	0.96	0.92	0.86
2		x	0.96	0.92	0.92
3			x	0.96	0.83
4				x	0.87
5					x

4. Discussion

The intense floral scent in *F. baldschuanica*, detectable by the human nose, is composed of many interesting volatile organic compounds (VOCs) that can influence or manipulate insect behavior. The first of these is β-farnesene, one of two naturally occurring stereoisomers of this compound. Both (α- and β-) isomers are also insect semiochemicals, i.e., organic compounds used by insects to convey specific chemical messages that modify behavior or physiology. β-Farnesene is the most common isomer of the pair. It is found in the coating of apples, and other fruits, and it is responsible for the characteristic green apple odor [30]. It is a constituent of various essential oils, it occurs both in gymnosperms such as *Larix leptolepis* [31], and in several families of angiosperms, e.g., in Fabaceae: *Medicago sativa* [32]; in Asteraceae, *Anthemis tinctoria*, *Chamomilla recutita*, *C. suaveolens*, *Leucanthemum vulgare* [33]; and *Matricaria perforate*, in Lamiaceae, *Mentha aquatica* var. *citrata* [34]; as well as in the Cannabaceae family, *Cannabis* spp. [35]. Several plants, including potato species, have been shown to synthesize this semiochemical as a natural insect repellent [36,37], e.g., transgenic plants of *Arabidopsis thaliana* emitted this compound as a repellent to the *Myzus persicae* (Hemiptera, Aphididae) [38]. Furthermore, this compound is also widespread in the animal kingdom. For example, increased amounts of β-farnesene have been found in the urine of dominant male mice (*Mus domesticus*), which probably plays a role in marking the territory [39]. Several insect pheromones, including β-farnesene, were found in the urine of female African elephants (*Loxodonta africana*) [40].

This substance fulfills many tasks, especially in insects, for example as a pheromone in marking the nests of solitary bees belonging to the genus *Andrena* (Andreninae, Andrenidae) [41], as a defense allomone, and as a trace pheromone of the worker ant species *Myrmecia nigriceps* [42] or as kairomone for finding the prey in some predatory ground beetles (Coleoptera: Carabidae) [37,43]. It acts as an alarm pheromone in termites [44] or a food attractant for the apple tree pest, the codling moth [45]. Moreover, β-farnesene has been reported in the floral scent of a number of male euglossine bee-pollinated orchids [46,47]. This compound is a component of the sex pheromone of the medfly fly, *Ceratitis capitate* and may also be a pheromone component in a beetle [48]. Interestingly, its derivatives (E,E)-farnesol has frequently been reported as a component of the secretions of the Dufour's glands of Andrenid bees, of the Nasonov glands of honey bee workers, of the labial glands of bumble bees, and of the mandibular glands of leaf-cutting ants [48]. Unfortunately, the role of this compound as an insect attractant in *Fallopia* spp. has yet to be proven.

Additionally, β-farnesene plays an important role in aphid behavior [36]. It is also released by greenflies as an alarm pheromone upon death to warn away other aphids [49]. This sesquiterpene is produced by many species of aphids and is a signal for nearby individuals to stop foraging and escape. Aphids are plant pests, they suck plant juices, feed on shoot juice, and usually feed on young, juicy apical shoots, as well as on young leaves, inflorescences, and flower buds, which can damage them. Alert pheromones, apart from repelling aphids, are often attractants to their natural enemies. We believe that the ability to produce this floral scent compound may be considered a beneficial adaptation of the pest elimination by *F. baldschuanica*. This hypothesis needs further examination. β-Farnesene is also reported as an oviposition stimulant [50] for the hoverfly *Episyrphus balteatus*, an insect that has been observed as a visitor and true pollinator of *F. baldschuanica* (Figure 3C).

A repellent for aphid nymphs of *Panaphis juglandis* and *Chromaphis juglandicola* [51] is limonene, another VOCs which has been identified in *F. baldschuanica*. The larvicidal activity effect also has another compound identified by us, i.e., γ-terpinene. This terpene is a component of essential oils of many plant species e.g., in the family Lamiaceae, in *Thymus vulgaris* and *Origanum* species it is also considered an effective repellent against mosquitoes [52]. An important scent compound emitted by *F. baldschuanica* is also hexanal, considered an insect attractant, among others for flies of Psilidae (Diptera) [53]. Additionally, other VOCs, i.e., limonene and β-ocimene that we detected in *F. baldschuanica,* have been reported as constituents of the volatile bouquet of several citrus species [54]. These compounds were identified as an ingredient of different infested fruit species that attracted other parasitoids, such as *Agathis bishopi* (Hymenoptera: Braconidae) and *Aphidius gifuensis* (Hymenoptera: Aphidiidae) [54]. In addition, the β-ocimene has very common plant volatiles released in significant amounts from the leaves and flowers of many plant species, and is a general attractant of a wide spectrum of pollinators [55], including the honeybees *Apis mellifera* and bumblebees (*Bombus* spp.) [56,57] that we have observed (Figure 3G–I, Table 2). This acyclic monoterpene can play several biological functions in plants, depending on the organ and the time of emission and potentially affecting floral visitors, and also by mediating defensive responses to herbivory [55]. Due to its attractive fragrance, β-ocimene may be one of the key compounds emitted by *F. baldschuanica* that lures pollinating insects, and also attracts natural enemies of the phytophagous. Besides, phytophagous insects can identify the β-ocimene, and use it as chemical cues to identify their host plants [58,59]. The presence of this floral aroma compound has not yet been reported in representatives of the genus *Fallopia*. The β-ocimene and limonene are also reported as predominant components of essential oils of species of many plant families [60]. The nonanal, another compound we identified in *F. baldschuanica*, attracts e.g., the observed by us as flower visitors, flesh flies (Sarcophagidae) [61], and was also attractive to the herbivorous beetle, *Hylastes opacus*

(Coleoptera, Scolytidae), in North America [62]. Unfortunately, we did not observe the mentioned taxa of beetles visiting *F. baldschuanica*, but we identified beetles from other families, i.e., Cantharidae—*Cantharis pellucida* and Cerambycidae—*Gaurotes virginea*.

It is worth mentioning that heptanal, α-pinene, and limonene, have been isolated from flowers of the genus *Ophrys* [63], taxon pollinated by the sphecid and scoliid wasps and solitary bees, including long-horned bees from genus *Eucera* (Hymenoptera, Apidae). Orchids, especially of the *Ophrys* genus, belong to plants that are highly specialized in attracting specific groups of pollinators. Thus, it is possible that the high frequency of *F. baldschuanica* visitors of Apidae (*Apis mellifera*, *Bombus* spp.) is the result of the emission of these three volatile chemicals. Interestingly, neotropical orchids pollinated by the *Euglossini* (Hymenoptera, Apidae) emit a very intense fragrance but do not produce nectar or pollen food. Their flowers are pollinated by male euglossine bees, who are attracted by volatile semiochemicals, e.g., α-pinene and ocimene [63], and thus the compounds emitted by *F. baldschuanica*. Although we did not observe the *Euglossini* on *F. baldschuanica*, because these insects exclusively occur in South or Central America, but in the center of Wroclaw city, the flowers of *F. baldschuanica* are frequently visited by other bee taxa (*Apis* spp., *Bombus terrestris*). However, it seems that the synergistic effect of all VOCs identified in *F. baldschuanica* best explains the frequency of all groups of visitors and potential pollinator insects that we have observed. Moreover, based on our observations, the emission of a strong, perceptible odor by *F. baldschuanica* over a long period of the day, i.e., from about 10.00 am to sunset, results in the possibility of visiting a large group of insects and thus increases the chance of pollinating more flowers in inflorescences. Additionally, the flowers of this plant produce nectar from mid-July to the end of October (until the first frosts). We believe that the relatively long period of nectar production, and thus food provision for many groups of insects, may pose a key factor in a plant's success and may ensure its invasion. *F. baldschuanica* produces smaller flowers of large amounts, constantly offering visiting insects access to food. Some researchers indicate that pollinators or other groups of beneficial insects (predators, parasitoids) recognize the host plant, not by single compounds, but by specific ratios of ubiquitous compounds [64,65]. Additionally, insects process the olfactory signals by the receptor neurons [61]. This should be taken into account in planning olfactory experiments on insect attractants. Field observations of the insects' behavior have shown that during intensive flowering, i.e., from mid-July to mid-October 2019 and 2020, the plant emits a very intense scent and is visited by a large number of insects, although its activity changed seasonally. In total, 24 species were recognized, which is high diversity for an urban ecosystem. All the species found are native to Poland and central Europe. Based on our research, we are not able to access if the occurrence of this plant species decreased the number of pollinators visiting native plant species and how it affects native populations. The recent study of Kovács-Hostyánszki et al. [66] addresses this question, indicating that the populations of alien plant species in the long-term negatively affect native plant species and pollinator communities. Surprisingly, invasive plants can increase the foraging resources of pollinators, but only for the short term [66]. On the other hand, knowing that *F. baldschuanica* is a food base for many pollinator species, it may be considered as a beneficial plant for pollinators in the high-urbanized area, characterized by low plant diversity. Nonetheless, in designing greenery, the invasiveness of *F. baldschuanica* should be considered and native species, which are equally attractive to pollinators, should be introduced.

To compare another related plant species, i.e., Japanese knotweed *Reynoutria japonica* (syn. *Fallopia japonica*), it was visited by 14 pollinators [21]. The species diversity indices were low, which is rather specific for urban areas. Urbanization generally reduced pollinator diversity when compared to natural ecosystems [67]. Adult insects (imagines) of flies (Diptera), mainly representatives of Syrphidae, Muscidae, Sarcophagida, and Calliphoridae, as well Hymenoptera (*Vespula vulgaris*, *Apis* sp., *Bombus* sp.), were the most active pollinators of *F. baldschuanica* from July to even until mid-October. Kovács-Hostyánszki et al. [66] found, that mainly hoverflies benefit from plant invasion, which is in line with our studies. The same authors indicate that the number of wild bees decreased and the number of honey bees increased after plant invasions. Wild bees are often more closely associated with particular plant species and therefore their population is declining after being impoverished by the ecosystem [68].

We also observed differences between pollinator populations in the five study sites, which is probably also a result of the studied habitat specificity. The urban ecosystems are characterized by a high degree of habitat heterogeneity with microclimates and micro-habitats variations [69]. However, in 2020, the species similarity between the five study sites was very high. We may therefore suspect that insect populations have mixed up over time. Interestingly, the results of our research confirm the general data on the pollination biology of related plant species, i.e., *R. japonica*, provided by Balough [70]. According to this author, the most frequent visitors of *R. japonica* flowers are syrphid flies (Diptera, Syrphidae) and muscid flies (Diptera, Muscidae). Additionally common are hymenopterans (Hymenoptera), beetles (Coleoptera), true bugs (Hemiptera, Rhynchota), moths, and butterflies (Lepidoptera) [70]. Among this rich list of insects, we have not only observed the bugs (Rhynchota) as visitors/potential pollinators of *F. baldschuanica*, but the mentioned bugs feed on plants, using the sucking and piercing mouthparts to extract plant sap. We have often observed these phytophagous insects near *F. baldschuanica* plants, but never on flowers. Additionally, in the study on other related species *R. japonica*, ants were classified as insect visitors [21].

5. Conclusions

Among the main fragrance components of *F. baldschuanica* floral scent, the most important are β-ocimene, heptanal, nonanal, α-pinene, 3-thujene, and the alarm pheromones, β-farnesene, and limonene. Emitting such strong attractants by this potential invasive plant explains the observed numerous groups of flower-visiting insects both Hymenoptera (*Apis* sp., *Vespula* sp.) and Diptera (Syrphidae, Calliphoridae, Muscidae). Based on the results obtained, we hypothesize that the chemical composition of floral aroma in *F. baldschuanica* is a key factor in this species' evolution because volatile organic compounds (VOCs) attract a large group of potential pollinators.

Supplementary Materials: The following supporting information can be downloaded at: https://www.mdpi.com/article/10.3390/insects13100904/s1, Supplementary Materials S1: Comparison of the identified organic compounds (VOCs) with the mass spectral library. Supplementary Materials S2: The description of ecological indices used for data analysis. Table S1: List of floral visitors and true pollinators of *Fallopia baldschuanica* in Wrocław, SW Poland. Table S2: The statistics of the Canonical Correspondence Analysis (CCA) for 2019. Table S3: The statistics of the Canonical Correspondence Analysis (CCA) for 2020. Figure S1: The CCA biplot shows the taxa abundance in relation to the date and time of observations in 5 sites in 2019. Figure S2: The CCA biplot shows the taxa abundance in relation to the date and time of observations in 5 sites in 2020.

Author Contributions: Conceptualization, A.J.-B., M.D., I.G., and M.J.K.; methodology, A.J.-B., M.D., I.G., and M.J.K.; validation, A.J.-B., M.D., I.G., and M.J.K.; formal analysis, A.J.-B., M.D., and I.G.; investigation, A.J.-B., M.D., and M.J.K.; data curation, A.J.-B.; writing—original draft preparation, A.J.-B., I.G., and M.J.K.; writing—review and editing, A.J.-B. and I.G.; visualization, A.J.-B. and I.G.; supervision, A.J.-B.; project administration, A.J.-B.; funding acquisition, A.J.-B. All authors have read and agreed to the published version of the manuscript.

Funding: This research was funded by University of Wrocław (ZBot/2020/73/501/MPK 2599180000/10110).

Institutional Review Board Statement: Not applicable.

Informed Consent Statement: Not applicable.

Data Availability Statement: The data sets generated and analyzed in the present study may be available from the corresponding author upon request.

Acknowledgments: The authors would like to thank Bogdan Wiśniowski (University of Rzeszów), Robert Żóralski, Łukasz Mielczarek (Krakow Municipal Greenspace Authority, Poland), hab. Paweł

59. Kariyat, R.R.; Mauck, K.E.; Balogh, C.M.; Stephenson, A.G.; Mescher, M.C.; De Moraes, C.M. Inbreeding in horsenettle (*Solanum carolinense*) alters night-time volatile emissions that guide oviposition by *Manduca sexta* moths. *Proc. R. Soc. B Biol. Sci.* **2013**, *280*, 20130020. [CrossRef] [PubMed]

60. Shi, T.; Yue, Y.; Shi, M.; Chen, M.; Yang, X.; Wang, L. Exploration of Floral Volatile Organic Compounds in Six Typical *Lycoris* taxa by GC-MS. *Plants* **2019**, *8*, 422. [CrossRef] [PubMed]

61. James, D.G. Further field evaluation of synthetic herbivore-induced plant volatiles as attractants for beneficial insects. *J. Chem. Ecol.* **2005**, *31*, 481–495. [CrossRef] [PubMed]

62. De Greet, P.; Poland, T.M. Attraction of *Hylastes opacus* (Coleoptera: Scolytidae) to nonanal. *Can. Entomol.* **2003**, *135*, 309–311. [CrossRef]

63. Metcalf, R.L.; Kogan, M. Plant volatiles as insect attractants. *Crit. Rev. Plant Sci.* **1987**, *5*, 251–301. [CrossRef]

64. Bruce, T.J.; Wadhams, L.J.; Woodcock, C.M. Insect host location: A volatile situation. *Trends Plant Sci.* **2005**, *10*, 269–274. [CrossRef]

65. McCormick, A.C.; Unsicker, S.B.; Gershenzon, J. The specificity of herbivore-induced plant volatiles in attracting herbivore enemies. *Trends Plant Sci.* **2012**, *17*, 303–310. [CrossRef]

66. Kovács-Hostyánszki, A.; Szigeti, V.; Miholcsa, Z.; Sándor, D.; Soltész, Z.; Török, E.; Fenesi, A. Threats and benefits of invasive alien plant species on pollinators, *Basic Appl. Ecol.* **2022**. [CrossRef]

67. Wenzel, A.; Grass, I.; Belavadi, V.V.; Tscharntke, T. How urbanization is driving pollinator diversity and pollination–A systematic review. *Biol. Conserv.* **2020**, *241*, 108321. [CrossRef]

68. Scheper, J.; Reemer, M.; van Kats, R.; Ozinga, W.A.; van der Linden, G.T.; Schaminée, J.H.; Siepel, H.; Kleijn, D. Museum specimens reveal loss of pollen host plants as key factor driving wild bee decline in The Netherlands. *Proc. Natl. Acad. Sci. USA* **2014**, *111*, 17552–17557. [CrossRef]

69. Cadenasso, M.L.; Pickett, S.T.A.; Grove, J.M. Dimensions of ecosystem complexity: Heterogeneity, connectivity, and history. *Ecol. Complex.* **2006**, *3*, 1–13. [CrossRef]

70. Balogh, L. Japanese, giant and bohemian knotweed. In *The Most Important Invasive Plants in Hungary*; Botta-Dukát, Z., Balogh, L., Eds.; Institute of Ecology and Botany, Hungarian Academy of Sciences: Vácrátót, Hungary, 2008; pp. 13–33.

Article

Physicochemical Properties of Sugarcane Cultivars Affected Life History and Population Growth Parameters of *Sesamia nonagrioides* (Lefebvre) (Lepidoptera: Noctuidae)

Seyedeh Atefeh Mortazavi Malekshah [1], Bahram Naseri [1,*], Hossein Ranjbar Aghdam [2], Jabraeil Razmjou [1], Seyed Ali Asghar Fathi [1], Asgar Ebadollahi [3] and Tanasak Changbunjong [4,*]

[1] Department of Plant Protection, Faculty of Agriculture and Natural Resources, University of Mohaghegh Ardabili, Ardabil 5619911367, Iran
[2] Iranian Research Institute of Plant Protection, Agricultural Research, Education and Extension Organization (AREEO), Tehran 1985813111, Iran
[3] Department of Plant Sciences, Moghan College of Agriculture and Natural Resources, University of Mohaghegh Ardabili, Ardabil 5697194781, Iran
[4] Department of Pre-Clinic and Applied Animal Science, Faculty of Veterinary Science, Mahidol University, Nakhon Pathom 73170, Thailand
* Correspondence: bnaseri@uma.ac.ir (B.N.); tanasak.cha@mahidol.edu (T.C.)

Citation: Mortazavi Malekshah, S.A.; Naseri, B.; Ranjbar Aghdam, H.; Razmjou, J.; Fathi, S.A.A.; Ebadollahi, A.; Changbunjong, T. Physicochemical Properties of Sugarcane Cultivars Affected Life History and Population Growth Parameters of *Sesamia nonagrioides* (Lefebvre) (Lepidoptera: Noctuidae). *Insects* **2022**, *13*, 901. https://doi.org/10.3390/insects13100901

Academic Editors: Gianandrea Salerno, Stanislav N. Gorb and Manuela Rebora

Received: 5 September 2022
Accepted: 29 September 2022
Published: 3 October 2022

Publisher's Note: MDPI stays neutral with regard to jurisdictional claims in published maps and institutional affiliations.

Simple Summary: The sugarcane stem borer, *Sesamia nonagrioides* (Lefebvre), is the most important pest of sugarcane in Iran and some other regions of the world. Variation in the resistance of six commercial sugarcane cultivars to *S. nonagrioides* was investigated using the oviposition preference, life history, and population growth parameters. Moreover, the physical and biochemical properties of the tested sugarcane cultivars were estimated to understand any possible correlation between the insect's parameters and the physiochemical features of the cultivars tested. The physicochemical properties of sugarcane cultivars significantly affected *S. nonagrioides* oviposition behavior, life history, and population parameters. Based on the obtained results, the resistant cultivar, SP70-1143, could be recommended for cultivation in sugarcane fields where the risk of *S. nonagrioides* damage is usually high.

Abstract: The use of resistant cultivars is an efficient management strategy against *S. nonagrioides*. The effects of different sugarcane cultivars, CP48-103, CP57-614, CP69-1062, CP73-21, SP70-1143, and IRC99-02 were evaluated on the oviposition preference (free-choice assay), life history, and life table parameters of *S. nonagrioides* at 27 ± 1 °C, 60 ± 5% RH and a photoperiod of 16: 8 (L: D) h. The longest and shortest developmental times were on cultivars SP70-1143 and CP48-103, respectively. The oviposition preference of *S. nonagrioides* was the highest on cultivars CP48-103 and CP69-1062, and negatively correlated with the shoot trichome density and shoot rind hardness of the cultivars. The highest intrinsic rate of increase of *S. nonagrioides* was on cultivar CP48-103 and the lowest was on cultivar SP70-1143. The shortest mean generation time was on CP48-103 and the longest was on SP70-1143. The results indicate that cultivars CP48-103 and CP69-1062 were susceptible, and cultivar SP70-1143 was partially resistant against *S. nonagrioides*. This information could be useful for developing integrated management programs of *S. nonagrioides*, such as the use of resistant cultivars to reduce the damage caused by this pest in sugarcane fields.

Keywords: sugarcane stem borer; life table; sugarcane cultivar; oviposition preference; biochemical traits

1. Introduction

The need for more food for the world's growing population has increased the quantity and quality of agricultural crops and their maintenance. Therefore, in order to optimally manage agricultural products, we must have correct information about production facilities, production methods, and monitor the production process until consumption [1–3]. Sugarcane, *Saccharum officinarum* L. (Poaceae), an industrial multi-usage plant, is a major source of sugar and raw materials for agro-industries [4]. This plant is extremely vulnerable to insect pest attacks, including stem borers belonging to *Sesamia* spp. [4].

The sugarcane stem borer, *Sesamia nonagrioides* (Lefebvre) (Lepidoptera: Noctuidae) is one of the most dangerous insect pests on sugarcane, corn, sorghum, millet, rice, grasses, melon, asparagus, palms, and banana [5,6]. It has a wide distribution in the Khuzestan and Fars provinces of Iran, southern Europe, North Africa, and the Middle East [7–9]. The sugarcane stem borer is a devastating pest on all phenological stages of the sugarcane in Khuzestan province where it has up to five generations per year [9]. The feeding of *S. nonagrioides* larvae at the tillering stage of sugarcane results in dead hearts and causes severe damage, after the formation of internodes, by reduction in stalk weight and sugar quality [10]. Both the quality and quantity of sugar extracted from sugarcane plants can be decreased when they are severely attacked by *S. nonagrioides* [11–13].

Owing to the development of insect pests' resistance to insecticides and the negative effects of insecticides on human health and the environment, the use of alternative methods of insect control is essential [14,15]. Host plant resistance is one of the suitable alternatives to insecticides in pest management programs both economically and environmentally [16–18]. Based on integrated pest management (IPM) programs, the use of sugarcane cultivars that are resistant to stem borers could be an efficient method, along with other controlling methods such as chemical, cultural, and biological control [10,19–21].

The quality of host plants can influence the life history, population growth, oviposition preference, and feeding performance of insect pests [22,23]. Plant physicochemical properties including primary and secondary metabolites, as well as trichome density, tissue hardness, and moisture content can determine the oviposition behavior and population performance of insect pests [24–28]. Therefore, study of these parameters, as resistance indices, could be useful in the identification of resistant cultivars for use in IPM programs [29–32].

Several studies have been conducted on the resistance mechanisms, biology, and life table parameters of *S. nonagrioides* on different host plants. For example, Ranjbar Aghdam and Kamali [33] studied the rearing of *Sesamia cretica* Lederer and *S. nonagrioides* on various host plants under laboratory conditions and showed that maize and sorghum are suitable hosts to *S. nonagrioides* and *S. cretica*, respectively. The resistance mechanisms of sugarcane cultivars to *Sesamia* spp. were investigated by Asgarianzadeh [34], who identified that the antixenosis and tolerance were more important than antibiosis in providing resistance. The effects of sugarcane and maize were assessed on the life table parameters of *S. nonagrioides*, and sugarcane was reported to be more susceptible than maize to this pest [35].

Due to the importance of sugarcane as a strategic agricultural crop in Iran and many other countries of the world [4], and the high economic damages of *S. nonagrioides* on it, the current study was conducted to estimate the oviposition preference, life history, and life table parameters of *S. nonagrioides* on six commercially cultivated cultivars of sugarcane in Iran. In addition, we assessed the associations between the physical and biochemical traits of examined cultivars and the pest's life history and life table parameters. The findings of this investigation could be helpful in an IPM program of *S. nonagrioides*, such as the use of genetically engineered cultivars that are resistant to this pest.

2. Materials and Methods

The experiments were performed in a growth chamber set at 27 ± 1 °C, $60 \pm 5\%$ RH, and a photoperiod of 16:8 (L:D) h in the Iranian Research Institute of Plant Protection, Agricultural Research, Education and Extension Organization (AREEO), Tehran, Iran. We used a completely randomized design for all the experiments.

2.1. Host Plant

The cut stems of 6 sugarcane cultivars, CP48-103, CP57-614, CP69-1062, and CP73-21 (Canal Point USA); SP70-1143 (Sao Paulo, Brazil); and IRC99-02 (cross-made in Cuba and selected in Iran) were collected from the sugarcane cultivars bank of Sugarcane and Byproduct Development Company, Khuzestan, Iran (48°46′53.81″ E and 31°50′23.79″ N). These cultivars were selected based on commerciality, cultivated area, and their reaction to the stem borers under field conditions. Some sugarcane cultivars such as SP70-1143 and CP57-614 were cultivated in Matana Agricultural Research Station, Egypt, showing the importance of these cultivars around the world [36]. About 100 plants from each cultivar were cultivated under field conditions. Stems between the elongation phase and the ripening phase (250–300 days after planting)) were cut every week and kept at 4 °C until their use in the experiments.

2.2. Physical and Biochemical Properties of Sugarcane Cultivars

To examine the relationship between sugarcane cultivars and *S. nonagrioides* population parameters and oviposition preference, some physical-biochemical traits of sugarcane cultivars were evaluated as follows:

2.2.1. Density of the Shoot Trichomes

Sugarcane shoots were selected randomly from 5 plants of each cultivar. The trichome density of each cultivar was estimated by counting the number of trichomes per cm^2 of the sheath margin (the upper, middle, and lower parts of the shoots) using a stereomicroscope (SZX16, Olympus) [37].

2.2.2. Moisture Content

The 5-cm cut stems of each cultivar (five replicates) were weighed based on the fresh weight (*FW*) using an electronic balance (A & D Co., LTD, Tokyo, Japan, d = 0.01 g), and kept in an oven set at 70 °C for 48 h. After reaching a constant weight, they were re-weighed (dry weight: *DW*), and the moisture content (%*MC*) in the stems of each cultivar was calculated as the following equation:

$$MC = \frac{FW - DW}{FW} \times 100$$

2.2.3. Rind and Pith Hardness

Using a fruit hardness tester (LutronFR-5120, Taiwan, tip size: 3 mm), the rind hardness of stems and shoots, and the pith hardness of stems were measured. The stems rind hardness was measured by piercing the middle part of the third and fourth internodes (as a target internode for the penetration of the stem borer larvae) at 300 days after planting, and the penetrating point was registered in Newton (N) as a hardness index. The pith hardness was determined in the cross-section (5-cm pieces) of the same internodes. The shoot rind hardness was measured by punching the initial and middle parts of the sugarcane shoots, as the target site for *S. nonagrioides* oviposition. Each test was conducted for each cultivar in five replications.

2.2.4. Determination of Silica

One gram of the pulverized sample of the stems and shoots of different sugarcane cultivars was mixed with 9 g of lithium tetraborate (weight ratio 1:9) into a platinum labware, and then the final mixture was melted at 900 °C and dissolved in concentrated nitric acid. Eventually, 2 mL of the solution was placed inside the set ICP-OES (Varian Vista MPX), and the amount of silica in the sample was measured based on the different wavelengths (250–288 nm) created on the metal in the raw material [38].

2.3. Determination of Primary and Secondary Metabolites

2.3.1. Sample Preparation

All experiments were performed in five replicates per cultivar. The biochemical metabolites of sugarcane cultivars were measured using the extract of sugarcane stems. After washing the stems in water (five stems from each cultivar), they were dried at room temperature for 4 h and chopped using an electric grinder.

2.3.2. Determination of Total Protein

The protein content of the sugarcane cultivars was evaluated according to Bradford [39] using bovine serum albumin (BSA) as a standard. About 0.1 g of the powdered stems was mixed with 100 mL of 100 mM phosphate buffer and immediately centrifuged (12,000× rpm at 4 °C for 15 min). Then, 100 µL of the resulting supernatant was taken and transferred to the test tubes containing 5 mL of Bradford reagent. The absorbance was measured at 595 nm.

2.3.3. Determination of Carbohydrate

The carbohydrate content was measured using the Anthrone reagent [40]. A quantity of 0.5 g of the powdered stems was mixed with 5 mL ethanol 80% and centrifuged (4000× rpm at 4 °C for 15 min). Then, 0.1 mL of the obtained supernatant was added to 3 mL of Anthrone reagent, and the absorbance was read at 620 nm.

2.3.4. Determination of Total Phenolics

The total phenolics were assessed according to the Folin–Ciocalteu method [41]. The mixture of 0.5 g of the powdered stems and 5 mL of methanol 80% was centrifuged (4000× rpm at 4 °C for 15 min). A total of 1 mL of the obtained supernatant was transferred to a test tube and its volume was increased to 50 mL by distilled water. Then, 0.25 mL of Folin–Ciocalteu and 1.25 mL of sodium carbonate 20% were added to the mixtures and mixed completely. The tubes were kept at room temperature in the dark for 40 min. The absorbance of each sample was measured at 725 nm using a spectrophotometer (Unico, UV/Vis2100, Fairfield, NJ, USA). The result was expressed as mg of gallic acid equivalent/100 g of fresh weight sample.

2.3.5. Determination of Total Flavonoids

The total flavonoids were determined using the aluminum chloride colorimetric assay method [42]. A total of 1 mg of powdered stems was mixed with 1 mL of 2% methanolic solution of aluminum chloride, and stored at room temperature for 15 min. The absorbance was read at 430 nm. The flavonoid content was expressed as mg of quercetin equivalents (QE)/g of fresh sample.

2.3.6. Determination of Tannins

The condensed tannin content in the stems was analyzed using the vanillin–HCL method [43]. About 0.5 mg of the powdered samples was mixed with 3 mL of vanillin (4%) (*w/v*, vanillin in methanol). Then, 1.5 mL of HCL was added to this extract, and after storage at room temperature in the dark for 15 min, the absorbance was measured at 500 nm. The tannin values of the samples were expressed as mg of catechin equivalent (CE)/100 g of fresh sample.

2.4. Rearing of Sesamia nonagrioides

The initial population of *S. nonagrioides* was established using about 10,000 eggs collected from sugarcane fields of Sugarcane and Byproducts Development Company, Khouzestan, Iran, in May 2019. The eggs were kept in 9 cm Petri dishes with a piece of wet cotton ball to prevent the eggs from desiccation. The hatched neonate larvae were reared on the cut stems of each sugarcane cultivar in transparent plastic containers (19 × 13 × 4 cm) with a hole covered by a cloth mesh for ventilation and kept in a growth chamber set at

27 ± 1 °C, 60 ± 5% RH, and a photoperiod of 16:8 (L:D) h. Every 2–3 days, the larvae were transferred to new containers with fresh-cut sugarcane stems. After 7–10 days, the larvae that were developed to the third instar were reared individually on 20-cm cut stems until pupation. Each 10 cut stems were placed in cylindrical plexiglass jars (25 × 17 cm, Venus Co., Tehran, Iran), which were covered at the top by a fine mesh net for ventilation. After pupation, the sex of individuals was distinguished based on the distance between the sexual hole and anus, which is longer in females than in males [44], and the pupae were sterilized using 5% sodium hypochlorite solution [33]. Male and female pupae were kept separately in disposable containers (7 cm diameter × 3 cm depth with a cap) and checked daily until adult emergence. After the emergence of the adult moths, 3–7 pairs of moths (with equal sex ratio) were transferred into cylindrical transparent plexiglass jars (17 cm diameter × 25 cm depth) for mating and oviposition. Each oviposition jar contained 3–5 sugarcane shoots to provide natural oviposition substrate and a piece of cotton ball soaked in 10% honey:water solution as a source of carbohydrate for feeding of the adult moths [33]. During the oviposition period, the eggs were gathered from the leaf sheath daily using a fine brush (No. 000), and the sugarcane shoots were replaced with new ones. The sterilization of the eggs was carried out using the procedure described by Ranjbar Aghdam and Kamali [33]. In order to nutritionally adapt the *S. nonagrioides* on each sugarcane cultivar, the insect was reared for two generations on each cultivar, and the eggs obtained from the third generation were used in the experiments.

2.5. Oviposition Preference of Sesamia nonagrioides

The oviposition preference (number of eggs laid) of *S. nonagrioides* was tested on six sugarcane cultivars according to the free-choice assay. Three detached sugarcane shoots (length 20 cm) from each cultivar were randomly arranged in wooden cages (80 × 80 × 60 cm) covered by a white mesh cloth. Five pairs of one-day-old male and female adults were released in the center of the cage, and after 72 h, the eggs laid under the leaf sheath were counted separately for each cultivar. This experiment was performed in five replications.

2.6. Life History and Survival of Sesamia nonagrioides

The life history and survival of *S. nonagrioides* were investigated using 200 eggs (<24 h old) as a cohort for each treatment (sugarcane cultivar), which were laid by the females reared on the same cultivar. All 200 eggs for each cultivar were transferred into sterilized glass Petri dishes (diameter 9 cm), and the number of hatched eggs and their incubation periods were recorded daily. Newly emerged larvae (<24 h old) were placed individually in transparent plastic containers (9 cm diameter × 5 cm depth) containing a piece of the sugarcane cut stem (3 cm diameter × 7 cm height). The cut stems of each cultivar were replaced by fresh ones every 3–4 days during the larval development until prepupal stage. The pre-pupae remained unchanged inside the stems until pupation. After pupation, the pupae were sexed and weighed using laboratory scales (Sartorius AG Germany GCA803S, $d = 0.001$ ct) at 24 h after pupation on each sugarcane cultivar and placed individually in new containers (7 cm diameter × 3 cm depth with a cap) until adult emergence. Larval and pupal containers were checked daily, and their duration and survival were recorded. The emerged adult moths were transferred to oviposition jars (17 cm diameter × 25 cm depth), and adult longevity and the number of eggs laid by each female (fecundity) were recorded daily. In order to determine egg hatching (fertility), the egg masses laid at different times were kept separately. Daily monitoring was continued until the death of the last female and male moths. The sex ratio was calculated as the following formula:

$$\text{Sex Ratio} = \frac{\text{females}}{\text{females} + \text{males}} \times 100$$

2.7. Life Table Parameters of Sesamia nonagrioides

The life table (population growth) parameters of *S. nonagrioides* were assessed based on the age-stage, two-sex life table theory [45,46] using TWOSEX-MSChart software [47]. The age-specific survival rate (l_x) was calculated as:

$$l_x = \sum_{j=1}^{k} s_{xj}$$

where k is the number of stages and s_{xj} is the probability that newly laid eggs will survive to age x and stage j. Age-specific fecundity (m_x) was measured using the following formula:

$$m_x = \frac{\sum_{j=1}^{k} s_{xj} f_{xj}}{\sum_{j=1}^{k} s_{xj}}$$

where f_{xj} (age-stage-specific fecundity) is the number of hatched eggs that were produced by an adult female at age x. The intrinsic rate of increase (r) was calculated as the following formula:

$$1 = \sum_{x=0}^{\infty} e^{-r(x+1)} l_x m_x$$

Other population growth parameters including the gross reproductive rate (GRR), net reproductive rate (R_0), finite rate of increase (λ), and mean generation time (T) were calculated as $GRR = \sum_{x=0}^{\infty} m_x$, $R_0 = \sum_{x=0}^{\infty} l_x m_x$, $\lambda = e^r$ and $T = (\ln R_0)/r$ [48,49].

2.8. Statistical Analysis

The physicochemical traits data, the weights of pupae, and the oviposition preference of *S. nonagrioides* were examined for normality with a Kolmogorov–Smirnov test using SPSS version 16.0 [50], and all the data were normally distributed. These data were analyzed by a one-way analysis of variance (ANOVA), and the statistical differences among the treatments were compared using Tukey's HSD test at a 5% probability level. The developmental time, adult pre-oviposition period (APOP: the time between the emergence of an adult female and the start of its oviposition), total pre-oviposition period (TPOP: the duration from egg to first oviposition), oviposition period, fecundity, and life table parameters were analyzed based on the age-stage, two-sex life table theory [45,46]. Additionally, the bootstrapping with 100,000 replications was used to estimate the variances and standard errors of life history and population parameters [51,52], and the paired bootstrap test was used to compare the means [53,54]. The relationship between the physicochemical features of the tested sugarcane cultivars and *S. nonagrioides* oviposition preference and population parameters was evaluated by Pearson correlation test using the average data of variables [55,56]. A dendrogram by Ward's method was drawn based on the oviposition preference, life history, and life table parameters of the insect reared on examined sugarcane cultivars using SPSS version 22.0 (IBM, Chicago, IL, USA).

3. Results

3.1. Properties of Sugarcane Cultivars

The physical and biochemical features of sugarcane cultivars used for the feeding of *S. nonagrioides* are presented in Table 1. The highest trichome density in sugarcane shoot was observed in cultivar SP70-1143, whereas the lowest density was determined in cultivars CP48-103, CP69-1062, and IRC99-02 ($F = 278.34$; df = 5, 29; $p < 0.001$). The stem of cultivar SP70-1143 (66.91%) had the lowest moisture content compared with other cultivars ($F = 352.57$; df = 5, 29; $p < 0.001$). The results showed significant differences in the shoot rind hardness (F = 1176.21; df = 5, 29; $p < 0.001$) and stem rind hardness (F = 6020.53; df = 5, 29; $p < 0.001$) among the various sugarcane cultivars. The stem and shoot rind hardness indices were the highest in cultivar SP70-1143 and the lowest in cultivar CP48-103. Among the tested sugarcane cultivars, the highest value of stem pith hardness was recorded in cultivar SP70-1143 (F = 977.49; df = 5, 29; $p < 0.001$).

Table 1. Mean ($\pm$SE) physical and biochemical traits (n = 5) in examined sugarcane cultivars.

Parameters	Cultivars					
	CP48-103	IRC99-02	CP69-1062	CP73-21	CP57-614	SP70-1143
Moisture content (%)	83.27 ± 0.20 b	86.13 ± 0.57 a	87.80 ± 0.16 a	77.92 ± 0.37 c	71.96 ± 0.57 d	66.77 ± 0.58 e
Trichome density (per cm^2)	1.40 ± 0.25 d	1.20 ± 0.2 d	1.40 ± 0.25 d	5.80 ± 0.20 c	9.20 ± 0.37 b	12.60 ± 0.40 a
Shoot rind hardness (Newton)	14.80 ± 0.37 f	17.80 ± 0.40 e	22.0 ± 0.32 d	32.60 ± 0.25 c	36.20 ± 0.49 b	46.40 ± 0.25 a
Stem rind hardness (Newton)	56.80 ± 0.58 f	86.80 ± 0.37 e	111.80 ± 0.58 d	120.0 ± 0.45 c	140.0 ± 0.45 b	163.0 ± 0.45 a
Stem pith hardness (Newton)	10.20 ± 0.37 c	9.00 ± 0.32 c	8.80 ± 0.20 c	19.0 ± 0.37 b	32.60 ± 0.25 a	33.20 ± 0.37 a
Total protein (mg/mL)	0.015 ± 0.002 d	0.069 ± 0.004 b	0.014 ± 0.002 d	0.048 ± 0.002 c	0.055 ± 0.004 c	0.096 ± 0.002 a
Carbohydrate content (mg/mL)	396.78 ± 1.30 b	394.36 ± 0.62 b	430.79 ± 1.48 a	394.29 ± 0.93 b	398.19 ± 1.15 b	399.67 ± 0.97 b
Total tannins (mg CE/100 g FW)	23.36 ± 0.32 e	31.664 ± 0.81 c	28.490 ± 0.44 d	30.988 ± 0.52 c	33.882 ± 0.23 b	37.972 ± 0.51 a
Total phenolics (mg GA/100 g FW)	8.374 ± 0.15 e	10.966 ± 0.17 c	9.69 ± 0.03 d	10.833 ± 0.17 c	12.726 ± 0.16 b	13.862 ± 0.12 a
Flavonoids content(mg QE/g FW)	58.57 ± 0.36 e	147.44 ± 3.55 b	108.68 ± 2.36 d	122.75 ± 1.49 c	174.97 ± 1.95 b	256.94 ± 2.64 a

Means in each row followed by different letters are significantly different ($p < 0.05$, Tukey's HSD test).

The highest amount of silica was observed in the stem (F = 389.66; df = 5, 17; $p < 0.001$) and shoot (F = 13,837.68; df = 5, 17; $p < 0.001$) of cultivar CP69-1062, while the lowest amount was observed in the stem and shoot of cultivars CP73-21 and SP70-1143 (Figure 1).

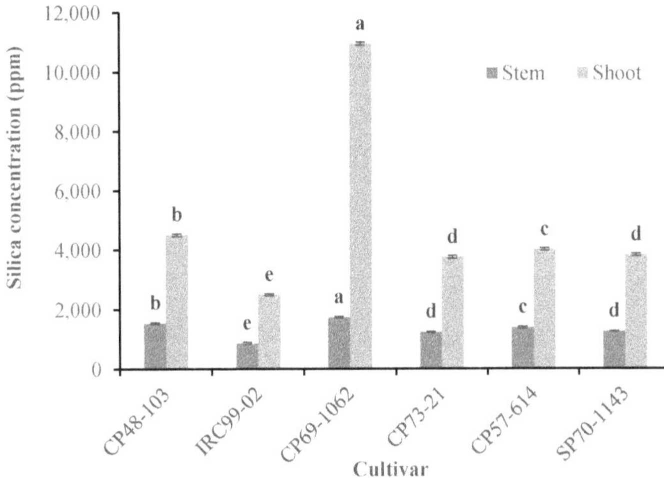

Figure 1. Mean ($\pm$SE) silica concentration (ppm) in the stem and shoot of tested sugarcane cultivars. Mean values followed by different letters are significantly different according to Tukey's HSD test ($p < 0.05$).

The total protein content (F = 139.18; df = 5, 24; $p < 0.001$) significantly varied among the tested sugarcane cultivars; the lowest values were in cultivars CP48-103 and CP69-1062, and the highest value was in cultivar SP70-1143. The carbohydrate content in cultivar CP69-1062 was significantly higher than in the other cultivars (F = 87.27; df = 5, 29; $p < 0.001$). The condensed tannins content was the highest in cultivar SP70-1143, whereas the lowest was in CP48-103 (F = 94.82; df = 5, 29; $p < 0.001$). A significant difference in the total phenolics (F = 197.10; df = 5, 29; $p < 0.001$) and total flavonoids (F = 872.13; df = 5, 29; $p < 0.001$) was detected in different sugarcane cultivars; with higher content in cultivar SP70-1143 and lower content in cultivar CP48-103.

3.2. Oviposition Preference of Sesamia nonagrioides

The free-choice oviposition preference of *S. nonagrioides* on the sugarcane cultivars is displayed in Figure 2. The maximum number of eggs laid was on cultivars CP48-103 and

CP69-1062, and the minimum number was on cultivar SP70-1143 ($F = 1043.05$; df = 5, 29; $p < 0.001$).

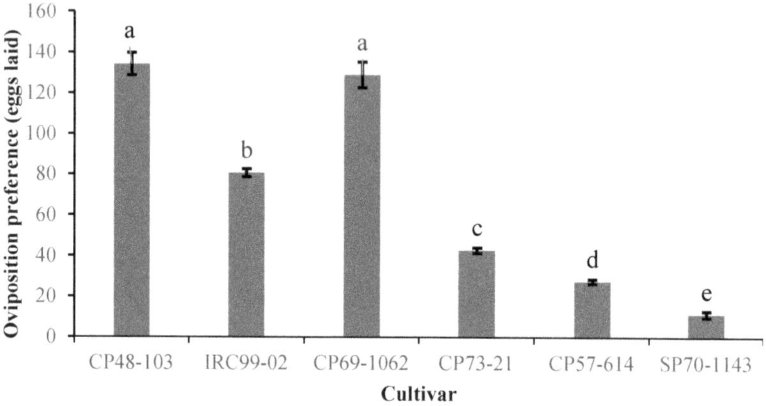

Figure 2. Mean (±SE) oviposition preference of *Sesamia nonagrioides* on different sugarcane cultivars. Mean values followed by different letters are significantly different according to Tukey's HSD test ($p < 0.05$).

3.3. Developmental Time and Survival of Sesamia nonagrioides

The outcomes of the effect of different sugarcane cultivars on the developmental time and survival (from egg to adult emergence) of *S. nonagrioides* are shown in Table 2. The average incubation period of *S. nonagrioides* on cultivar CP48-103 was significantly shorter than on the others ($p < 0.05$). The longest larval period was on cultivar SP70-1143 (50.09 ± 2.12 days), and the shortest was on cultivar CP48-103 (31.63 ± 1.02 days) ($p < 0.05$). No significant differences ($p > 0.05$) were observed for pre-pupal period among the sugarcane cultivars. The longest and shortest pupal period was seen on cultivars SP70-1143 (11.19 ± 0.07 days) and CP69-1062 (10.52 ± 0.05 days) ($p < 0.05$). The mean developmental time (from egg to adult emergence) on cultivar CP48-103 was about 25.50 days shorter than on cultivar SP70-1143 ($p < 0.05$). The highest immature (egg, larva, prepupa and pupa) survival was on cultivars CP48-103 ($81.25 \pm 3.07\%$) and CP69-1062 ($76.25 \pm 3.36\%$), and the lowest on SP70-1143 ($19.34 \pm 2.56\%$).

Table 2. Mean (±SE) developmental time (day) and survival rate (%) of *Sesamia nonagrioides* immature stages on examined sugarcane cultivars.

Cultivars	Egg Incubation	Larval Period	Pre-Pupal Period	Pupal Period	Developmental Time	Immature Survival
CP48-103	6.01 ± 0.01 b	31.63 ± 1.02 e	2.03 ± 0.01 a	10.66 ± 0.05 cd	50.33 ± 1.07 e	81.25 ± 3.07 a
IRC99-02	6.04 ± 0.01 ab	33.00 ± 0.73 de	2.04 ± 0.02 a	10.70 ± 0.08 cd	51.68 ± 0.75 de	55.61 ± 3.93 b
CP69-1062	6.06 ± 0.02 a	34.26 ± 0.51 d	2.03 ± 0.01 a	10.52 ± 0.05 d	52.73 ± 0.53 d	76.25 ± 3.36 a
CP73-21	6.06 ± 0.02 a	36.70 ± 0.51 c	2.05 ± 0.02 a	10.73 ± 0.08 bc	55.38 ± 0.55 c	58.13 ± 3.89 b
CP57-614	6.07 ± 0.02 a	40.79 ± 0.88 b	2.09 ± 0.03 a	10.96 ± 0.08 b	60.15 ± 0.99 b	48.13 ± 3.95 b
SP70-1143	6.10 ± 0.03 a	50.09 ± 2.12 a	2.09 ± 0.04 a	11.19 ± 0.07 a	75.81 ± 2.19 a	19.34 ± 2.56 c

Means in each column followed by different letters are significantly different ($p < 0.05$, paired bootstrap test).

3.4. Weight of Male and Female Pupae

Figure 3 shows the weight of the male and female pupae of *S. nonagrioides* on the examined sugarcane cultivars. The mean weight of the female pupae significantly varied from 110.2 mg on cultivar SP70-1143 to 186.4 mg on cultivar CP69-1062 ($F = 78.06$; df = 5, 164; $p < 0.001$). Furthermore, the male pupal weight on cultivar SP70-1143 (100.4 mg) was significantly lower than the other cultivars ($F = 5.37$; df = 5, 179; $p < 0.001$).

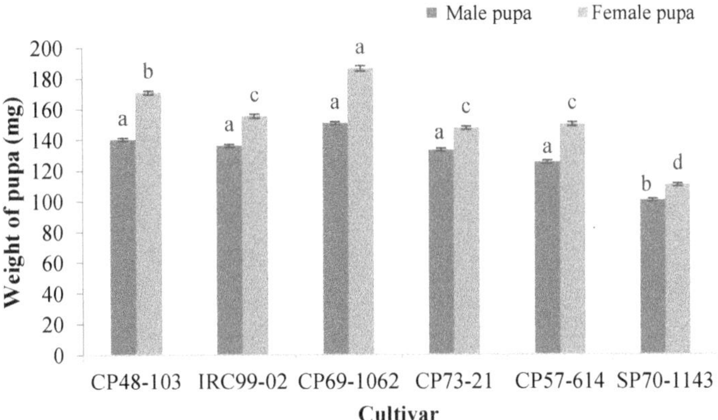

Figure 3. Mean (±SE) weight of pupal stage (mg) of *Sesamia nonagrioides* on examined sugarcane cultivars. Mean values followed by different letters are significantly different according to Tukey's HSD test ($p < 0.05$).

3.5. Longevity and Reproductive Variables of Sesamia nonagrioides

Adult longevity showed significant differences on the examined sugarcane cultivars ($p < 0.05$). The longevity of both male and female adults of *S. nonagrioides* was shortest on cultivar SP70-1143, while the longest female longevity was observed on cultivars CP48-103 and CP69-1062 (Table 3). The pre-oviposition and oviposition periods, fecundity and fertility of *S. nonagrioides* on the tested cultivars are presented in Table 3. Various sugarcane cultivars had a significant influence on the adult pre-oviposition period (APOP) and total pre-oviposition period (TPOP) of *S. nonagrioides*. The longest and shortest APOP was on cultivars SP70-1143 and CP69-1062, respectively ($p < 0.05$). The longest TPOP was observed on cultivar SP70-1143 and the shortest was on cultivars CP69-1062 and CP48-103 ($p < 0.05$). Moreover, the longest oviposition period was on cultivars CP48-103 and CP69-1062, and the shortest was on cultivar SP70-1143 ($p < 0.05$). The fecundity and fertility of *S. nonagrioides* were the highest on cultivars CP48-103 and CP69-1062, and the lowest on cultivar SP70-1143 ($p < 0.05$).

Table 3. Mean (±SE) duration (day) of pre-oviposition and oviposition period, fecundity (eggs), fertility (percentage of hatched eggs), and adult longevity (day) of *Sesamia nonagrioides* reared on examined sugarcane cultivars.

Cultivars	APOP	TPOP	Oviposition Period	Fecundity (Eggs/Female)	Fertility (%)	Male Adult Longevity	Female Adult Longevity
CP48-103	0.97 ± 0.02 b	52.69 ± 1.52 d	4.70 ± 0.12 a	228.79 ± 7.12 a	94.71 ± 0.16 a	6.88 ± 0.13 a	6.04 ± 0.12 a
IRC99-02	1.00 ± 0.01 b	54.64 ± 1.15 dc	3.56 ± 0.10 c	188.60 ± 6.81 c	88.25 ± 0.21 c	6.60 ± 0.13 ab	4.84 ± 0.12 b
CP69-1062	0.85 ± 0.04 c	54.00 ± 0.79 d	4.79 ± 0.12 a	228.76 ± 7.17 a	93.50 ± 0.14 a	6.57 ± 0.14 ab	5.86 ± 0.12 a
CP73-21	1.00 ± 0.01 b	57.18 ± 0.71 bc	3.79 ± 0.11 c	213.88 ± 7.72 ab	89.62 ± 0.19 b	6.44 ± 0.14 b	4.93 ± 0.13 b
CP57-614	0.97 ± 0.02 b	60.50 ± 1.54 b	4.14 ± 0.10 b	197.38 ± 11.15 bc	82.89 ± 0.14 d	6.64 ± 0.08 ab	5.05 ± 0.14 b
SP70-1143	1.19 ± 0.08 a	78.89 ± 2.97 a	2.95 ± 0.16 d	67.76 ± 9.09 d	74.75 ± 0.72 e	5.99 ± 0.12 c	4.61 ± 0.10 c

APOP, adult pre-ovipositional period (from the emergence of an adult female to the start of its oviposition); TPOP, total pre-ovipositional period (from egg to first oviposition). Means in each column followed by different letters are significantly different ($p < 0.05$, paired bootstrap test).

3.6. Survival Rate and Fecundity Curves

The age-stage-specific survival rate (s_{xj}) of *S. nonagrioides* on various sugarcane cultivars is presented in Figure 4. The survival rates of *S. nonagrioides* at similar ages and developmental stages on CP48-103 and CP69-1062 were higher than on the other tested cultivars. Furthermore, the survival curves of male and female adults were further outspread on these cultivars, indicating an increase in oviposition period and the insect population on these cultivars.

The age-specific survival rate (l_x), age-specific fecundity (m_x), and the age-stage-specific fecundity (f_{xj}) of *S. nonagrioides* on various cultivars are plotted in Figure 5. The convexity of the l_x curve was highest on cultivars CP48-103 and CP69-103, while the highest concavity of this curve was seen on cultivar SP70-1143. The m_x curve indicated that the highest number of eggs laid occurred on day 72 (98 eggs per individual per day) on CP69-1062. However, the reproduction of female moths performed much earlier on the CP48-103 cultivar (on the thirty-sixth day) than the other cultivars. Moreover, females lived longer on this cultivar. The highest f_{xj} of *S. nonagrioides* on CP48-103, CP69-1062, CP73-21, IRC99-02, CP57-614, and SP70-1143 was122, 123, 85, 110, 83, and 45 eggs per female per day, respectively. These curves showed that the survival and fecundity rates, as well as oviposition period, were higher on CP69-1062 and CP48-103 cultivars than the other cultivars.

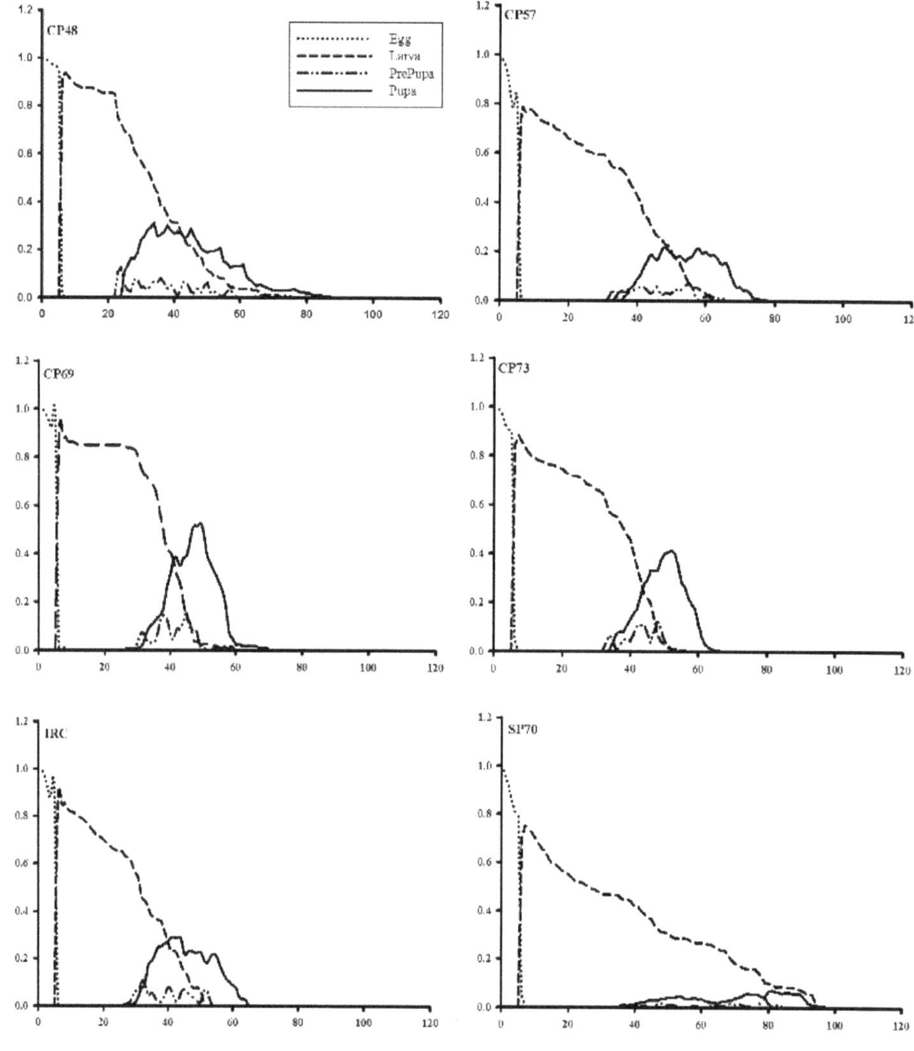

Figure 4. *Cont.*

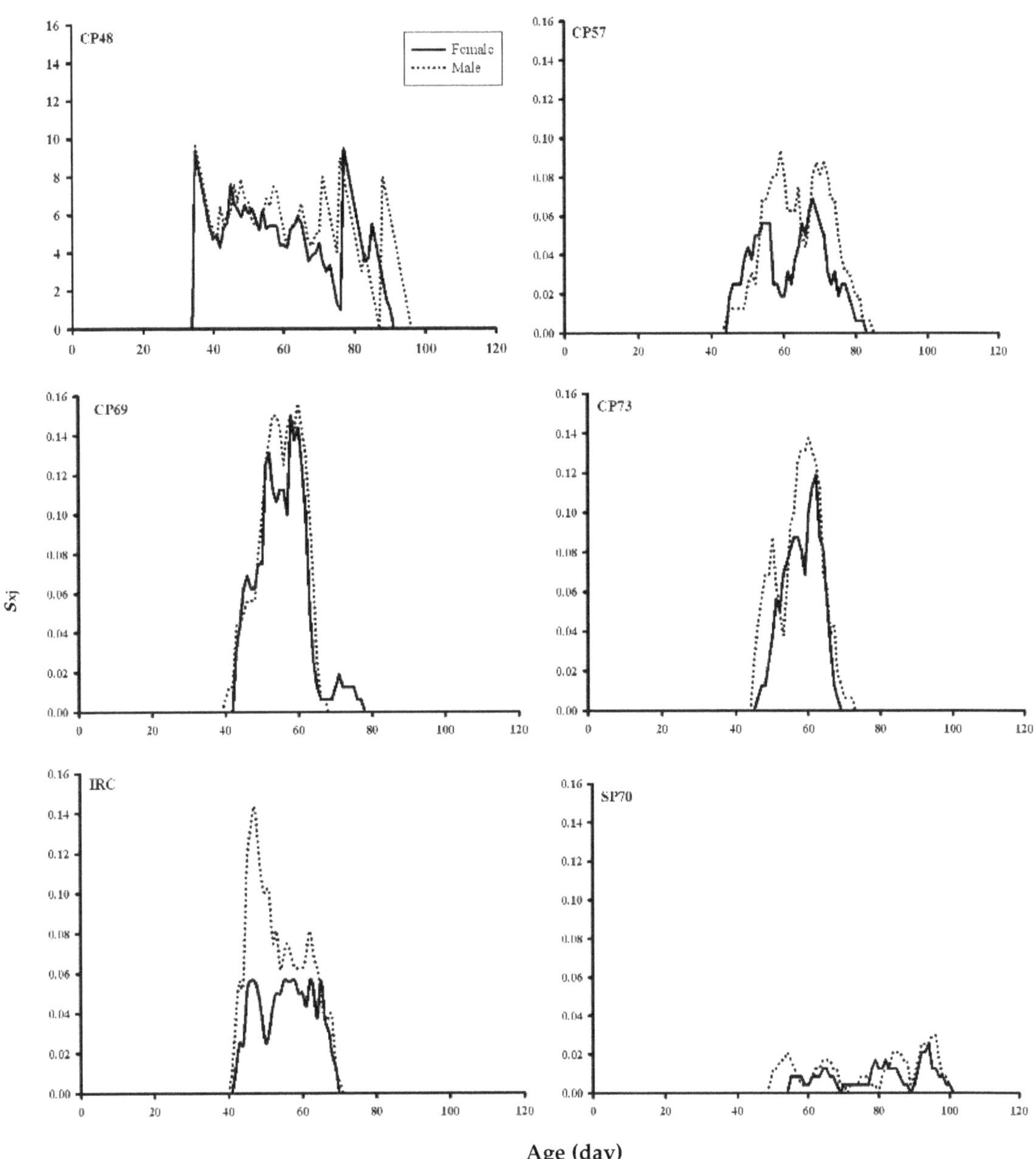

Figure 4. Age-stage survival rate (s_{xj}) of *Sesamia nonagrioides* on examined sugarcane cultivars.

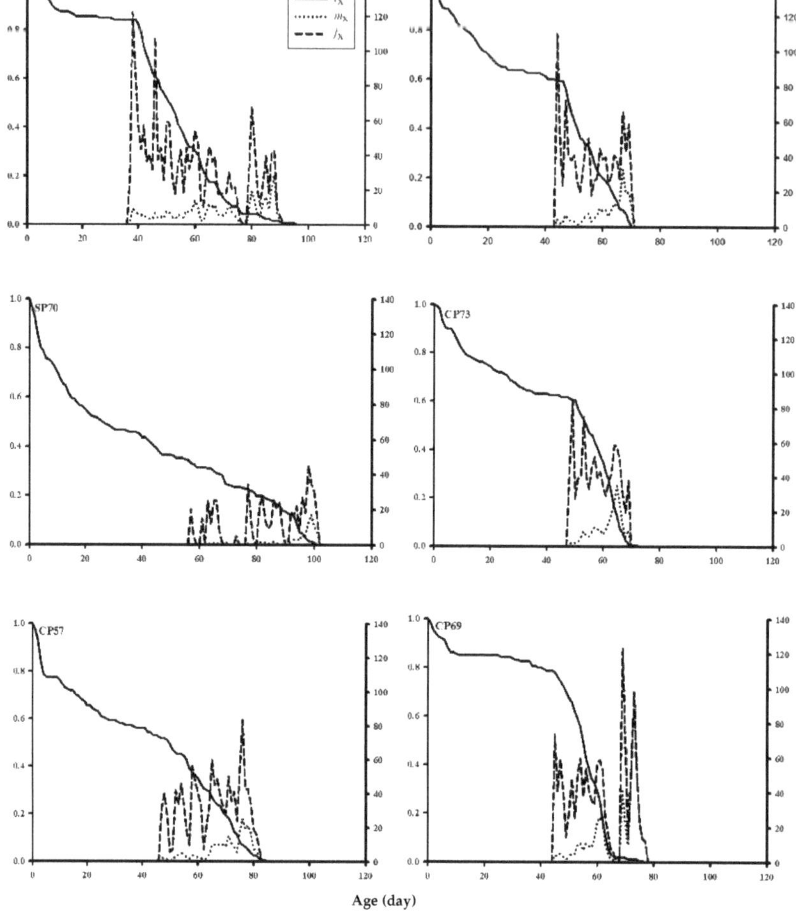

Figure 5. Age-specific survival rate (l_x), age-specific fecundity (m_x), and age-stage-specific fecundity (f_{xj}) of *Sesamia nonagrioides* on examined sugarcane cultivars.

The sex ratio of *S. nonagrioides* was also affected by the host plant cultivars. The results show that the values of sex ratio were 55.74, 51.64, 48.42, 44.21, 46.75, and 42.01% on cultivars CP48-103, CP69-1062, CP73-21, IRC99-02, CP57-614, and SP70-1143, respectively.

3.7. Life Table Parameters of Sesamia nonagrioides

The population growth parameters of *S. nonagrioides* reared on different sugarcane cultivars are given in Table 4. There were significant differences in all computed population parameters on the examined cultivars. The lowest and highest gross reproductive rate (*GRR*) was observed on cultivars SP70-1143 and CP69-1062, respectively ($p < 0.05$). The highest net reproductive rate (R_0) was observed on cultivars CP48-103 and CP69-1062 and the lowest on cultivar SP70-1143 ($p < 0.05$). The intrinsic rate of increase (r) and finite rate of increase (λ) were the highest when the insect was reared on cultivar CP48-103, and the lowest when it was reared on cultivar SP70-1143 ($p < 0.05$). Among sugarcane cultivars, the mean generation time (T) significantly differed from 47.06 days on cultivar CP48-103 to 82.63 days on cultivar SP70-1143 ($p < 0.05$).

Table 4. Age-stage, two-sex life table parameters (mean $\pm$ SE) of *Sesamia nonagrioides* reared on examined sugarcane cultivars.

Cultivars	*GRR* (offspring)	R_0 (offspring)	*r* (day^{-1})	λ (day^{-1})	*T* (day)
CP48-103	388.90 $\pm$ 61.86 ab	99.76 $\pm$ 9.74 a	0.0977 $\pm$ 0.0036 a	1.1027 $\pm$ 0.0040 a	47.06 $\pm$ 1.17 e
IRC99-02	230.55 $\pm$ 40.67 c	45.97 $\pm$ 6.62 b	0.0706 $\pm$ 0.0034 c	1.0731 $\pm$ 0.0037 c	54.11 $\pm$ 1.35 d
CP69-1062	478.28 $\pm$ 79.19 a	90.10 $\pm$ 9.28 a	0.0818 $\pm$ 0.0021 b	1.0853 $\pm$ 0.0023 b	54.91 $\pm$ 0.77 d
CP73-21	209.08 $\pm$ 34.27 c	58.81 $\pm$ 7.85 ab	0.0697 $\pm$ 0.0025 c	1.0722 $\pm$ 0.0027 c	58.25 $\pm$ 0.79 c
CP57-614	256.23 $\pm$ 52.24 bc	44.41 $\pm$ 6.97 b	0.0607 $\pm$ 0.0028 d	1.0626 $\pm$ 0.0030 d	62.21 $\pm$ 1.51 b
SP70-1143	83.06 $\pm$ 37.00 d	5.98 $\pm$ 1.46 c	0.0212 $\pm$ 0.0030 e	1.0215 $\pm$ 0.0031 e	82.63 $\pm$ 3.36 a

Means in each column followed by different letters are significantly different ($p < 0.05$, paired bootstrap test). *GRR*, gross reproductive rate; R_0, net reproductive rate; *r*, intrinsic rate of increase; λ, finite rate of increase; *T*, mean generation time.

3.8. Correlation Analysis

Table 5 indicates the values of correlation coefficients between *S. nonagrioides* life history, life table parameters, and oviposition preference and different physicochemical characteristics of the tested sugarcane cultivars. The life history and life table parameters of *S. nonagrioides* were not significantly correlated with the carbohydrate content of the cultivars. Significant positive correlations were found between the developmental time and condensed tannins, total phenolics, and flavonoids content of the sugarcane cultivars. The survival rate of immature stages, fecundity, weight of pupae (male and female), oviposition preference, R_0, r, and λ of *S. nonagrioides* were negatively correlated with the protein content, tannins content, total phenolics, and flavonoids content of the cultivars. In contrast, the *T* values of *S. nonagrioides* were positively correlated with the protein content, tannins content, total phenolics, and flavonoids content of the sugarcane cultivars.

Table 5. Pearson correlation coefficients (*r*) of population parameters and oviposition preference of *Sesamia nonagrioides* reared on examined sugarcane cultivars with some biochemical and physical traits of these cultivars.

Parameters	Total Protein	Carbohydrate Content	Condensed Tannins	Total Phenolics	Flavonoids Content	Moisture Content	Trichome Density	Shoot Rind Hardness	Stem Rind Hardness	Stem Silica	Shoot Silica
Developmental time	0.776	−0.137	0.841 *	0.870 *	0.920 **	−0.892 *	0.921 **	0.915 **	0.852 *	−0.117	−0.220
Survival rate	−0.953 **	0.354	−0.954 **	−0.961 **	−0.978 **	0.848 *	−0.881 *	−0.873 *	−0.828 *	0.471	0.470
Fecundity	−0.872 *	0.233	−0.814 *	−0.813 *	−0.914 **	0.754	−0.770	−0.755	−0.690	0.349	0.348
Weight of female pupa	−0.921 **	0.554	−0.829 *	−0.852 *	−0.866 *	0.874 *	−0.864 *	−0.829 *	−0.693	0.511	0.629
Weight of male pupa	−0.872 *	0.458	−0.817 *	−0.868 *	−0.890 *	0.926 **	−0.910 *	−0.855 *	−0.725	0.358	0.533
Female longevity	−0.941 **	0.518	−0.890 *	−0.854 *	−0.821 *	0.628	−0.671	−0.694	−0.670	0.762	0.633
Male longevity	−0.776	0.016	−0.842 *	−0.783	−0.867 *	0.667	0.747	−0.831 *	−0.825 *	0.242	0.117
Oviposition preference	−0.845 *	0.486	−0.898 *	−0.927 **	−0.843 *	0.876 *	−0.904 *	−0.897 *	−0.832 *	0.463	0.554
R_0	−0.983 **	0.405	−0.961 **	−0.945 **	−0.961 **	0.778	−0.815 *	−0.814 *	−0.783	0.345	0.534
r	−0.902 **	0.192	−0.954 **	−0.947 **	−0.986 **	0.826 *	−0.880 *	−0.899 *	−0.880 *	0.594	0.310
λ	−0.904 **	0.193	−0.958 **	−0.950 **	−0.987 **	0.824 *	−0.879 *	−0.901 *	−0.882 *	0.353	0.312
T	0.815 *	−0.094	0.900 *	0.903 *	0.953 **	−0.847 *	0.900 *	0.924 **	0.895 *	−0.180	−0.194

* and ** show significant correlations at $p < 0.05$ and $p < 0.01$, respectively. R_0, net reproductive rate; *r*, intrinsic rate of increase; λ, finite rate of increase; *T*, mean generation time.

There was a significant negative correlation between R_0 values with the trichome density and shoot hardness index of sugarcane cultivars. The oviposition preference, immature survival, *r* and λ values of *S. nonagrioides* were negatively correlated with shoot trichome density and the hardness index of the shoot and stem, whereas these variables were positively correlated with the moisture content of the cultivars. Positive correlations were discovered between the developmental time and *T* values of *S. nonagrioides* with shoot trichome density and the hardness index of the shoot and stem, while there were negative correlations between moisture content and *S. nonagrioides* developmental time and *T* values (Table 5).

Significant correlations were found for the weight of male and female pupae with the life table variables and fecundity of *S. nonagrioides* (Table 6). There were positive correlations between the pupal weight and fecundity, R_0, r and λ of *S. nonagrioides*, whereas *T* value was negatively correlated with pupal weight.

Table 6. Pearson correlation coefficients (*r*) between population parameters and pupal weight of *Sesamia nonagrioides* reared on examined sugarcane cultivars.

Parameters	Intrinsic Rate of Increase		Net Reproductive Rate		Finite Rate of Increase		Mean Generation Time		Fecundity		Female Longevity		Male Longevity	
	r	*p*	*r*	*p*	*r*	*p*	*r*	*p*	*r*	*p*	*r*	*p*	*r*	*p*
Weight of female pupa	0.910	0.012	0.926	0.008	0.908	0.012	−0.877	0.022	0.901	0.014	0.843	0.035	-	-
Weight of male pupa	0.921	0.009	0.897	0.015	0.918	0.01	−0.914	0.011	0.93	0.007	-	-	0.788	0.063

3.9. Cluster Analysis

A dendrogram based on the oviposition preference and population growth parameters of *S. nonagrioides* on various sugarcane cultivars is shown in Figure 6. The tested sugarcane cultivars were grouped into three clusters: A (cultivars IRC99-02, CP73-21 and CP57-614); B (cultivars CP48-103 and CP69-1062); and C (cultivar SP70-1143).

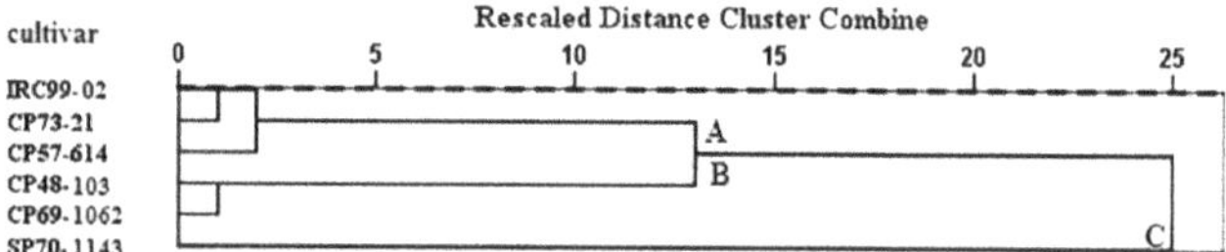

Figure 6. Dendrogram of sugarcane cultivars classified based on the oviposition preference and population parameters of *Sesamia nonagrioides* fed with these cultivars (Ward's method).

4. Discussion

The findings of this research revealed significant effects of the sugarcane cultivars on the oviposition preference, growth, development, survival, and reproduction of *S. nonagrioides*. In this study, the plant secondary metabolites, moisture content, shoot trichome density, and shoot hardness had a notable effect on the oviposition choice of *S. nonagrioides* because these variables were significantly correlated with the oviposition preference. Previous investigations indicated that trichome density [37,57], concentrations of secondary metabolites [32], and tissue hardness [58,59] significantly affected host selection by ovipositing female insects. The outcomes of the oviposition preference and population performance of *S. nonagrioides* demonstrated the suitability of cultivars CP48-103 and CP69-1062 for the offspring development and population increase of the pest. This agrees with the preference–performance theory proposed by Jaenike [60], who expressed that ovipositing female insects choose the best-quality host plants for the survival and optimal growth of their progeny.

The population dynamics of herbivorous insects can be influenced by environmental factors, natural enemies, and insecticides [61–65], as well as host plants [66–69]. The development of resistant cultivars serves as an effective approach in IPM to reduce losses due to insect attacks [70,71]. In the present study, *S. nonagrioides* developed rather quickly on cultivar CP48-103, demonstrating its better nutritive quality than the other tested cultivars. Differences in the duration of immature stages of herbivorous insects on various plant cultivars might be relevant to the nutritional quality and quantity or biochemical attributes of the host plants [72,73]. Moreover, in this study, associations between the hardness index and moisture content of the tested cultivars and the developmental time of *S. nonagrioides* showed that these physical traits could be important factors responsible for the development of immature stages. In some cultivars of sugarcane, thickened stems prevent the penetration of larvae or slow their penetration into the stem by increasing the layers of epidermal cells [59]. The shortest larval period of *S. nonagrioides* on cultivar CP48-103 may be related to lower concentrations of secondary metabolites such as tannins, phenolics, and flavonoids in this cultivar. Moreover, it may be due to the lower hardness index of rind and stem and the higher moisture content of this cultivar, which leads to the easier penetration and feeding of larvae inside the stem. Positive correlations between the developmental time of *S. nonagrioides* and the concentration of condensed tannins, total phenolics and flavonoids, and hardness index of tested cultivars indicated that, among physicochemical features, these factors play the most important role in the developmental rate of *S. nonagrioides*. Furthermore, developmental time was negatively correlated with the amount of moisture in the stems of sugarcane cultivars, which revealed its role in the development of the immature stages of this pest. The slow development of the immature stages of *S. nonagrioides* on cultivar SP70-1143 can be attributed to the higher concentrations of secondary metabolites, lower moisture content, and higher hardness index in this cultivar. In our study, the larval and pupal periods of *S. nonagrioides* on the tested cultivars were longer than those reported by Ranjbar Aghdam [74] and Sedighi [35]. These discrepancies can be due to the differences in the genetic pattern of plant cultivars and the pest, as well as the experimental conditions.

Body weight is one of the major biological indices of the insect population, which is correlated to the type of diet consumed [75]. The high pupal weight of the individuals reared on cultivar CP69-1062 demonstrated that this cultivar is more favorable than the others. In this study, negative correlations between pupal weight and the concentration of tannins, total phenolics, and flavonoids suggested that the larval and pupal growth are affected by the quality of food consumed or the amount of secondary compounds [29,76]. Moreover, the nutritional effects of unsuitable cultivars are reflected in weight of the pupae [32,77]. Therefore, the high pupal weight on cultivar CP69-1062 might be related to the low contents of tannins, total phenolics, and flavonoids in this cultivar. No significant correlation between pupal weight and the carbohydrate content of sugarcane cultivars points to the more important effect of secondary compounds than carbohydrate on the growth of pupae [32,77]. According to the results of this study, the females reared on cultivars CP69-1062 and CP48-103 had higher longevity and fecundity. Therefore, these cultivars were suitable hosts for the growth and development of the pest, which led to the emergence of adults with higher potential reproduction. These results can be related to the lower concentration of tannins, total phenolics, and flavonoids in these cultivars than the other cultivars. This is confirmed by the negative correlation observed between *S. nonagrioides* female longevity and fecundity and the secondary metabolites content of the tested cultivars. Furthermore, the lowest female longevity and fecundity of *S. nonagrioides* on cultivar SP70-1143 might be attributed to the high contents of tannins, total phenolics, and flavonoids in this cultivar, which revealed its poor nutritional quality for this pest. In the other words, the immature survival and development of *S. nonagrioides* on cultivar SP70-1143 are negatively influenced by the secondary compounds, leading to reduced adult longevity and reproductive capacity. The pupal weight is one of the main indices of fecundity performance (number of eggs laid) in female adults. The high pupal weight expresses the better nutrition of the insect in the larval stage and consequently increases adult longevity and reproductive potential. Cultivars CP69-1062 and CP48-103 had suitable conditions for the growth and nutrition of *S. nonagrioides* larvae, which led to the increase in the pupal weight, fecundity, and fertility of this pest. Moreover, no significant correlation between physical traits and adult longevity and fecundity suggested that these traits have no significant effect on the adult life history parameters. The longest immature period and lowest pupal weight of *S. nonagrioides* on cultivar SP70-1143 can be attributable to the low amount of moisture, high trichome density, and shoot and stem hardness in this cultivar. Previously, it has been reported that the quality and quantity of nutrients in the host plants are major factors influencing the adult longevity and reproduction of herbivorous insects [28,32,78]. Therefore, the quality of food and the presence of secondary metabolites in host plants are key factors affecting the survival and growth rate of herbivorous insects, which in turn will affect the reproductive potential of the adults [79].

The highest survival rate and shortest developmental time of *S. nonagrioides* on cultivar CP48-103 are likely due to its better nutritional quality, which results in increased values of the growth index [80,81]. Differences in nutrient quality or physicochemical characteristics among sugarcane cultivars can influence *S. nonagrioides* larval growth. In this study, cultivar CP48-103 showed low levels of tannins, total phenolics, and flavonoids, low hardness of the shoot and stem, low density of trichomes, and high amount of moisture. Hence, it is a suitable cultivar for the development and survival of this pest. Moreover, the high mortality of the pest on cultivars SP70-1143, CP73-21, and CP57-614 may be related to the high shoot hardness, and high amounts of tannins, phenolics, and flavonoids in these cultivars. Although no significant correlations were found, in our study, between *S. nonagrioides* population parameters and the tested properties of the sugarcane cultivars, previous works showed that in cultivars with high silica content, the inner surface of the mandibles of *S. nonagrioides* larvae could be damaged during the larval feeding process due to the increased hardness of stem tissues, and many of these larvae die due to decreased digestibility or starvation [82–84]. Sugarcane cultivars can absorb silica and amass it in the plant leaves and stalks, leading to an increase in plant resistance to pests [84,85].

The intrinsic rate of increase (r) is an important index for expressing the level of plant resistance to insects, which is greatly affected by the nutritional quality of host plant [86]. The highest and lowest r values were on cultivars CP48-103 and SP70-1143, respectively; hence, these cultivars were the most susceptible and resistant, respectively. The sugarcane stem borers reared on cultivar CP48-103 had greater survival, fecundity, and fertility, and shorter developmental time, which resulted in higher r value. Furthermore, the increase in the R_0 and GRR values of *S. nonagrioides* fed on cultivars CP48-103 and CP69-1062 may be attributed to the better nutritive quality of these cultivars than the other tested cultivars (e.g., low tannins content, total phenolics, flavonoids, low shoot and stem stiffness, and high moisture content in these two cultivars). Furthermore, the lowest values of r, R_0, GRR, and λ of *S. nonagrioides* on cultivar SP70-1143 can be related to high amounts of tannins, total phenolics, flavonoids, high shoot and stem stiffness, and low moisture content in this cultivar. Therefore, the mentioned physicochemical characteristics can be the important resistance factors of cultivar SP70-1143 to this pest. In our study, there is a negative correlation between the r, R_0, GRR, and λ of *S. nonagrioides* and secondary metabolites content, trichome density, and shoot and stem hardness, as well as a positive correlation with the carbohydrate and moisture content of the tested cultivars. These results agree with the reports of Abedi [32], who noted that the population parameters of *Ectomyelois ceratoniae* Zeller (Lepidoptera: Pyralidae) significantly correlated with the primary and secondary metabolites of host plants. Our findings regarding the life table parameters on cultivar CP69-1062 were different from Sedighi's [35] findings, who reported that r, R_0, GRR, and λ values were high for *S. nonagrioides* reared on this cultivar. This discrepancy may be due to either genetic variations in studied populations or differences in the rearing conditions of the tested insects.

The cluster analysis of the tested sugarcane cultivars indicated that SP70-1143 classified in cluster C was the most resistant cultivar. Cluster B consisted of cultivars CP48-103 and CP69-1062 as susceptible cultivars, and other cultivars classified in cluster A were partially resistant cultivars. The *S. nonagrioides* reared on cultivars CP48-103 and CP69-1062 showed higher immature survival and fecundity than that reared on other cultivars, which led to the significant increase in the values of r, R_0, and λ on these cultivars. In addition, the low values of r, R_0, and λ in *S. nonagrioides* fed on cultivar SP70-1143 indicated that this cultivar is an unsuitable host for the population increase of this pest. Hence, it could be recommended for cultivation in sugarcane fields where the risk of *S. nonagrioides* damage is high.

5. Conclusions

The results of this study confirm that the quality of sugarcane cultivars influenced the developmental rate and population parameters of *S. nonagrioides*. The *S. nonagrioides* reared on cultivars CP48-103 and CP69-1062 showed faster development and higher survival and fecundity than that reared on the other cultivars, leading to the highest intrinsic rate of increase. Therefore, these can be considered as the most suitable cultivars for *S. nonagrioides*, which would result in higher infestations in the sugarcane field [87]. However, cultivar SP70-1143 was the least suitable (most resistant) cultivar for *S. nonagrioides*, which could be recommended for cultivation in the infested sugarcane farms, where the damage of *S. nonagrioides* is usually high. Since CP57-614, CP73-21, and SP70-1143 are reported as resistant sugarcane cultivars to *S. cretica* [77], they can be recommended for use in transgenic expression for resistance to *S. nonagrioides* and *S. cretica*. The outcomes obtained in this research may improve the use of resistant cultivars in the better management of sugarcane stem borers in IPM programs, which, in turn, lead to the reduction in the loss caused by this pest in sugarcane fields. Since volatiles and secondary chemicals derived from non-living plants could be different from those emitted by living plants, to generalize the results of this study to field conditions, further works about the oviposition preference and population parameters of *S. nonagrioides* on living sugarcane cultivars will be necessary.

Author Contributions: Conceptualization, S.A.M.M. and B.N.; methodology, S.A.M.M., B.N. and H.R.A.; software, S.A.M.M. and B.N.; formal analysis, B.N.; investigation, S.A.M.M. and B.N.; data curation, S.A.M.M., B.N., H.R.A., J.R. and S.A.A.F.; writing—original draft preparation, S.A.M.M., B.N., H.R.A., J.R. and S.A.A.F.; writing—review and editing, S.A.M.M., B.N., A.E. and T.C.; supervision, B.N.; funding acquisition, T.C. All authors have read and agreed to the published version of the manuscript.

Funding: This research received no external funding.

Institutional Review Board Statement: Not applicable.

Data Availability Statement: The data presented in this study are available on request from the corresponding author.

Acknowledgments: This research received financial and technical support from the University of Mohaghegh Ardabili and the Iranian Research Institute of Plant Protection, which is greatly appreciated.

Conflicts of Interest: The authors declare no conflict of interest.

References

1. Miao, R.; Xueli, Q.; Guo, M.; Musa, A.; Jiang, D. Accuracy of space-for-time substitution for vegetation state prediction following shrub restoration. *J. Plant Ecol.* **2018**, *11*, 208–217. [CrossRef]
2. Munyasya, A.N.; Koskei, K.; Zhou, R.; Liu, S.; Indoshi, S.N.; Wang, W.; Zhang, X.C.; Cheruiyot, W.K.; Mburu, D.M.; Nyende, A.B.; et al. Integrated on-site & off-site rainwater-harvesting system boosts rainfed maize production for better adaptation to climate change. *Agric. Water Manag.* **2022**, *269*, 107672. [CrossRef]
3. Yang, Y.; Chen, X.; Liu, L.; Li, T.; Dou, Y.; Qiao, J.; Wang, Y.; An, S.; Chang, S.X. Nitrogen fertilization weakens the linkage between soil carbon and microbial diversity: A global meta-analysis. *Glob. Change Biol.* **2022**, *28*, 6446–6461. [CrossRef]
4. Khalid, Z.R.; Sundaisa, A.B.; Arif, M. Comparative studies on quantity and quality of pests' incursion sugarcane grown at rahim Yar khan Pakistan. *Agri. Res. Tech.* **2020**, *24*, 556276. [CrossRef]
5. Tsitsipis, J.A. *Contribution toward the Development of an Integrated Control Method for the Corn Stalk borer Sesamia nonagrioides (Lef.), Pesticides and Alternatives*; Casida, J.E., Ed.; Elsevier: Amsterdam, The Netherlands, 1990; pp. 217–228.
6. Stamopoulos, D.K. *Insects of Stored Products, Crops and Vegetables*, 2nd ed.; Ziti Press: Thessaloniki, Greece, 1999.
7. Fantinou, A.A.; Perdikis, D.; Chatzoglou, C.S. Development of immature stages of *Sesamia nonagrioides* (Lepidoptera: Noctuidae) under alternating and constant temperatures. *Environ. Entomol.* **2003**, *32*, 1337–1342. [CrossRef]
8. Eizaguirre, M.; Fantinou, A.A. Abundance of *Sesamia nonagrioides* (Lef.) (Lepidoptera: Noctuidae) on the Edges of the Mediterranean Basin. *Psyche* **2012**, 1–7. [CrossRef]
9. Askarianzadeh, A.; Moharramipour, S.; Kamali, K.; Fathipour, Y. Evaluation of damage caused by stalk borers, *Sesamia* spp. (Lepidoptera: Noctuidae), on sugarcane quality in Iran. *Entomol. Res.* **2008**, *38*, 263–267. [CrossRef]

10. Nikpay, A.; Nejadian, E.S.; Goldasteh, S.; Farazmand, H. Efficacy of silicon formulations on sugarcane stalk borers, quality characteristics and parasitism rate on five commercial varieties. *Proc. Natl. Acad. Sci. India Sect. B Biol. Sci.* **2017**, *87*, 289–297 [CrossRef]

11. Goebel, F.R.; Achadian, E.; Kristini, A.; Sochib, M.; Adi, H. Investigation of crop losses due to moth borers in Indonesia. *Proc. Aust. Soc. Sugar Cane Technol.* **2011**, *33*, 1–9.

12. McGuire, P.J.; Dianpratiwi, T.; Achadian, E.; Goebel, F.R.; Kristini, A. Extension of better control practices for moth borers in the Indonesian sugar industry. *Proc. Aust. Soc. SugarCane Technol.* **2012**, *34*, 1–10.

13. Sattar, M.; Mehmood, S.S.; Khan, M.R.; Ahmad, S. Influence of egg parasitoid *Trichogramma chilonis* Ishii on sugarcane stem borer (*Chilo infuscatellus* Snellen) in Pakistan. *Pak. J. Zool.* **2016**, *48*, 989–994.

14. Wright, D.J.; Verkert, R.H.J. Integration of chemical and biological control systems for arthropods; evaluation in a multitrophic context. *Pestic. Sci.* **1995**, *44*, 207–218. [CrossRef]

15. Saeb, H.; Nouri-Ganbalani, G.; Rajabi, G. Comparison of the resistance of some rice genotypes of Guilan province to the striped stem borer, *Chilo suppressalis* Walker and investigating the role of silica in resistance. *J. Agric. Sci.* **2002**, *7*, 17–25.

16. Tolmay, V.L.; Vander Westhuizen, M.C.; Van Derenter, C.S. A week screening method for mechanism of host plant resistance to *Diglyphus noxia* in wheat accessions. *Euphytica* **1999**, *107*, 79–89. [CrossRef]

17. Golizadeh, A.; Kamali, K.; Fathipour, Y.; Abbasipour, H. Life table of the diamondback moth, *Plutella xylostella* (L.) (Lepidoptera: Plutellidae) on five cultivated brassicaceous host plants. *J. Agr. Sci. Technol.-Iran.* **2009**, *11*, 115–124.

18. Soufbaf, M.; Fathipour, Y.; Karimzadeh, J.; Zalucki, M.P. Bottom-up effect of different host plants on *Plutella xylostella* (Lepidoptera: Plutellidae): A life-table study on canola. *J. Econ. Entomol.* **2010**, *103*, 2019–2027. [CrossRef]

19. Mesbah, H.A.; Fahmy, I.S.; EI-Deeb, A.S.; Gaber, A.A.; Nour, A.H. The susceptibility of ten sugarcane varieties to infestation with borers and mealy bugs at Alexandria, Egypt. *Bull. Entomol. Soc. Egypt* **1980**, *6*, 403–411.

20. Beuzelin, J.; Mészáros, A.; Akbar, W.; Reagan, T. Sugarcane planting date impact on fall and spring sugarcane borer (Lepidoptera: Crambidae) infestations. *Fla. Entomol.* **2011**, *94*, 242–252. [CrossRef]

21. Nikpay, A.; Kord, H.; Goebel, F.-R.; Sharafizadeh, P. Assessment of natural parasitism of sugarcane moth borers *Sesamia* spp. by *Telenomus busseolae*. *Sugar Tech.* **2014**, *16*, 325–328. [CrossRef]

22. Scriber, J.M.; Slansky, F. The nutritional ecology of immature insects. *Annu. Rev. Entomol.* **1981**, *26*, 183–211. [CrossRef]

23. Sarfraz, M.; Dosdall, L.M.; Keddie, B.A. Diamondback moth-host plant interactions: Implications for pest management. *Crop. Prot.* **2006**, *25*, 625–636. [CrossRef]

24. Adango, E.; Onzo, A.; Hanna, R.; Atachi, P.; James, B. Comparative demography of the spider mite *Tetranychus ludeni* (Acari: Tetranychidae) on two host plants in West Africa. *J. Insect Sci.* **2006**, *6*, 1–9. [CrossRef]

25. Naseri, B.; Borzoui, E.; Majd, S.; Mansouri, S.M. Influence of different food commodities on life history, feeding efficiency, and digestive enzymatic activity of *Tribolium castaneum* (Coleoptera: Tenebrionidae). *J. Econ. Entomol.* **2017**, *110*, 2263–2268. [CrossRef] [PubMed]

26. El-Rodeny, W.M.; Abeer, A.S.; Salwa, M.M.; Amany, M.M. Comparative resistance as function of physical and chemical properties of selected faba bean promising lines against *Callosobruchus maculatus* pos harvest. *J. Plant Prod. Mansoura Univ.* **2018**, *9*, 609–617.

27. Locatelli, D.P.; Castorina, G.; Sangiorgio, S.; Consonni, G.; Limonta, L. Susceptibility of maize genotypes to *Rhyzopertha dominica* (F.). *J. Plant Dis. Prot.* **2019**, *126*, 509–515. [CrossRef]

28. Naseri, B.; Majd-Marani, S. Assessment of eight rice cultivars flour for feeding resistance to *Tribolium castaneum* (Herbst) (Coleoptera: Tenebrionidae). *J. Stored Prod. Res.* **2020**, *88*, 101650. [CrossRef]

29. Borzoui, E.; Naseri, B.; Rahimi Namin, F. Different diets affecting biology and digestive physiology of the Khapra beetle, *Trogoderma granarium* Everts (Coleoptera: Dermestidae). *J. Stored Prod. Res.* **2015**, *62*, 1–7. [CrossRef]

30. Golizadeh, A.; Abedi, Z. Comparative performance of the Khapra beetle *Trogoderma granarium* Everts (Coleoptera: Dermestidae) on various wheat cultivars. *J. Stored Prod. Res.* **2016**, *69*, 159–165. [CrossRef]

31. Golizadeh, A.; Abedi, Z.; Borzoui, E.; Golikhajeh, N.; Jafary, M. Susceptibility of five sugar beet cultivars to the black bean aphid, *Aphis fabae* Scopoli (Hemiptera: Aphididae). *Neotrop. Entomol.* **2016**, *45*, 427–432. [CrossRef]

32. Abedi, Z.; Golizadeh, A.; Soufbaf, M.; Hassanpour, M.; Jafari-Nodoushan, A.; Akhavan, H.R. Relationship between performance of carob moth, *Ectomyelois ceratoniae* Zeller (Lepidoptera: Pyralidae) and phytochemical metabolites in various pomegranate cultivars. *Front. Physiol.* **2016**, *10*, 1425. [CrossRef]

33. Ranjbar Aghdam, H.; Kamali, K. In vivo rearing of *Sesamia cretica* and *Sesamia nonagrioides* botanephaga. *J. Entomol. Soc. Iran.* **2002**, *22*, 63–78.

34. Askarianzadeh, A.R.; Moharamipour, S.; Kamali, K.; Fathipour, Y. Evaluation of resistance to stem borers (*Sesamia* spp.) in some sugarcane cultivars at tillering stage. *J. Seed Plant.* **2006**, *22*, 117–128.

35. Sedighi, L.; Ranjbar Aghdam, H.; Imani, S.; Shojai, M. Age-stage two-sex life table analysis of *Sesamia nonagrioides* (Lep.: Noctuidae) reared on different host plants. *Arch. Phytopathol. Plant Prot.* **2017**, *50*, 438–453. [CrossRef]

36. Abo Elenen, F.M.; Eid Mehareb, M.; Ghonema, M.A.; El-Bakry, A. Selection in sugarcane germplasm under the Egyptian conditions. *J. Agri. Res.* **2018**, *3*, 000162.

37. Tabari, M.A.; Fathi, S.A.A.; Nouri-Ganbalani, G.; Moumeni, A.; Razmjou, J. Antixenosis and antibiosis resistance in rice cultivars against *Chilo suppressalis* (Walker) (Lepidoptera: Crambidae). *Neotrop. Entomol.* **2016**, *46*, 1–9. [CrossRef]

38. Jafari Petroudy, S.R.; Resalati, H.; Rasouli Garmaroudi, E. The effect of type pulping process on the desilication of rice straw black liquor. *J. Wood Forest Sci. Technol.* **2015**, *22*, 45–60.

39. Bradford, M.M. A rapid and sensitive method for the quantitation of microgram quantities of protein utilizing the principle of protein-dye binding. *Anal. Biochem.* **1976**, *72*, 248–254. [CrossRef]

40. Haghighi, R.; Mirmohammady Maibody, S.A.M.; Sayed Tabatabaei, B.E. Evaluation of enzymatic and non-enzymatic changes in flowering process of saffron (*Crocus sativus* L.). *J. Plant Proc. Funct.* **2020**, *9*, 297–308.

41. Galvão, M.A.M.; de Arruda, A.O.; Bezerra, I.C.F.; Ferreira, M.R.A.; Soares, L.A.L. Evaluation of the Folin-Ciocalteu method and quantification of total tannins in stem barks and pods from *Libidibia ferrea* (Mart. ex Tul) L. P. Queiroz. *Braz. Arch. Biol. Technol.* **2018**, *61*, e18170586. [CrossRef]

42. Huang, D.J.; Chun-Der, L.; Hsien-Jung, C.; Yaw-Huei, L. Antioxidant and antiproliferative activities of sweet potato (*Ipomoeabatatas* [L.] Lam Tainong 57') constituents. *Bot. Bull. Acad. Sin.* **2004**, *45*, 179–186.

43. Broadhurst, R.B.; Jones, W.T. Analysis of condensed tannins using acidified vanillin. *J. Sci. Food Agric.* **1978**, *29*, 788–794. [CrossRef]

44. Sreng, I. La pheromonesexuelle de *Sesamia* spp. Lepidoptera- Noctuidae. Ph.D. Thesis, Universite de Dijon, Dijon, France, 1984.

45. Chi, H.; Liu, H. Two new methods for the study of insect population ecology. *Bull. Inst. Zool. Academia Sin.* **1985**, *24*, 225–240.

46. Chi, H. Life-table analysis incorporating both sexes and variable development rates among individuals. *Environ. Entomol.* **1988**, *17*, 26–34. [CrossRef]

47. Chi, H. TWOSEX-MS Chart: A Computer Program for the Age-Stage, Two-Sex Life Table Analysis. 2020. Available online: http://140.120.197.173/Ecology/ (accessed on 13 January 2020).

48. Carey, J.R. *Applied Demography for Biologists with Special Emphasis on Insects*; Oxford University Press, Inc.: New York, NY, USA, 1993.

49. Huang, Y.B.; Chi, H. Life tables of *Bactrocera cucurbitae* (Diptera: Tephritidae): With an invalidation of the jackknife technique. *J. Appl. Entomol.* **2013**, *137*, 327–339. [CrossRef]

50. Efron, B.; Tibshirani, R.J. *An Introduction to the Bootstrap*; Chapman and Hall: New York, NY, USA, 1993.

51. Yu, J.Z.; Chi, H.; Chen, B.H. Comparison of the life tables and predation rate of *Harmonia dimidiate* (F.) (Coleoptera: Coccinellidae) fed on *Aphis gossypii* Glover (Hemiptera: Aphididae) at different temperatures. *Biol. Control* **2013**, *64*, 1–9. [CrossRef]

52. Reddy, G.V.; Chi, H. Demographic comparison of sweet potato weevil reared on a major host, *Ipomoea batatas*, and an alternative host, *I. triloba*. *Sci. Rep.* **2015**, *5*, 11871. [CrossRef]

53. Bahari, F.; Fathipour, Y.; Talebi, A.A.; Alipour, Z. Long-term feeding on greenhouse cucumber affects life table parameters of two-spotted spider mite and its predator *Phytoseiulus persimilis*. *Syst. Appl. Acarol.* **2018**, *23*, 2304–2317. [CrossRef]

54. Kitch, L.W.; Shade, R.E.; Murdock, L.L. Resistance to the cowpea weevil (*Callosobruchus maculatus*) larva in pods of cowpea (*Vigna unguiculata*). *Entomol. Exp. Appl.* **1991**, *60*, 183–192. [CrossRef]

55. Naseri, B.; Majd-Marani, S. Different cereal grains affect demographic traits and digestive enzyme activity of *Rhyzopertha dominica* (F.) (Coleoptera: Bostrichidae). *J. Stored Prod. Res.* **2022**, *95*, 101898. [CrossRef]

56. Majd-Marani, S.; Naseri, B.; Nouri-Ganbalani, G.; Borzoui, E. Maize hybrids affected nutritional physiology of the Khapra beetle, *Trogoderma granarium* Everts (Coleoptera: Dermestidae). *J. Stored Prod. Res.* **2018**, *77*, 20–25. [CrossRef]

57. Smith, C.M. *Plant Resistance to Arthropods: Molecular and Conventional Approaches*; Springer Press: Dordrecht, The Netherlands, 2005; p. 413. [CrossRef]

58. Chang, H.; Shih, C.Y. A study on the leaf mid-rib structure of sugarcane as related with resistance to the top borer (*Scirpophaganivella* F.). *Taiwan Sugar Exp. Sta. Rep.* **1959**, *19*, 53–56.

59. Martin, G.A.; Richard, C.A.; Hensley, S.D. Host resistance to *Diatraea saccharalis* (F.): Relationship of sugarcane node hardness to larval damage. *Environ. Entomol.* **1975**, *4*, 687–688. [CrossRef]

60. Jaenike, J. On optimal oviposition behavior in phytophagous insects. *Theor. Popul. Biol.* **1978**, *14*, 350–356. [CrossRef]

61. Greenberg, S.M.; Sétamou, M.; Sappington, T.W.; Liu, T.X.; Coleman, R.J.; Armstrong, J.S. Temperature-dependent development and reproduction of the boll weevil (Coleoptera: Curculionidae). *Insect Sci.* **2005**, *12*, 449–459. [CrossRef]

62. Barteková, A.; Praslička, J. The effect of ambient temperature on the development of cotton bollworm (*Helicoverpa armigera* Hübner, 1808). *Plant Prot. Sci.* **2006**, *42*, 135–138. [CrossRef]

63. Yang, T.C.; Chi, H. Life tables and development of *Bemisia argentifolii* (Homoptera: Aleyrodidae) at different temperatures. *J. Econ. Entomol.* **2006**, *99*, 691–698. [CrossRef] [PubMed]

64. Atlihan, R.; Chi, H. Temperature-dependent development and demography of *Scymnus subvillosus* (Coleoptera: Coccinellidae) reared on *Hyalopterus pruni* (Homoptera: Aphididae). *J. Econ. Entomol.* **2008**, *101*, 325–333. [CrossRef] [PubMed]

65. Wang, X.G.; Johnson, M.W.; Daane, K.M.; Nadel, H. High summer temperatures affect the survival and reproduction of olive fruit fly (Diptera: Tephritidae). *Environ. Entomol.* **2009**, *38*, 1496–1504. [CrossRef]

66. Yin, J.; Sun, Y.; Wu, G.; Ge, F. Effects of elevated CO_2 associated with maize on multiple generations of the cotton bollworm. *Entomol. Exp. Appl.* **2010**, *136*, 12–20. [CrossRef]

67. Goldasteh, S.; Talebi, A.A.; Rakhshani, E.; Gholdasteh, S. Effect of four wheat cultivars on life table parameters of *Schizaphis graminum* (Hemiptera: Aphididae). *J. Crop Prot.* **2012**, *1*, 121–129.

68. Goodarzi, M.; Fathipour, Y.; Talebi, A.A. Antibiotic resistance of canola cultivars affecting demography of *Spodoptera exigua* (Lepidoptera: Noctuidae). *J. Agric. Sci. Technol.* **2012**, *17*, 23–33.

69. Tazerouni, Z.; Talebi, A.A.; Fathipour, Y.; Soufbaf, M. Bottom-up effect of two host plants on life table parameters of *Aphis gossypii* (Hemiptera: Aphididae). *J. Agric. Sci. Technol.* **2016**, *18*, 1–12.

70. Sachan, J.N. Progress in host–plant resistance work in chickpea and pigeonpea against *Helicoverpa armigera* (Hubner). In Summary Proceedings of the First Consultative Group Meeting on the Host Selection Behavior of *Helicoverpa armigera*, Patancheru, India, 5–7 March 1990; pp. 19–22.

71. Jallow, M.F.A.; Cunningham, J.P.; Zalucki, M.P. Intra-specific variation for host plant use in *Helicoverpa armigera* (Hubner) (Lepidoptera: Noctuidae): Implications for management. *Crop Prot.* **2004**, *23*, 955–964. [CrossRef]

72. Soufbaf, M.; Fathipour, Y.; Zalucki, M.P.; Hui, C. Importance of primary metabolites in canola in mediating interactions between a specialist leaf-feeding insect and its specialist solitary endoparasitoid. *Arthropod-Plant Interact.* **2012**, *6*, 241–250. [CrossRef]

73. Nouri-Ganbalani, G.; Borzoui, E.; Shahnavazi, M.; Nouri, A. Induction of resistance against *Plutella xylostella* (L.) (Lep.: Plutellidae) by jasmonic acid and mealy cabbage aphid feeding in *Brassica napus* L. *Front. Physiol.* **2018**, *9*, 859. [CrossRef] [PubMed]

74. Ranjbar Aghdam, H. *Mass Rearing of the Sugarcane Stem Borers, Sesamia spp. Using Semi Artificial Diet*; Final Project Report No. 4–16–16–92169; Iranian Research Institute for Plant Protection: Tehran, Iran, 2016; No. 50897.

75. Li, Y.; Hill, C.B.; Hartman, G.L. Effect of three resistant soybean genotypes on the realized fecundity, mortality and maturation of soybean aphid (Homoptera, Aphididae). *J. Econ. Entomol.* **2004**, *97*, 1106–1111. [CrossRef]

76. Liu, Z.D.; Li, D.M.; Gong, P.Y.; Wu, K.J. Life table studies of the cotton bollworm, *Helicoverpa armigera* (Hübner) (Lepidoptera: Noctuidae), on different host plants. *Environ. Entomol.* **2004**, *33*, 1570–1576. [CrossRef]

77. Babamir-Satehi, A.; Habibpour, B.; Ranjbar Aghdam, H.; Hemmati, S.A. Interaction between feeding efficiency and digestive physiology of the pink stem borer, *Sesamia cretica* Lederer (Lepidoptera: Noctuidae), and biochemical compounds of different sugarcane cultivars. *Arthropod Plant Interact.* **2022**, *16*, 309–316. [CrossRef]

78. Razmjou, J.; Naseri, B.; Hemati, S.A. Comparative performance of the cotton bollworm, *Helicoverpa armigera* (Hubner) (Lepidoptera: Noctuidae) on various host plants. *J. Pest Sci.* **2014**, *87*, 29–37. [CrossRef]

79. Awmack, C.S.; Leather, S.R. Host plant quality and fecundity in herbivorous insects. *Annu. Rev. Entomol.* **2002**, *47*, 817–844. [CrossRef]

80. Greenberg, S.M.; Sappington, T.W.; Legaspi, B.C.; Liu, T.X.; Setamou, M. Feeding and life history of *Spodoptera exigua* (Lepidoptera: Noctuidae) on different host plants. *Ann. Entomol. Soc. Am.* **2001**, *94*, 566–575. [CrossRef]

81. Farahani, S.; Talebi, A.A.; Fathipour, Y. Life cycle and fecundity of *Spodoptera exigua* (Lep.: Noctuidae) on five soybean varieties. *J. Entomol. Soc. Iran.* **2011**, *30*, 1–12.

82. Panda, N.; Pradhan, B.; Samalo, A.P.; Rao, P.S.P. Note on the relationship of some biochemical factors with the resistance in rice varieties to yellow rice borer. *Indian J. Agric. Sci.* **1975**, *45*, 499–501.

83. Reynolds, O.L.; Keeping, M.G.; Meyer, J.H. Silicon-augmented resistance of plants to herbivorous insects: A review. *Ann. Appl. Biol.* **2009**, *155*, 171–186. [CrossRef]

84. Ma, J.F. Role of silicon in enhancing the resistance of plants to biotic and abiotic stresses. *Soil Sci. Plant Nutr.* **2004**, *50*, 11–18. [CrossRef]

85. Meyer, J.H.; Keeping, M.G. Impact of silicon in alleviating biotic stress in sugarcane in South Africa. *Sugar Technol. Assoc.* **2005**, *23*, 14–18.

86. Razmjou, J.; Moharramipour, S.; Fathipour, Y.; Mirhoseini, S.Z. Demographic parameters of cotton aphid, *Aphis gossypii* Glover (Homoptera: Aphididae) on five cotton cultivars. *Insect Sci.* **2006**, *13*, 205–210. [CrossRef]

87. Taherkhani, K.; Moazen-Rezamahaleh, H. *Sugarcane Stem Borers and Their Managements*, 1st ed.; Iranian Sugarcane Training Institute (ISCRTI): Ahvaz, Iran, 2012; p. 27.

Article

Quantifying the Influence of Pollen Aging on the Adhesive Properties of *Hypochaeris radicata* Pollen

Steven Huth [1,*], Lisa-Maricia Schwarz [2,3] and Stanislav N. Gorb [1]

1 Zoological Institute, Kiel University, Am Botanischen Garten 1-9, 24118 Kiel, Germany
2 Grassland Ecology & Grassland Management, Department of Plant Nutrition, Institute of Crop Science and Resource Conservation (INRES), University of Bonn, Karlrobert-Kreiten-Str. 13, 53115 Bonn, Germany
3 Biodiversity Research and Systematic Botany, Institute of Biochemistry and Biology, University of Potsdam, Maulbeerallee 1, 14469 Potsdam, Germany
* Correspondence: shuth@zoologie.uni-kiel.de

Simple Summary: Pollination is the transfer of pollen from a plant's male part (anther) to the corresponding female part (stigma). It is a fundamental biological process that ensures plant reproduction. Most studies investigate pollination from a biological perspective, but the underlying physical processes are poorly understood. Many plants rely on insects to transport pollen and the forces with which pollen adhere to insects and floral surfaces are fundamental for successful pollination. We quantified pollen adhesion by measuring the forces necessary to detach *Hypochaeris radicata* (catsear, a common insect-pollinated plant) pollen from glass and studied for the first time how the adhesion forces change with pollen aging. Our results show that newly formed adhesion bonds between *H. radicata* pollen and glass are stronger for fresh pollen than for old ones. On the other hand, when *H. radicata* pollen age in contact with glass, the adhesion between pollen and glass strengthens over time. These effects are probably caused by the viscous liquid covering most pollen (pollenkitt) changing its viscoelastic properties as it dries.

Citation: Huth, S.; Schwarz, L.-M.; Gorb, S.N. Quantifying the Influence of Pollen Aging on the Adhesive Properties of *Hypochaeris radicata* Pollen. *Insects* 2022, 13, 811. https://doi.org/10.3390/insects13090811

Academic Editor: Bertrand Schatz

Received: 6 July 2022
Accepted: 23 August 2022
Published: 6 September 2022

Publisher's Note: MDPI stays neutral with regard to jurisdictional claims in published maps and institutional affiliations.

Abstract: Although pollination is one of the most crucial biological processes that ensures plant reproduction, its mechanisms are poorly understood. Especially in insect-mediated pollination, a pollen undergoes several attachment and detachment cycles when being transferred from anther to insect and from insect to stigma. The influence of the properties of pollen, insect and floral surfaces on the adhesion forces that mediate pollen transfer have been poorly studied. Here, we investigate the adhesive properties of *Hypochaeris radicata* pollen and their dependence on pollen aging by quantifying the pull-off forces from glass slides using centrifugation and atomic force microscopy. We found that the properties of the pollenkitt—the viscous, lipid liquid on the surface of most pollen grains—influences the forces necessary to detach a pollen from hydrophilic surfaces. Our results show that aged *H. radicata* pollen form weaker adhesions to hydrophilic glass than fresh ones. On the other hand, when a pollen grain ages in contact with glass, the adhesion between the two surfaces increases over time. This study shows for the first time the pollen aging effect on the pollination mechanism.

Keywords: pollination; pollen adhesion; pollenkitt; atomic force microscopy; cryogenic scanning electron microscopy; centrifugation

1. Introduction

Pollination is a highly efficient and species-specific biological particle transport system that forms the basis for plant reproduction and distribution [1,2]. It is fundamental for our planet's ecosystem and for human food production, and it plays an important role in the mediation of atmospheric phenomena [3–5]. Despite the fact that the principle of pollination is common knowledge, remarkably little is known about the details of pollen transfer mechanisms [6]. Most studies have focused on pollination ecology, chemical recognition mechanisms, germination or other biochemical aspects of pollination [1,3,7–11]. However, adhesion processes play a crucial role during pollen transfer from the anther to the stigma. Especially with insect-dependent pollination, pollen undergo multiple attachment and detachment cycles to surfaces with various chemical and structural features [5]. In order to facilitate efficient pollen transfer, floral and pollen surfaces must present a specific set of properties that passively influence adhesion forces. These adhesion forces cause the pollen to be transferred from the flower anther to the insect body and eventually from the insect to the flower stigma. Therefore, adhesive mechanisms, as well as the influence of floral and insect surface properties on adhesion forces, are an essential part of the pollination process.

This has only been recently addressed by very few studies [2,5,6,12]. These studies have shown that pollen adhesion dynamics are affected by an interplay of surface structures and liquids that are often present on floral surfaces. For instance, pollenkitt—the surface liquid covering pollen grains—is widely accepted to increase pollen adhesion [10,13,14]. On the other hand, it has recently been shown that pollenkitt may reduce adhesion to wet surfaces [6], suggesting that its role during pollination is more complex than widely assumed. It has also been stated that the pollenkitt of most pollen partially dries out during the transfer process from the anther to the stigma [10,15].

To the best of our knowledge, the adhesion properties of pollen as well as the influence of pollen aging and dehydration have not yet been quantified experimentally. It is plausible to hypothesize that pollenkitt, which is the most outer layer of pollen grains, is influenced by dehydration and therefore by the time that has passed since the pollen presentation on the anther. Here, we study how the adhesion properties of *Hypochaeris radicata* pollen change over time by quantifying the forces necessary to detach pollen from glass surfaces using centrifugation and atomic force microscopy (AFM). Centrifugation allows the quantification of pollen adhesion for many pollen grains in one experiment and therefore offers the opportunity to draw statistically relevant conclusions. AFM experiments, on the other hand, quantify the adhesion of single pollen grains. These experiments are repeatable and therefore allow us to examine one and the same pollen individual at different experimental and environmental conditions with highly comparable results. This opens the possibility to realize very detailed studies on the influence of a variety of factors on pollen adhesion. Therefore, such a detailed AFM analysis is complementary to the statistically relevant centrifugation approach. We also employed cryo-scanning electron microscopy (cryo-SEM) to observe pollen grains in their native conditions with their surface liquids present.

In this study, we examine the change in morphology of *H. radicata* pollen grains due to aging and the impact this has on their adhesive bonds' properties as well as their ability to form new adhesions to hydrophilic surfaces. We also studied the dependence of pull-off forces necessary to detach an individual *H. radicata* pollen grain from a hydrophilic glass surface on both the contact force and the detachment speed.

A deeper understanding of the pollination process will provide insights into bioinspired passive microparticle manipulation through the altering of surface properties, which is relevant for artificial pollination, drug delivery [16], respiratory medicine [17], and the engineering of biomimetic microgripping systems [2].

2. Materials and Methods

2.1. Pollen Sample Preparation

The flowering stems of *Hypochaeris radicata* (Asteraceae) were collected from open spaces around Kiel University (Kiel, Germany) between July and November. To preserve the functionality of the collected flowering stems in the laboratory, they were kept in water. Ray flowers that had pushed their styles through the anthers were chosen and by carefully brushing an eyelash over the styles, the pollen grains were removed and poured onto cleaned glass coverslips. Using an ultrasonic bath, glass slides were cleaned for 5 min in 70% ethanol and subsequently for 5 min in double-distilled water. They were blown dry with compressed air. To analyze pollen without pollenkitt, the latter was removed by washing the pollen grains ten times in a mixture of chloroform and methanol (3:1). The solvent containing the pollenkitt was filtered out between washing procedures.

2.2. Centrifugation

For centrifugation assays, glass slides were cut into pieces fitting a 2.5 mL Eppendorf tube, cleaned and populated with pollen grains. The bottom of each glass piece was marked for target regions using 8 mm × 6 mm grids. Only target regions that contained at least 20 pollen grains were chosen for analysis. For each target area, an image of the pollen particles was taken using a Leica M205 microscope (Leica Microsystems GmbH, Wetzlar, Germany) before and after the sample was centrifuged for 3 min at 1000 rpm in a Heraeus Sepatech Biofuge A table centrifuge (Heraeus Holding GmbH, Hanau, Germany). The experiment was repeated 13 times with increasing rotation speeds (1000 rpm, 2000 rpm, 3000 rpm, … 13,000 rpm). The number of single pollen grains on each of the resulting images was counted using a home-written Matlab (MathWorks, Natick, US) algorithm. Clumps of pollen grains were excluded.

The centrifugal force is defined as $F = m\omega^2 r$ with m being the mass of a single pollen, ω the angular velocity and r the radius of rotation. Therefore, the mass of pollen particles was necessary to obtain the centrifugal force. To determine the pollen mass, we first weighed approximately 1000 pollen grains with a UMX2 Ultra-microbalance (Mettler Toledo, Columbus, USA). These grains were subsequently spread on a glass slide to ensure that they were positioned in a single layer. We then took a light microscopy image of the monolayer. The customized Matlab particle counting algorithm was used to determine the exact number of pollen grains per sample, which was, in turn, used to calculate the average mass of a single pollen grain. We repeated the experiment ten times for fresh and 7-day old *H. radicata* pollen, respectively.

2.3. Cantilever Preparation

We calibrated AFM cantilevers (HQCSC37/NO_AL, MikroMasch OÜ, Tallinn, Estonia) with a Nanowizard I AFM (JPK, Berlin, Germany). The sensitivity was determined by collecting 10 force–distance curves with a setpoint of 1 V at a speed of 10 µm/s on each of 16 different positions on a cleaned glass slide. Subsequently, the spring constant was calibrated ten times using the well-established thermal noise method and the mean value was used for the experiments. The cantilevers used in this study had spring constants ranging from 0.3 N/m to 0.5 N/m.

A single *Hypochaeris radicata* pollen particle on a glass slide was chosen and a small amount of 2-component epoxy adhesive (UHU Schnellfest, UHU GmbH & Co. KG, Bühl, Germany) was deposited next to it. A calibrated cantilever was approached so that the tip was dipped into the glue. Afterwards, residual glue was removed by dipping the cantilever onto the glass once or twice before the cantilever was brought into contact with the pollen particle. The pollen and cantilever were retracted from the glass slide and the glue was allowed to cure for 30 min before the experiments were conducted.

2.4. Atomic Force Microscopy

In order to quantify the influence of the setpoint force and cantilever speed on the pull-off forces of a *Hypochaeris radicata* pollen from glass (Figures 4 and 6), a pollen-cantilever was used to collect 10 force–distance curves on each of 9 positions on a cleaned glass slide with the respective experimental parameters. For 0.1 µm/s, only 5 curves were recorded per position. The sampling rates were synchronized so that each curve consisted of 1024 data points. The results were plotted as boxplots using a home-written Python algorithm. Statistical tests were carried out in Python (Python Software Foundation, Beaverton, OR, USA). Since the residuals of our data were not normally distributed, we conducted Kruskal–Wallis tests and Dunn's post hoc test with Bonferroni correction. For this purpose, we tested the data in groups. For the influence of setpoint force, we tested groups of data collected at the same cantilever speed. Similarly, the influence of cantilever speed was tested by analyzing data with same setpoint force.

The long-time measurements (Figure 5) were realized on a clean glass slide with 1 nN setpoint force. Curves were collected at several positions to exclude artifacts from non-specific pollen–glass interactions. A list of positions was repeatedly measured, recording one curve per position. Per cycle, the mean of all positions was calculated and plotted. For the measurement carried out at 10 µm/s, the curves were collected at 9 different positions with a pause of 300 s after each curve. For 0.1 µm/s, curves were collected at 9 different positions without any pause and for 1 µm/s, curves were collected at 6 different positions with 600 s pause after each curve.

2.5. Cryo-Scanning Electron Microscopy (Cryo-SEM)

Opened florets or AFM cantilevers carrying a pollen particle were fixated on SEM stubs. These stubs were subsequently frozen at −140 °C in the Gatan Alto 2500 cryo-preparation system (Gatan, Pleasanton, CA, USA) and sputter coated with 10 nm Au-Pd under frozen conditions so that surface liquids did not evaporate and thus could be visualized. Images were taken at −120 °C and 3 kV accelerating voltage with a Hitachi S 4800 scanning electron microscope (Hitachi High Technology, Tokyo, Japan).

3. Results and Discussion

We imaged fresh *Hypochaeris radicata* pollen with and without pollenkitt using cryo-SEM (Figure 1a,b and Figure 1c,d, respectively). Moreover, cryo-SEM images of a *H. radicata* pollen, glued to an AFM cantilever and stored at room temperature for more than 7 days were taken (Figure 1e,f). A pollenkitt layer covers the entire surface of fresh *H. radicata* pollen grains and only the tips of the pollen spines are visible (Figure 1a,b). As the pollenkitt is removed, the pollen surface morphology changes considerably: the pollen show deeper cavities, the rims of which are formed by elevated areas that carry the fully exposed spines (Figure 1c,d). Apparently, most of the pollenkitt is deposited in these cavities, covering the porous surface of the pollen's outer wall (exine). Old and dehydrated *H. radicata* pollen also show these cavities (Figure 1e,f). However, the pores, despite being clearly visible, are still covered by a thin layer of pollenkitt. Moreover, the aged pollenkitt is more apparent at the cavities' edges (marked with white arrows) and between the spines, which are not as exposed as they are for pollen grains without pollenkitt. Thus, the pollenkitt on aged pollen still covers the pollen surface, but has been dehydrated and reduced to a very thin layer that renders the morphology to be dominated by the exine.

To study the influence of pollen aging on pollen adhesion, we quantified the adhesive properties of fresh and aged *Hypochaeris radicata* pollen with a centrifugation assay. Pollen adhesion was quantified by measuring the force necessary to detach pollen particles from glass via the centrifugation of a glass slide with adhering *H. radicata* pollen particles at different centrifugation speeds and therefore centrifugal forces. After each centrifugation, the amount of pollen that detached due to the respective centrifugal force was determined. It is important to note that higher centrifugation speeds do not only correspond to higher forces, but also to an extended duration of mechanical load as the centrifuge increases its speed gradually from 0 rpm. Especially for viscous samples, this might result in lower detachment forces compared to instantaneous force application. However, since we employ this method to compare the adhesion of different pollen particles and as these considerations apply to all our samples equally, centrifugation is applicable and our comparative results are valid. The advantage of this type of experiment is that a single experiment results in the adhesion analysis of many pollen particles at the same conditions.

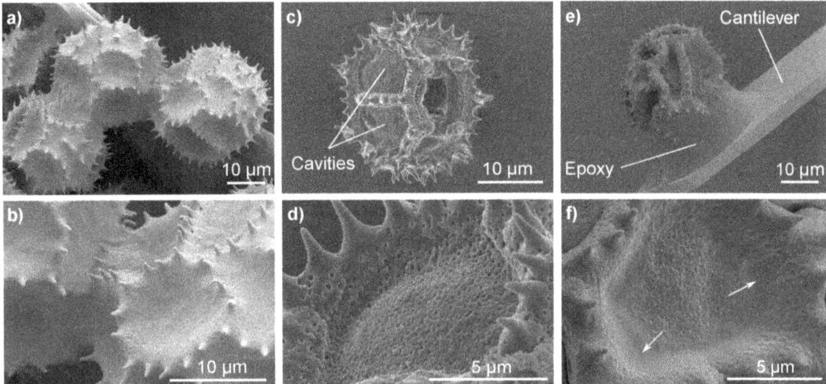

Figure 1. Cryo-scanning electron microscopy (cryo-SEM) images of fresh *Hypochaeris radicata* pollen with pollenkitt present (**a**,**b**) and with pollenkitt removed (**c**,**d**) as well as 7-day-old *H. radicata* pollen (**e**,**f**). The white arrows in (**f**) indicate remnants of pollenkitt on the surface of aged pollen. The influence of the pollen's aging and dehydration process on the pollenkitt are clearly visible. In (**e**), a pollen glued to an atomic force microscopy (AFM) cantilever is presented, showing that the pollen grain surface is left pristine by the gluing process.

While fresh pollen were directly deposited on the glass slide after being collected from the plant stylus, aged pollen were collected fresh and kept at room temperature for seven days before they were deposited on the glass slide. In both cases, centrifugation experiments were carried out directly after the pollen deposition, which means that short-term adhesive interactions were quantified. Furthermore, the long-term adhesive properties of *H. radicata* pollen were quantified by depositing fresh pollen on a glass slide and allowing them to adhere to it for seven days before the centrifugation experiments were carried out.

The results of the centrifugation experiment are presented in Figure 2. The adhesion of fresh *H. radicata* pollen that aged in contact with glass was much stronger than that of fresh and old pollen that were tested immediately after being deposited on glass. Most fresh and old pollen grains detached from the glass surface at a force range between 10 nN and 100 nN. In both latter cases, less than 30% of pollen remained attached after experiencing a centrifugal force of 100 nN. For the pollen grains aged in contact, on the other hand, there was no critical centrifugal force that removed the majority of pollen grains, but they detached more gradually and at higher forces. After applying a centrifugal force of 831 nN, 40% of these pollen still remained adhering to the glass, while less than 20% of fresh and old pollen withstood this force. Moreover, the results suggest that old *H. radicata* pollen adhere less strongly to glass than fresh ones. The difference is not as pronounced as it is for fresh pollen aged in contact, but for each centrifugal speed, the mean detachment force of old pollen was lower than that of fresh pollen.

One possible explanation of these observations is a change of pollenkitt properties, when the pollen dehydrate. Pollenkitt is largely regarded as a sticky emulsion of water and oil [14] and should thus become denser, more viscous and more adhesive when dehydrated [18]. On the one hand, a stickier pollenkitt may lead to an increase in pollen adhesion, however, the increased viscosity of the pollenkitt, on the other hand, slows down or even inhibits the formation of fluid bridges, which might be crucial for the fast and efficient formation of strong pollen adhesion [6]. This latter effect reduces the contact area—and therefore adhesive interactions—between the pollen particle and the substrate. The contact area is further reduced due to the fact that the pollenkitt layer shrinks with time because of the partial pollenkitt evaporation (Figure 1). Thus, short-term pollen adhesion, which has been analyzed for fresh and old pollen grains immediately after the contact formation with the substrate, is weakened when the pollenkitt evaporates partially. The reason for this is that the higher viscosity of old pollenkitt increases the time that is needed for the pollenkitt to flow onto the substrate to form fluid bridges. Long-term adhesion, which has been quantified for fresh pollen grains after they spent seven days in contact with the substrate, is increased, presumably due to the fact that fluid bridges that had already formed were solidified and strengthened when the pollenkitt became more condensed and stickier after partial evaporation.

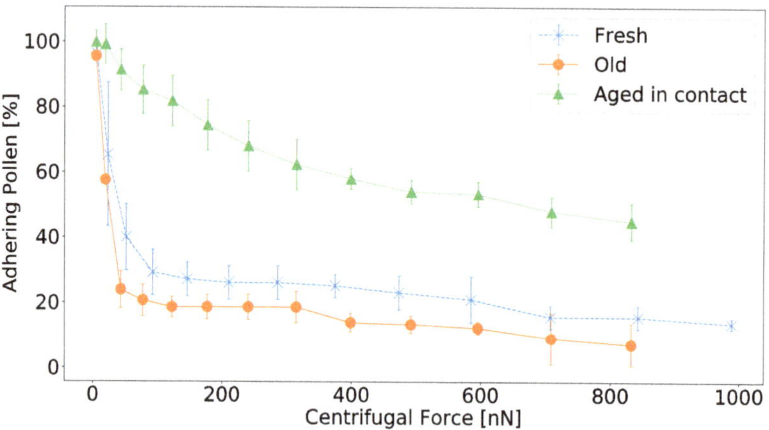

Figure 2. Quantification of *Hypochaeris radicata* pollen adhesion to glass using centrifugation. The amount of adhering pollen grains—relative to the number of pollen prior to centrifugation—is computed versus the centrifugal force. The adhesive properties were analyzed for pollen grains that were (1) fresh (blue crosses, n = 115); (2) old (orange circles, n = 165) when being deposited on the glass slide; and (3) were deposited freshly, but aged in contact with the glass prior to centrifugation (green triangles, n = 122). The mean values of four individual experiments are presented for each group of pollen and errorbars correspond to the standard deviation. Pollen that aged in contact with the glass adhered strongest to it. Fresh pollen grains had stronger adhesion than old ones.

To study the effects of pollen aging on the adhesive properties of pollen grains in more detail, we attached single *H. radicata* pollen grains to AFM cantilevers. Figure 1e) shows an exemplary image of an old *H. radicata* pollen glued to an AFM cantilever. It can be seen that the epoxy glue is only present between the pollen and the cantilever with the rest of the pollen surface left pristine during the gluing process. Pollen adhesiveness was quantified by bringing a pollen attached to a calibrated cantilever into contact with glass and by recording cantilever deformation while removing the pollen from the glass surface after the contact was formed. An example of a resulting force–distance curve is presented in Figure 3a. The minimum of the retract segment of the curves corresponds to the force necessary to detach the pollen from the glass slide ("pull-off force"). As this is the force that overcomes the adhesive forces between pollen and glass, we used the pull-off force to quantify pollen adhesion.

In some rare cases, electrostatic forces, visible as long-range interactions in the curves, were recorded. In these cases, we removed the surface charges from pollen and glass slide by employing an ionizing blow-off gun. Figure 3b presents the approach segments of an AFM curve recorded before and after the deionization procedure. The results clearly show that the long-range interactions can be present, but they were effectively removed by deionization.

The advantage of AFM-based adhesion quantification is that the adhesion of one and the same pollen particle can be repeatedly measured and thus, the influence of different experimental and environmental conditions can be analyzed using a single individual. Therefore, many errors originating from biological individuality do not affect such experiments and relevant qualitative conclusions can be drawn even with a relatively low number of individuals. Here, we chose to combine AFM studies with the centrifugation approach in order to have very detailed comparative results on the one hand and a high amount of individuals for quantitative results on the other. In order to ensure the repeatability of the AFM experiments, we chose the setpoint forces, which correspond to the force that the pollen experiences in contact with the glass slide, sufficiently low so that the pollen is not deformed during the experiments. As long as the pollen is undeformed, the part of the force–distance curve that is recorded during pollen–glass contact (marked with the dashed oval in Figure 3a) remains linear. For each experiment, we collected force–distance curves at 6–9 different positions on the glass slide to avoid artifacts from unspecific interactions between pollen grains and glass substrate. Furthermore, we repeatedly measured the pull-off forces of a single pollen particle 320 times and could not find a significant change in pollen adhesion (Figure 3c)

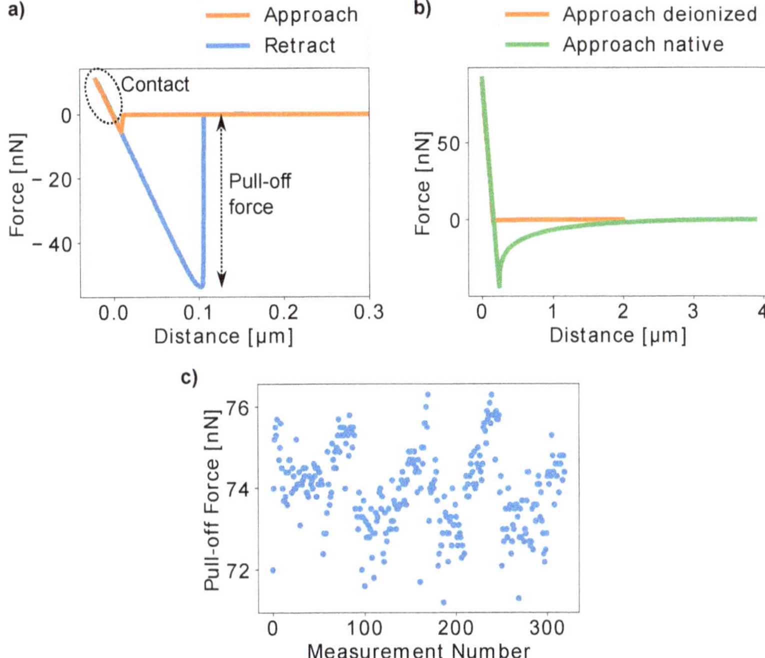

Figure 3. (**a**) An exemplary force–distance curve recorded with *Hypochaeris radicata* pollen on a glass slide. The curve's linear behavior during the contact between pollen and glass indicates that the pollen was not deformed during the measurement, which ensures the repeatability of the experiment. The minimum of the retract part of the curve serves as a quantification of the adhesive forces between the pollen and the glass substrate ("pull-off force"). (**b**) The approach segments of force–distance curves before and after deionizing pollen and glass show that electrostatic interactions, which are sometimes present, can be removed from the setup using deionization. (**c**) 320 curves were recorded with one *H. radicata* pollen. The pull-off forces did not change significantly, showing that our experimental procedure is stable and repeatable.

First, we tested the influence of the setpoint force and cantilever speed on the measured pull-off forces of single *H. radicata* pollen from glass in order to optimize these experimental parameters for our measurements. Figure 4a,b present the results obtained for two individual pollen grains. Moreover, we plotted the pull-off forces versus the measurement number for each pollen grain (Figure 4c,d) to ensure that our results were not influenced by any artifacts due to repeated measurements. For both individual pollen grains, the pull-off forces increased with the cantilever speed (pull-off rate), which is probably caused by the viscosity of the pollenkitt. Interestingly, at low cantilever speeds of 0.1 µm/s, both pollen grains had similar pull-off forces of approximately 50 nN. At higher cantilever speeds, the pull-off forces increased more strongly for the first pollen than for the second one. Thus, lower pull-off rates resulted in more comparable measurements between individual pollen grains.

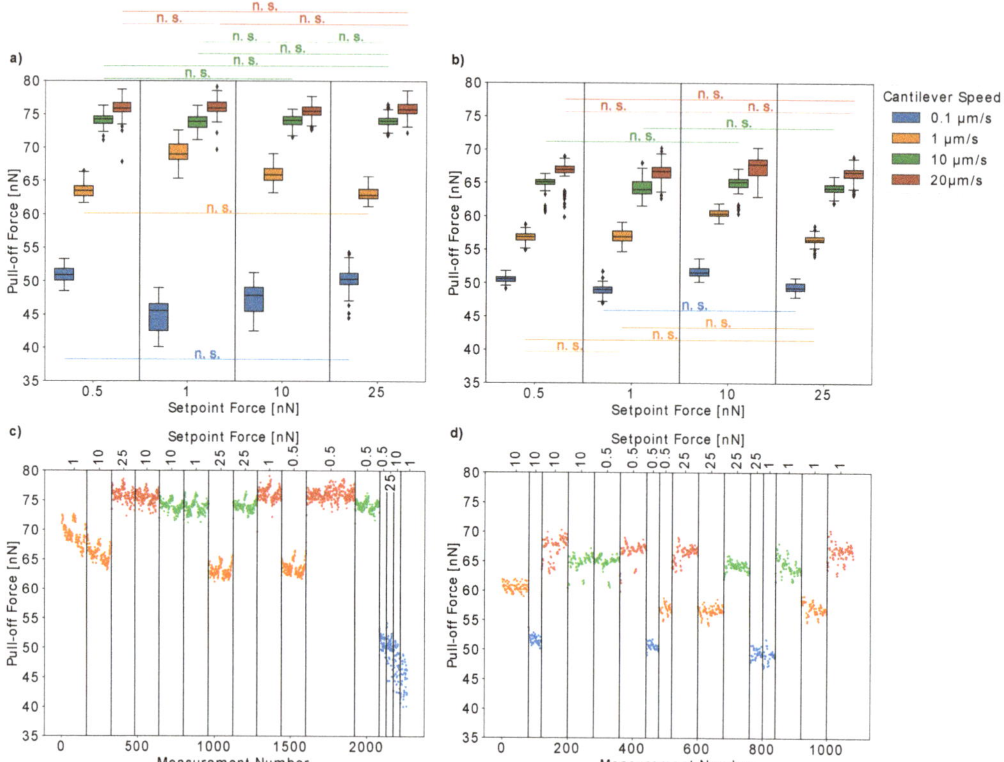

Figure 4. The influence of the setpoint force (contact force) and cantilever speed on the measured pull-off forces were analyzed for two fresh *Hypochaeris radicata* pollen. (**a**,**b**) The results presented as box plots. (**c**,**d**) The pull-off forces versus measurement number as an experimental control for the repeatability of measurements. For clarity, we only marked the boxes that are not significantly different from each other. Both individual pollen grains show an increase in pull-off forces with an increasing cantilever speed. The dependency on the setpoint force is not very pronounced. (**c**) The pull-off forces of the pollen presented in (**a**) decrease with a measurement number for low cantilever speeds, which could indicate that pollen aging influences our results. (**d**) The results obtained from the pollen presented in (**b**) were repeatable throughout the entire experiment. n.s.: no significance.

On the other hand, the pull-off forces of the first pollen continuously decreased over time when measured at low cantilever speeds of 0.1 µm/s or 1 µm/s (Figure 4c). At higher speeds, the measurements are repeatable. This did not only result in broader distributions of pull-off force values, but it is also a possible explanation for the fact that the influence of the setpoint forces on the pull-off forces is much more pronounced for low pull-off rates (Figure 4b). At higher cantilever speeds of 10 µm/s and 25 µm/s, the pull-off forces are not strongly influenced by setpoint forces between 0.5 nN and 25 nN.

This might be due to the fact that pollen aging influences our results, which should be more pronounced for slow cantilever speeds, as more time passes between consecutive measurements. Another possible explanation could be that slow cantilever speeds give the pollenkitt more time to form contact with the glass, which would increase the probability of some pollenkitt residues to remain on the glass when the pollen is removed. However, pollenkitt mostly consists of lipids, and is therefore a hydrophobic substance [14]. Glass, on the other hand, is hydrophilic and thus, an increased amount of pollenkitt on the glass should rather strengthen adhesion instead of weakening it. The second pollen did not show such a behavior, so this effect either might be due to the individual differences between pollen grains, their possible different ages when collected from the flower, or because the limits of repeatability were reached for the first pollen. In any case, it is important to double-check the results with such consecutive plots and to choose the pull-off rate carefully when conducting experiments. Lower pull-off rates ensure comparability between individual pollen grains but higher pull-off rates, on the other hand, decrease the risk of artifacts when a single pollen grain is measured repeatedly.

In order to study the influence of pollen age on the adhesive properties of a pollen grain, we repeatedly measured the pull-off force of three *H. radicata* pollen from glass for several hours. We carried out the experiments with different cantilever speeds to further study the mechanisms that make pollen adhesion dependent on pull-off rate. The results show that all three pollen decrease their adhesion over time (Figure 5). After the first pollen showed a continuous decrease in pull-off forces up to 16.2% from 108.5 nN to 90.8 nN over 12 h (Figure 5a), we increased the measurement time to 53 h for the next pollen (Figure 5b). For this pollen, the pull-off forces increased from 48.3 nN to 51.5 nN during the first hour and then decreased for 25% to 38.5 nN after 35 h, after which it did not change strongly anymore. The data for the third pollen show the decrease in pull-off forces by 10.4% from 65.7 nN to 58.9 nN in 26 h, after which the force decreased further after the measurement site on the glass slide was changed after 48 h (Figure 5c). Figure 5d shows the pull-off forces of all three pollen grains over time.

In addition to the fact that all pollen decreased their adhesiveness, other observations are worth noticing: For instance, we used a higher measurement frequency for the second pollen (Figure 5b) in order to test if the measured decrease in pull-off forces originates from the aging of pollen or from the pollen surface wear from measuring repeatedly. Apart from an initial increase, the pull-off forces of this pollen decreased over time in a comparable manner as for the other individual pollen grains. As this is the case despite the higher amount of collected force–distance curves, we conclude that our results indeed mainly originate from the effects of pollen aging instead of the pollen surface wear.

The initial increase in pull-off forces by 3.2 nN recorded for the second pollen (Figure 5b) might be caused by natural fluctuations of the results, which are due to the complexity of the pollen–glass interaction. The results presented in Figure 3c also show force fluctuations of this magnitude. Another possible explanation might be that very young pollen increase their adhesiveness for a short time, before the decrease begins. Gluing a pollen to a cantilever includes a curing time of 30 min. Moreover, individual pollen grains might have already different ages when they are glued to the cantilever, because the time at which a blossom already blooms before pollen collection varies as well. A combination of these two factors might lead to the situation that most individual pollen grains have already passed the time of increasing their adhesiveness before we started our experiment. The proof of this hypothesis would be a very interesting topic for future studies.

It is striking that the decrease in pull-off forces ceases for the second and third pollen after 35 h and 26 h, respectively. This might either be caused by the pollen itself, which might stop changing its adhesive properties, or it might be due to an assimilation of hydrophobic pollenkitt residuals on the hydrophilic glass slide, which would increase pollen adhesion. Indeed, the pull-off forces of the second and third pollen increased slightly after having reached a minimum. If the pollen continues to decrease its adhesiveness, these two effects could cancel each other out, which would explain the plateaus in our force data. The fact that the adhesion of the third pollen decreased further after we changed to a different position on the glass slide, is a confirmation of that hypothesis. On the other hand, the second pollen should show a similar effect much earlier, as we quantified its adhesion over time with a higher measurement frequency.

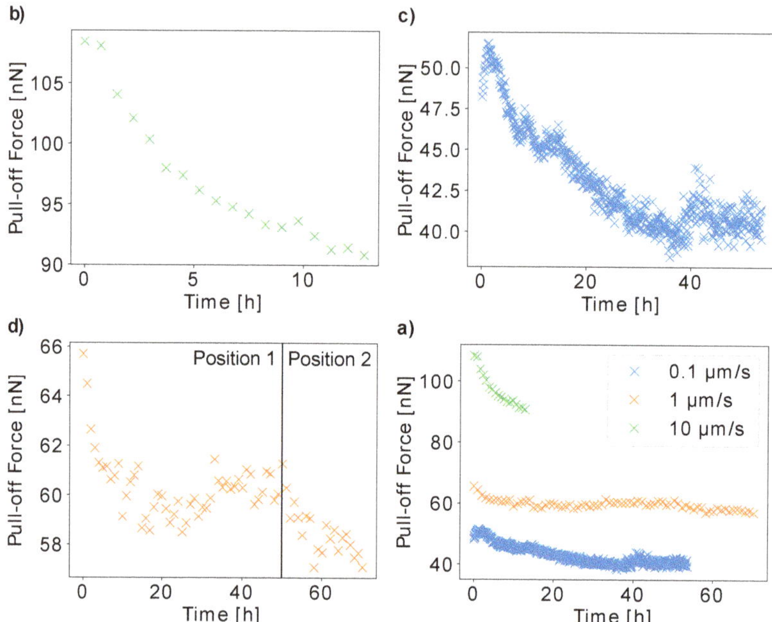

Figure 5. The pull-off forces of three single *Hypochaeris radicata* pollen grains from glass substrates measured repeatedly for 12 h (**a**) or more than 50 h (**b,c**) at different pull-off rates. (**d**) The pull-off forces plotted over time for all three pollen. The adhesiveness is decreasing over time for each pollen grain. The pollen measured with a cantilever speed of 10 µm/s has the highest pull-off forces, and the quickest decrease in adhesiveness.

Figure 5d shows the different pull-off force ranges for the three individual pollen grains presented in Figure 5a–c. This is in accordance with the speed dependency which we presented in Figure 4. Interestingly, the relatively high adhesion of pollen a, measured at 10 µm/s, decreased the fastest. A possible explanation for this result is that the pollenkitt becomes more viscous over time. Even though a higher pollenkitt viscosity should theoretically increase the pull-off forces for high cantilever speeds, it also increases the time that the pollenkitt needs to form fluid bridges with the glass slide. As higher cantilever speeds result in less contact time, an increased viscosity could actually weaken short-term adhesion.

It is also very interesting that the third pollen measured at 1 µm/s only very slightly decreases its adhesion. Possible reasons are individual pollen characteristics or a seasonal effect, as the pollen grains were harvested at different times of the year (August and November). Whether the season does influence pollen adhesiveness remains unclear and would be a very interesting topic for future studies.

We further tested the hypothesis that pull-off forces are mostly affected by the change of pollenkitt viscosity when the pollen particle ages by measuring the influence of setpoint forces and cantilever speeds for two *H. radicata* pollen older than seven days (Figure 6).

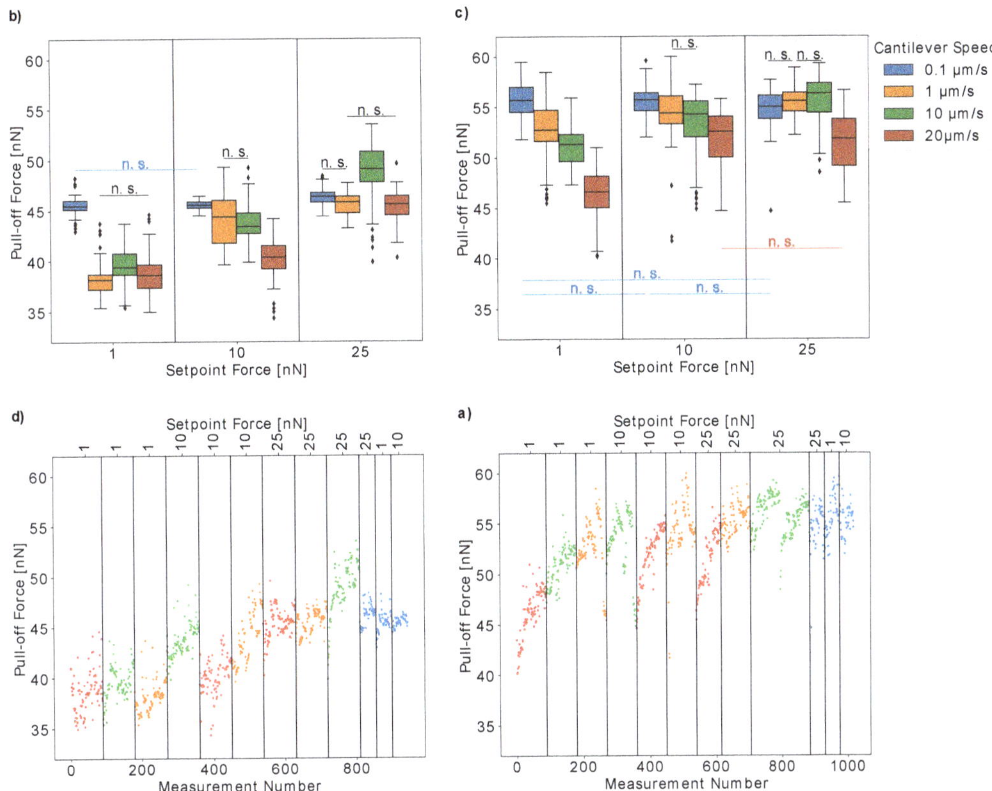

Figure 6. The influence of setpoint force and cantilever speed on pull-off forces for two individual pollen grains of *Hypochaeris radicata* that were older than seven days. (**a,b**) Box plots of the pull-off forces. (**c,d**) Pull-off forces versus measurement number. We marked only boxes that were not significantly different. In contrast to fresh pollen, the old pollen show a decrease in pull-off forces with increasing cantilever speed. For speeds above 1 µm/s, the pull-off forces increased with an increasing setpoint force. The pull-off forces for old pollen have a tendency to increase with the measurement number (**c,d**), which was not the case for fresh pollen. n.s.: no significance.

Strikingly, the influence of these two experimental parameters is very different than for fresh pollen. With few exceptions, higher cantilever speeds now result in lower pull-off forces. This fact is in accordance with our hypothesis that the higher viscosity of the pollenkitt increases with time, since this extends the time that the pollenkitt needs to form fluid bridges with the glass. Higher cantilever speeds result in a lower contact time and thus, the pollen adhesion decreases. Interestingly, for speeds above 0.1 µm/s, the pull-off forces of both aged pollen increase with setpoint force. Higher contact forces might increase the contact area between pollenkitt and glass, thus supporting the formation of fluid bridges. However, it is surprising that no similar effect was observed for fresh pollen (Figure 4). Another possible explanation is that during this experiment, pull-off forces increased when measured repeatedly, as shown by Figure 6c,d. Thus, higher setpoint forces, which were measured later, present higher values. This is an interesting observation in itself, as this was not observed for fresh pollen. Another possible reason for the different behavior of the pollen presented in Figures 4 and 6 might be that the former were harvested in July and August, while the latter were collected in November. It remains unclear whether flowers produce dissimilar pollen at different seasons and this would be a very interesting topic for future studies.

It would be interesting to repeat the experiments on hydrophobic surfaces, as pollen grains have to be transferred from hydrophilic plant surfaces to hydrophobic insect surfaces and vice versa during pollination. The pollen age (or the state of pollen evaporation) might be a factor regulating the adhesion forces during such transfer events.

During this study, we removed the surface charges from the pollen. However, it has been hypothesized that electrostatic interactions play an important role during pollination, especially for wind pollinated species [19–22]. Thus, in future studies, it would be very interesting to analyze the contribution of electrostatic forces to the adhesion of pollen in plant species with different pollination biology.

Another point is that the precise pollen age is not easily determined, as the exact moment the stylus pushes through the anther to present the pollen is hard to define. However, our results suggest that pollen properties might already change within an hour (Figure 5). Moreover, as it takes 30 min for the epoxy to cure when a pollen grain is glued to a cantilever, our method might not be able to quantify the initial adhesive properties of pollen grains. Novel AFM techniques, such as fluidic force microscopy, can potentially be employed to instantly attach a pollen grain to a cantilever with pressurized airflow to overcome this issue.

Furthermore, in order to understand the mechanics of pollination, the adhesion of pollen to more biologically relevant surfaces such as anther, stigma and insects, has to be studied in the future. As adhesion is mediated by a complex interplay of physical forces, which in turn are influenced by a variety of properties of several surfaces, these processes need to be disentangled by systematic research. The study of pollen adhesiveness to glass as a model surface is a first important step and the knowledge gained from this and similar future studies is crucial for the interpretation of more complex experimental designs.

4. Conclusions

After we visualized the different appearances of fresh and old *Hypochaeris radicata* pollen using cryo-SEM, we quantified the adhesion of *H. radicata* pollen and the influence of pollen age on pollen adhesiveness with a combination of centrifugation assays and AFM measurements. Our data indicate that pollen decrease their short-term adhesion and increase their long-term adhesion to hydrophilic surfaces over time. We found evidence that the viscosity of the pollenkitt influences the adhesive properties of the pollen. Our data suggest that the change of pollenkitt viscosity over time—which is most likely due to the evaporation of some volatile components—dominates the alteration of pollen adhesiveness. Long-term adhesion is strengthened by this effect, while the formation of new fluid bridges, which are a key factor for initial adhesion formation, is inhibited. This study is the first to investigate the influence of pollen aging on the adhesive properties of pollen grains and is a further step to understand the mechanics of pollination.

Author Contributions: Conceptualization, S.N.G. and S.H.; methodology, S.H. and L.-M.S.; software, S.H.; validation, S.H., L.-M.S. and S.N.G.; formal analysis, S.H. and L.-M.S.; investigation, S.H. and L.-M.S.; resources, S.N.G.; data curation, S.H. and L.-M.S.; writing—original draft preparation, S.H.; writing—review and editing, S.H., L.-M.S. and S.N.G.; visualization, S.H.; supervision, S.N.G.; project administration, S.N.G.; funding acquisition, S.N.G. All authors have read and agreed to the published version of the manuscript.

Funding: This research received no external funding.

Institutional Review Board Statement: Not applicable.

Informed Consent Statement: Not applicable.

Data Availability Statement: The data presented in this study are openly available in FigShare at https://doi.org/10.6084/m9.figshare.20238342.v1.

Acknowledgments: We would like to thank Shuto Ito for the cryo-SEM pictures of pollen without pollenkitt and for their support with the centrifugation assays.

Conflicts of Interest: The authors declare no conflict of interest.

References

1. Fægri, K.; Van der Pijl, L. *The Principles of Pollination Ecology*; Pergamon: Amsterdam, The Netherlands, 1979.
2. Ito, S.; Gorb, S.N. Attachment-based mechanisms underlying capture and release of pollen grains. *J. R. Soc. Interface* **2019**, *16*, 20190269. [CrossRef] [PubMed]
3. Willmer, P. *Pollination and Floral Ecology*; Princeton University Press: Princeton, UK, 2011. [CrossRef]
4. James, R.R.; Pitts-Singer, T.L. *Bee Pollination in Agricultural Ecosystms*; Oxford University Press: Oxford, UK, 2008.
5. Lin, H.; Qu, Z.; Meredith, J.C. Pressure sensitive microparticle adhesion through biomimicry of the pollen–stigma interaction. *Soft Matter* **2016**, *12*, 2965–2975. [CrossRef] [PubMed]
6. Ito, S.; Gorb, S.N. Fresh "pollen adhesive" weakens humidity-dependent pollen adhesion. *ACS Appl. Mater. Interfaces* **2019**, *11*, 24691–24698. [CrossRef] [PubMed]
7. Fattorini, R.; Glover, B.J. Molecular mechanisms of pollination biology. *Annu. Rev. Plant Biol.* **2020**, *71*, 487–515. [CrossRef] [PubMed]
8. Lone, A.H.; Ganie, S.A.; Wani, M.S.; Munshi, A.H. Wind Pollination: A Review. *Int. J. Mod. Plant Anim. Sci.* **2015**, *3*, 45–53.
9. Ollerton, J. Pollinator diversity: Distribution, ecological function, and conservation. *Annu. Rev. Ecol. Evol. Syst.* **2017**, *48*, 353–376. [CrossRef]
10. Pacini, E. From anther and pollen ripening to pollen presentation. *Plant Syst. Evol.* **2000**, *222*, 19–43. [CrossRef]
11. Corbet, S.A.; Chen, F.F.; Chang, F.F.; Huang, S.Q. Transient dehydration of pollen carried by hot bees impedes fertilization. *Arthr.-Plant Interact* **2020**, *14*, 207–214. [CrossRef]
12. Amador, G.J.; Matherne, M.; Waller, D.; Mathews, M.; Gorb, S.N.; Hu, D.L. Honey bee hairs and pollenkitt are essential for pollen capture and removal. *Bioinspir. Biomimetics* **2017**, *12*, 26015. [CrossRef] [PubMed]
13. Lin, H.; Gomez, I.; Meredith, J.C. Pollenkitt wetting mechanism enables species-specific tunable pollen adhesion. *Langmuir* **2013**, *29*, 3012–3023. [CrossRef] [PubMed]
14. Pacini, E.; Hesse, M. Pollenkitt–its composition, forms and functions. *Flora* **2005**, *200*, 399–415. [CrossRef]
15. Zinkl, G.M.; Zwiebel, B.I.; Grier, D.G.; Preuss, D. Pollen-stigma adhesion in Arabidopsis: A species-specific interaction mediated by lipophilic molecules in the pollen exine. *Development* **1999**, *126*, 5431–5440. [CrossRef] [PubMed]
16. Champion, J.A.; Katare, Y.K.; Mitragotri, S. Particle shape: A new design parameter for micro- and nanoscale drug delivery carriers. *J. Control. Release* **2007**, *121*, 3–9. [CrossRef]
17. Nel, A. Air pollution-related illness: Effects of particles. *Science* **2005**, *308*, 804 – 806. [CrossRef] [PubMed]
18. Lin, H.; Lizarraga, L.; Bottomley, L.A.; Carson Meredith, J. Effect of water absorption on pollen adhesion. *J. Colloid Interface Sci.* **2015**, *442*, 133–139. [CrossRef] [PubMed]
19. Vaknin, Y.; Gan-Mor, S.; Bechar, A.; Ronen, B.; Eisikowitch, D. The role of electrostatic forces in pollination. *Plant Syst. Evol.* **2000**, *222*, 133–142. [CrossRef]
20. Armbruster, W.S. Evolution of floral form: Electrostatic forces, pollination, and adaptive compromise. *New Phytol.* **2001**, *152*, 181–183. [CrossRef]
21. Bowker, G.E.; Crenshaw, H.C. Electrostatic forces in wind-pollination–Part 1: Measurement of the electrostatic charge on pollen. *Atmos. Environ.* **2007**, *41*, 1587–1595. [CrossRef]
22. Bowker, G.E.; Crenshaw, H.C. Electrostatic forces in wind-pollination—Part 2: Simulations of pollen capture. *Atmos. Environ.* **2007**, *41*, 1596–1603. [CrossRef]

Insects **2022**, *13*, 338

Article

Full-Length Transcriptome Sequencing-Based Analysis of *Pinus sylvestris* var. *mongolica* in Response to *Sirex noctilio* Venom

Chenglong Gao [1,2], Lili Ren [1,2], Ming Wang [1,2], Zhengtong Wang [1,2], Ningning Fu [1,2], Huiying Wang [3] and Juan Shi [1,2,*]

1   Beijing Key Laboratory for Forest Pest Control, Beijing Forestry University, Beijing 100083, China; gaocl0907@bjfu.edu.cn (C.G.); lily_ren@bjfu.edu.cn (L.R.); mingming66@bjfu.edu.cn (M.W.); wangzhengtong@bjfu.edu.cn (Z.W.); ning_fu@bjfu.edu.cn (N.F.)
2   Sino-France Joint Laboratory for Invasive Forest Pests in Eurasia, INRAE-Beijing Forestry University, Beijing 100083, China
3   Jilin Forestry Survey and Design Institute, Changchun 130000, China; wanghuiying2677@163.com
*   Correspondence: bjshijuan@bjfu.edu.cn; Tel.: +86-130-1183-3628; Fax: +86-10-6233-6423

Simple Summary: *Sirex noctilio*, as a devastating international forestry quarantine pest whose venom can cause a series of physiological changes in the host plants, such as needle wilting, yellowing, decreased transpiration rate and increased respiration rate, etc. In this study, a full-length reference transcript of *Pinus sylvestris* var. *mongolica* was constructed by combining second- and third-generation transcriptome sequencing technologies. We also identified the specific expression genes and transcription factors of *P. sylvestris* var. *mongolica* under *S. noctilio* venom and wounding stress. *S. noctilio* venom mainly induced the expression of genes related to ROS, GAPDH and GPX, and mechanical damage mainly induced the photosynthesis—related genes. The results provide a better understanding of the molecular regulation of pine trees in response to *S. noctilio* venom.

Citation: Gao, C.; Ren, L.; Wang, M.; Wang, Z.; Fu, N.; Wang, H.; Shi, J. Full-Length Transcriptome Sequencing-Based Analysis of *Pinus sylvestris* var. *mongolica* in Response to *Sirex noctilio* Venom. *Insects* **2022**, *13*, 338. https://doi.org/10.3390/insects13040338

Academic Editors: Gianandrea Salerno, Manuela Rebora and Stanislav N. Gorb

Received: 19 January 2022
Accepted: 25 March 2022
Published: 30 March 2022

Publisher's Note: MDPI stays neutral with regard to jurisdictional claims in published maps and institutional affiliations.

Abstract: *Sirex noctilio* is a major international quarantine pest that recently emerged in northeast China to specifically invade conifers. During female oviposition, venom is injected into the host together with its symbiotic fungus to alter the normal *Pinus* physiology and weaken or even kill the tree. In China, the Mongolian pine (*Pinus sylvestris* var. *mongolica*), an important wind-proof and sand-fixing species, is the unique host of *S. noctilio*. To explore the interplay between *S. noctilio* venom and Mongolian pine, we performed a transcriptome comparative analysis of a 10-year-old Mongolian pine after wounding and inoculation with *S. noctilio* venom. The analysis was performed at 12 h, 24 h and 72 h. PacBio ISO-seq was used and integrated with RNA-seq to construct an accurate full-length transcriptomic database. We obtained 52,963 high-precision unigenes, consisting of 48,654 (91.86%) unigenes that were BLASTed to known sequences in the public database and 4309 unigenes without any annotation information, which were presumed to be new genes. The number of differentially expressed genes (DEGs) increased with the treatment time, and the DEGs were most abundant at 72 h. A total of 706 inoculation-specific DEGs (475 upregulated and 231 downregulated) and 387 wounding-specific DEGs (183 upregulated and 204 downregulated) were identified compared with the control. Under venom stress, we identified 6 DEGs associated with reactive oxygen species (ROS) and 20 resistance genes in Mongolian pine. Overall, 52 transcription factors (TFs) were found under venom stress, 45 of which belonged to the AP2/ERF TF family and were upregulated. A total of 13 genes related to the photosystem, 3 genes related photo-regulation, and 9 TFs were identified under wounding stress. In conclusion, several novel putative genes were found in Mongolian pine by PacBio ISO seq. Meanwhile, we also identified various genes that were resistant to *S. noctilio* venom, such as GAPDH, GPX, CAT, FL2, CERK1, and HSP83A, etc.

Keywords: *Pinus sylvestris* var. *mongolica*; full-length transcriptome; *Sirex noctilio*; venom; wounding

Key Contribution: The study is the first to construct the full-length transcriptome of *P. sylvestris* var. *mongolica*, and we identified and analyzed 706 DEGs specifically induced by *S. noctilio* venom combined with NGS technology. The work carried out in this study will hopefully help researchers to better understand the interaction mechanism between Siricidae insects and host plants.

1. Introduction

Mongolian pine (*Pinus sylvestris* var. *mongolica*), a geographical variety of Scots pine (*P. sylvestris*), is naturally distributed in the Daxinganling Mountains and the Honghuaerji of Hulunbuir in northeast China, as well as in some parts of Russia and Mongolia [1]. It exhibits a high tolerance to cold, drought, and low soil fertility, thereby exhibiting strong adaptability and rapid growth [2,3]. Owing to these characteristics, Mongolian pine is currently the main coniferous plant that is utilized for windbreaking and sand fixation in the 3-North area of China, and thus plays a role in environmental protection and ecological construction. However, *Sirex noctilio* Fabricius (Hymenoptera: Siricidae), an invasive pest, was first identified in Daqing, Heilongjiang Province, northeast China, which specifically invaded *P. sylvestris* var. *mongolica* and caused tremendous economic losses and ecological damage [4]. To date, it has been spread to several provinces in China [5,6].

S. noctilio is a wood-wasp species native to Eurasia and North Africa. In the native range, the females attack only weak or dying *Pinus* spp. and are thus considered to be a secondary pest of negligible economic or ecological impact [7–10]. By contrast, *S. noctilio* is a major pest in invasive sites that causes damage and even kills a large number of healthy pines; for example, it has destroyed 70% of *P. radiata* in Uruguay, 30% of *P. taeda* and *P. elliotii* in northeast Argentina, and 75% *P. ponderosa* and *P. contorta* in southwest Argentina [11]. Thus, *S. noctilio* has a variety of hosts, which can cause great economic losses to the forestry of the invasive sites.

The females inject venom and symbiotic fungus together with eggs as they oviposit into the xylem of the host pines [12–15]. The synergetic effect of the venom and fungus is lethal to the host, although the venom alone can weaken trees by causing needle wilt and yellowing [16,17]. Studies have shown that the venom induces a series of physiological changes in the host that weaken the host's defense response and hence contribute to the growth of symbiotic fungus and the development of eggs [15]. These physiological changes include those related to needle dry weight, starch accumulation, peroxidase and amylase activity, respiration rate, and leaf pressure [18,19]. In addition, a few studies have used molecular methods to investigate the effects of the venom on the host plant. For example, in a study, the gene expression of pine tissue responding to *S. noctilio* venom was determined using a 26,496-feature loblolly pine cDNA microarray [20].

With the advancement of molecular-biology techniques, RNA-seq has become an indispensable tool for the analysis of defense gene expression and regulation in the transcriptome. The technique has been extensively used to study the response of conifers such as *P. halepensis* [21], *P. tecunumanii* [22], and *Picea sitchensis* [23] to biotic and abiotic stresses. However, for species lacking a reference genome, most genes obtained by RNA-seq were assembled from short reads. In this study, we obtained the full-length transcriptome of *P. sylvestris* var. *mongolica* for the first time by using the PacBio sequencing approach. Furthermore, we systematically identified and analyzed the functions and signaling pathways of differentially expressed genes (DEGs), and the molecular mechanisms of needles in response to venom and wounding stress. The study results may deepen our understanding of the molecular mechanism of the *P. sylvestris* var. *mongolica* resistance to *S. noctilio* venom, as well as help in identifying resistance-related genes and promoting genetic improvement.

2. Materials and Methods

2.1. Venom and Plant-Material Collection and Inoculation Trial

S. noctilio females obtained from the Biosafety Laboratory in Beijing Forestry University (BFU, Beijing, China) were immediately stored at −80 °C after their emergence from pine (*P. sylvestris* var. *mongolica*) logs infested by *S. noctilio*. Frozen wood wasps were dissected on ice to isolate the venom sac under a microscope (×40) (Leica M205C, Heidelberg, Germany). The pooled venom sac was added into a 1.5 mL centrifuge tube with four steel balls (diameter = 0.4 mm), and the mixture was shaken three times (2 min each time, frequency = 40/s). Then, the mixture was diluted to a 20 mg/mL concentration in a solution with deionized water. The resultant solution was centrifuged at 16,000× g and 4 °C for 30 min to remove cell debris, and the supernatant was stored at −20 °C until inoculation.

The inoculation experiments were conducted in approximately 10-year-old planted *P. sylvestris* var. *mongolica*, which were derived from a single seed lot and located in Ergetu Pine Plantation of Ulanhot CT, northeast China. Nine healthy pine trees with similar heights (174 ± 9 cm) and diameters at breast height (3.82 ± 0.22 cm) were used in the experiment that involved three treatments, namely control (non-inoculated), inoculated, and wounded. For inoculated and wounded samples, two 1.5 cm-deep holes were drilled at the 20 cm and 25 cm heights of the trunk with a cordless drill (6 mm diameter) and angled down by 45 degrees in order to hold a total of 2 mL of 20 mg/mL *S. noctilio* venom or water. Each hole was wrapped with parafilm strips. 3–5 needles were cut with sterile scissors in 4 directions (east, south, north and west) on the annual shoots at the top of each pine tree. The leaf needles were collected at three time points (0 h, 24 h, and 72 h) and immediately placed in liquid nitrogen and brought back to the laboratory for storage at −80 °C until RNA extraction.

2.2. PacBio Iso-Seq Library Preparation and Sequencing

A total of 27 samples (3 treatments: control, inoculation, wounding × 3 time points: 0 h, 24 h, 72 h × 3 biological replicates) were used for RNA sequencing. Total RNA was extracted from needles by using the E.Z.N.A. Plant RNA Kit (OMEGA bio-tek, Norcross, GA, USA), according to the manufacturer's protocol, and the RNA quality and concentration were measured using the NanoDrop 2000 Spectrophotometer (NanoDrop Products, Rockville, MD, USA). The RNA integrity number was assessed using Agilent 2100 Bioanalyzer (Agilent Technologies, Wilmington, DE, USA). The quality assessment of the 27 samples used for sequencing is shown in Table S1. Equal amounts of RNA from 27 samples (1 µg per sample) were pooled together to form total RNA, and then the SMART library was prepared using 5 µg total RNA. After PCR amplification, the products were used to construct the SMRTbell template library using the Iso-Seq protocol. Then, the libraries were prepared for sequencing by annealing a sequencing primer and binding polymerase to the primer-annealed template.

2.3. Iso-Seq Data Analysis

The raw polymerase reads were processed using the SMRT Pipe analysis workflow with default parameters. Next, CCSs were obtained through conditional screening (full passes of 1 and quality of 0.9) from the subread files. The CCSs were further classified into FL and non-full-length (nFL) transcripts irrespective of the presence of the 5′ primer sequence, 3′ primer sequence, and a poly-A tail. Full-length non-chimera (FLnc) reads were subjected to isoform-level clustering, followed by arrow polishing with the nFL sequence (hq_quiver_min_accuracy 0.99, bin_by_primer false, bin_size_kb 1, qv_trim_5p 100, qv_trim_3p 30). These polished sequences were further corrected, and redundant sequences were removed using the CD-HIT software (-c 0.99 -G 0 -aL 0.00 -aS 0.99 -AS 30 -M 0 -d 0 -p 1) to obtain the non-redundant, non-chimeric, and full-length transcripts for subsequent analysis (https://github.com/weizhongli/cdhit/wiki/3, accessed on 12 March 2022).

2.4. Illumina Library Preparation and Sequencing

The Illumina library was constructed using the TruseqTM RNA sample prep Kit (Illumina, San Diego, CA, USA). Briefly, polyadenylated mRNA was randomly disrupted into fragments. Under the action of reverse transcriptase, first-strand cDNA was generated using random hexamer primers (Invitrogen, Carlsbad, CA, USA), followed by second-strand cDNA synthesis. The purified fragmented cDNA was subjected to end-repair, followed by A-tailing. The final cDNA library was obtained by PCR enrichment and quantified using the TBS-380 (Tuner Biosystems, Sunnyvale, CA, USA). Finally, Illumina NovaSeq 6000 was used for sequencing.

2.5. Gene Functional Annotation

Full-length transcripts were annotated by performing BLASTX searches against six public databases, namely NR, KEGG, COG, Pfam, and Swiss-Prot, using the Blast2GO program for GO annotation based on NR annotation. The cut off E-value $< 10^{-5}$ was used in BLAST analysis against these databases.

2.6. Quantification of Unigene Expression Levels and Analysis of DEGs

To determine the difference in the gene expression of *P. sylvestris* var. *mongolica* under different treatments, we identified the gene-expression level of each sample by using RSEM software with the default parameters [24]. All clean data generated by Illumina sequencing were mapped to the full-length transcripts, and the read count of each gene was obtained in *P. sylvestris* var. *mongolica*. All the read-count values were converted into the TPM value to calculate the expression of each gene [25].

DEG analysis between the venom-stressed and control samples and between the wounding-stressed and control samples was performed using the DESeq R package (1.10.1) [26]. Significant DEGs were assigned with thresholds based on the false-discovery rate (FDR) < 0.05 and log2 | fold change | ≥ 1. GO and KEGG enrichment analyses for all DEGs were performed using the Python goatools package and KOBAS software. The potential TFs of DEGs were identified using the BLAST method with the PlantTFDB 4.0 database.

3. Results

3.1. Full-Length Transcriptome Sequencing

The RNA of all samples was mixed for third-generation sequencing (TGS) by using the PacBio Sequel platform. A total of 6,588,055 subreads were generated by filtering the raw data with an average length of 3625 bp and an N50 value of 4581 bp. Then, we obtained 849,415,352 bp circular consensus sequences (CCS) (199,841 reads of insert) with an average length of 4250 bp through conditional screening (full passes ≥ 1, quality > 0.90). The number of full-length (FL) and full-length nonchimeric (FLnc) reads was 163,130 and 162,036, respectively, with the corresponding average length of the two reads being 4232 and 4184 (Table 1).

Table 1. Summary of the transcriptome data from the PacBio platform.

Type	Number
Subread Reads	6,588,055
Average Subread Length	3625
Subread N50	4581
Total CCS	199,841
Average CCS Length (bp)	4250
FL Reads	163,130
FL Mean Length (bp)	4232
FLNC Reads	162,036
FLNC Mean Length (bp)	4184
Total Unigenes	52,963
Average Unigene Length (bp)	1801
TF number	1017

The advantage of the TGS technology by using the PacBio platform is that it provides long read lengths, although its single-base-error rate is high. To further improve the accuracy, we also sequenced 27 samples of the needles by using the Illumina NovaSeq 6000 platform (Illumina, Sam Diego, CA, USA) with 300 bp pair-end reads. Redundant and similar sequences were removed using CD-HIT software. Finally, 52,963 unigenes with an average length of 1801 bp were obtained and considered the reference transcriptome, and the length distribution of unigenes is shown in Figure S1.

3.2. Functional Annotation of the Full-Length Reference Transcriptome

To determine the possible functions of unigenes in *Pinus*, we analyzed a total of 52,963 unigenes by using six databases, namely NCBI nonredundant protein (NR), Gene Ontology (GO), Kyoto Encyclopedia of Genes and Genomes (KEGG), Clusters of Orthologous Groups of proteins (COG), Swiss-Prot, and Protein Family (Pfam). Overall, 48,654 (91.86%) unigenes were annotated to known sequences in public database, whereas 20,172 (38.09%) unigenes were simultaneously annotated in six databases. As shown in Figure 1A, the number of unigene hits was the highest (48,228; 91.06% and 44,087; 83.24%) in the Nr database, followed by those in the GO database (41,580; 78.51%), Swiss-Prot database (39,332; 74.26%), Pfam database (37,874; 71.51%), and KEGG database (25,157; 47.50%). According to the sequence alignment in the NR database, 24,848 (51.52%) sequences had significant homology against *P. sitchensis*, followed by those against *Amborella trichopoda* (3462, 7.18%), *Nelumbo nucifera* (1489, 3.09%), *Cinnamomum micranthum* (1364, 2.83%), *Macleaya cordata* (775, 1.61%), *P. taeda* (736, 1.53%), *Vitis vinifera* (611, 1.27%), *P. tabuliformis* (608, 1.26%), *P. sylvestris* (600, 1.24%), and *Elaeis guineensis* (570, 1.18%). Nevertheless, 28.47% of the sequences were homologous to those of other species (Figure 1B).

For the GO analysis, 41,580 unigenes were divided into 52 categories, namely molecular function (MF, 33,774), cellular component (CC, 27,629), and biological process (BP, 22,787). Of the total, 16 categories belonged to MF, 14 categories belonged to CC, and 22 categories belonged to BP (Figure 2A). To further profile the pathways in which the unigenes are involved, we conducted an analysis based on the KEGG database. A total of 25,157 unigenes were classified into five metabolic pathways (first category) and were found to be mainly associated with carbohydrate metabolism (3057, 12.15%), followed by energy metabolism (2445, 9.72%) and translation (2151, 8.55%) (Figure 2B).

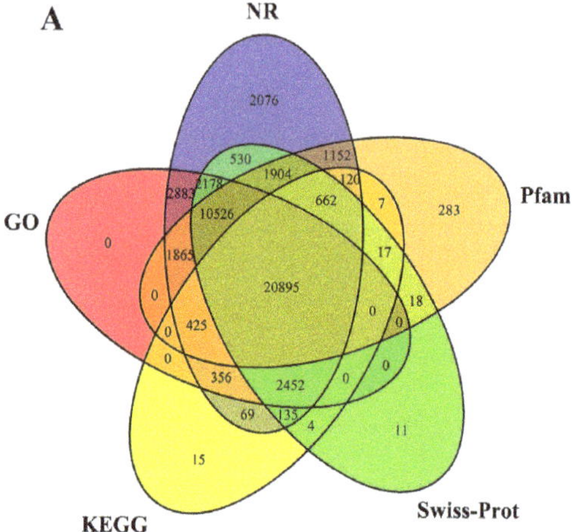

Figure 1. *Cont.*

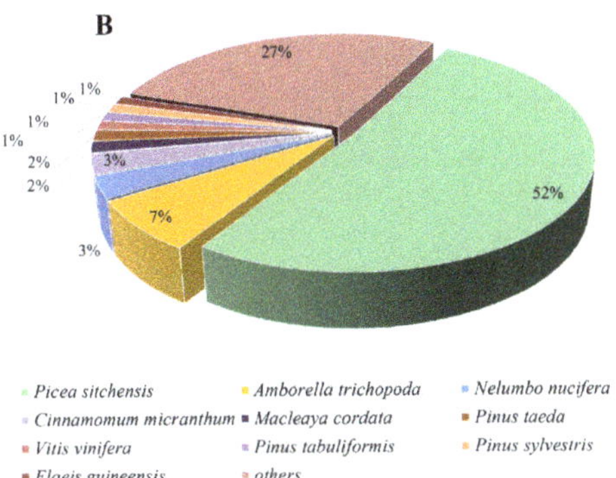

Figure 1. Unigene annotation of PacBio sequencing. (**A**) Overlap between the number of all unigenes according to five databases. (**B**) Distribution of unigene annotations based on the NR database for the species-distribution statistics.

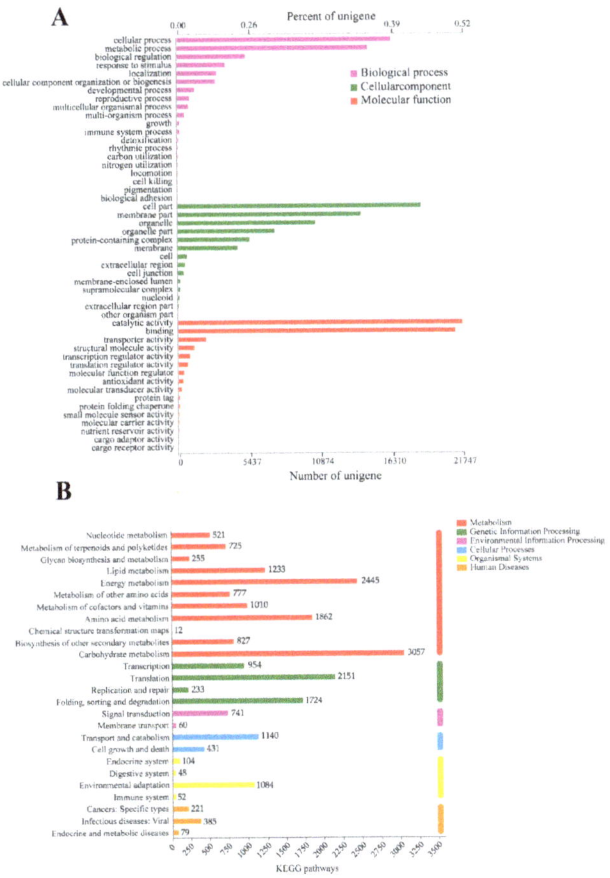

Figure 2. GO and KEGG classification of unigenes. (**A**) GO functional classification of all unigenes. (**B**) KEGG classification of all unigenes.

Furthermore, we also predicted 1017 unigenes, which were assigned to the 31 TF family. The three most abundant unigenes were MYB (170), AP2/ERF (130) and NAC (87) (Figure S2).

3.3. Identification of DEGs

Libraries obtained from the control, wounded, and inoculated samples were mapped to the reference transcripts from the PacBio ISO-seq. The matched rate of all the clean reads was >80% (Table S2). The difference in the expression levels between two groups was determined based on the transcripts-per-million (TPM) value. Samples that were inoculated with venom, wounded, and those without any treatment were replaced with PI, PW and CK, respectively. For inoculation, we identified a total of 266 DEGs (139 upregulated and 127 downregulated) between PI_0 hand PI_24 h, 995 DEGs (802 upregulated and 193 downregulated) between PI_24 h and PI_72 h, and 1253 DEGs (905 upregulated and 348 downregulated) between PW_0 hand PI_72 h. For wounding treatment, we identified 301 DEGs (111 upregulated and 190 downregulated) between PW_0 h and PW_24 h, 365 DEGs (258 upregulated and 107 downregulated) between PW_24 h and PW_72 h, and 1,193 DEGs (602 upregulated and 591 downregulated) between PW_0 h and PW_72 h. These results show that the number of DEGs increased with an increase in the treatment time. For example, the number of DEGs after 72 h of inoculation was 1253, which was approximately five times higher than that after 24 h of inoculation (266) (Figure 3).

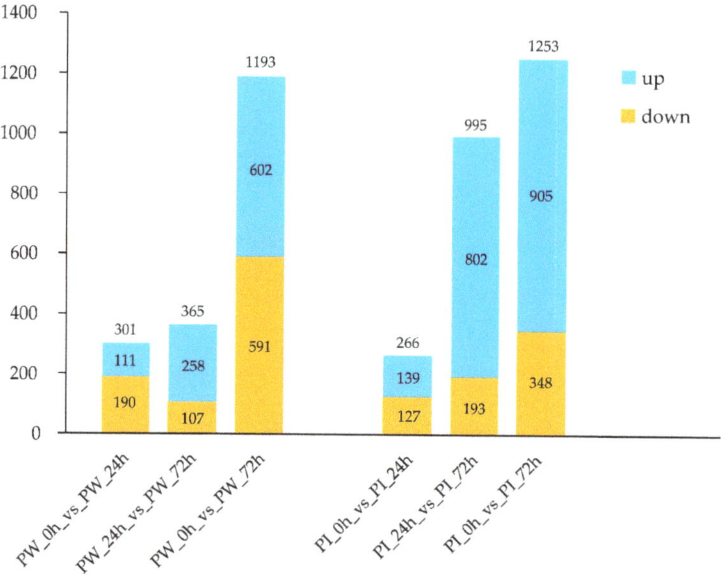

Figure 3. Number of DEGs induced by wounding and inoculation at different times.

3.4. Analysis of DEGs in Wounded and Inoculated Mongolian Pine at 72 h

Among the three time points, the specific-treatment time point was 72 h, at which the DEGs were most abundant. Therefore, the 72 h time point was used for subsequent treatments. Compared the CK_72 h, 825 DEGs were induced by *S. noctilio* venom, 506 DEGs were induced by wounding, and 119 DEGs were co-induced by venom and wounding.

3.4.1. Analysis of Inoculation-Specific DEGs

Among the 825 DEGs between control and inoculation, 559 genes were upregulated, and 266 genes were downregulated. A total of 706 inoculation-specific DEGs (474 upregulated and 232 downregulated) were identified after removing the DEGs common to wounding and inoculation (Table S3).

To investigate the molecular mechanisms that regulate interactions and coordination of pine under *S. noctilio* venom-specific presentation, a GO enrichment analysis (top 20) of these 706 DEGs was performed, which indicated that inoculation affected 12 MF categories, as well as seven BP and one CC. The most significantly enriched MF categories were "ADP binding" (31 upregulated and 3 downregulated), "transcription-regulator activity" (46 upregulated and 1 downregulated), and "DNA-binding transcription-factor activity" (45 upregulated and 1 downregulated). Only one CC category was found to be significantly enriched, namely "nucleus" (68 upregulated and 12 downregulated), whereas "signal transduction" (45 upregulated and 1 downregulated), "xyloglucan metabolic process", and "hemicellulose metabolic process" were the most significantly enriched BP categories. Notably, all six genes enriched in the "xyloglucan metabolic process" and "hemicellulose metabolic process" categories were upregulated and identical (Figure 4A). Of these, five genes were annotated as "xyloglucan endotransglucosylase/hydrolase 2", whereas one gene was annotated as "unknown" in the NR database.

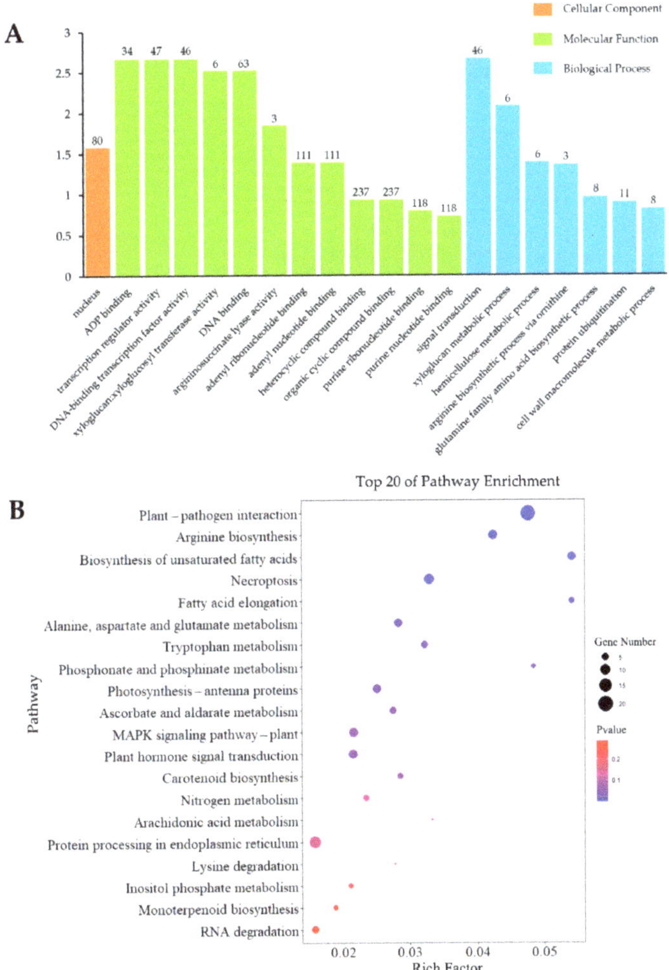

Figure 4. GO (**A**) and KEGG (**B**) analyses of DEGs induced by inoculation specificity.

Furthermore, the KEGG-pathway enrichment analyses of 706 DEGs were performed, and these DEGs were assigned to 80 KEGG pathways. The top 20 pathways were screened as having the most intensive response activities. The most significantly enriched pathways were "plant–pathogen interaction", "arginine biosynthesis", "biosynthesis of unsaturated fatty acids", "necroptosis", and "fatty-acid elongation" (Figure 4B).

3.4.2. Analysis of Wounding-Specific DEGs

A total of 506 DEGs were identified by comparing the gene-expression levels under wound treatment with those under control treatment. A total of 387 DEGs were differentially expressed specifically after wounding; of these, 183 DEGs were upregulated and 204 DEGs were downregulated. GO enrichment analysis (top 20) of the 387 DEGs showed that wounding affected 4 MF, 13 CC, and 3 BP categories. Seven exactly identical genes were assigned to the most significantly enriched MF categories, namely "inositol 3-alpha-galactosyltransferase activity", "glucuronosyltransferase activity", and "UDP-galactosyltransferase activity". Six DEGs were assigned to the most significantly enriched CC categories, namely "photosystem II antenna complex", "PSII-associated light-harvesting complex II", and "light-harvesting complex". The significantly enriched BP categories were "photosynthesis", light harvesting", "photosystem II assembly", and "nonphotochemical quenching" (Figure 5A).

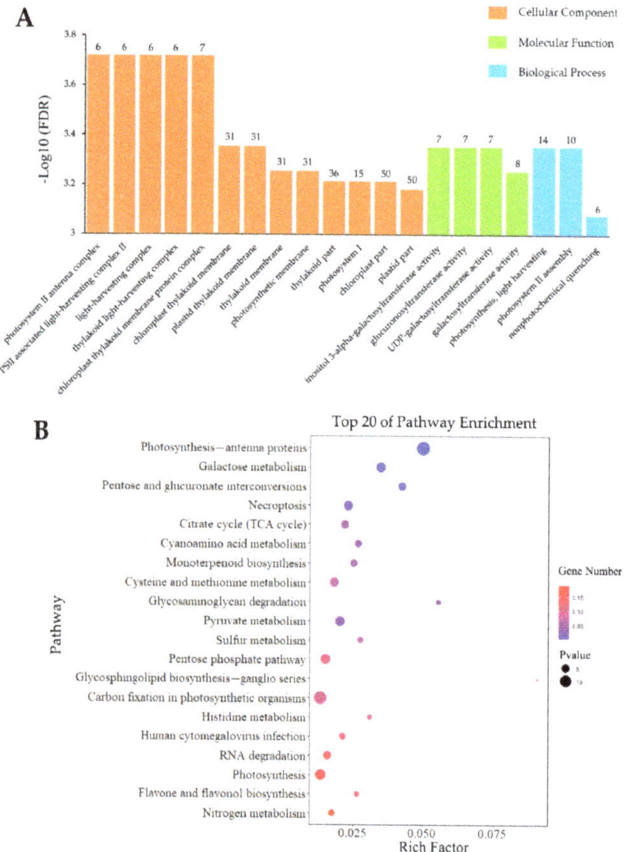

Figure 5. GO (**A**) and KEGG (**B**) analysis of DEGs induced by wounding specificity.

The wound-specific DEGs were also subjected to KEGG enrichment analysis and were assigned to 83 metabolism pathways. The top 20 significantly enriched pathways are shown in Figure 5B and were divided into four categories, namely "metabolism (M)", "cellular processes (CP)", "genetic information processing (GIP)", and "organismal systems (OS)". The most significantly enriched M categories were "photosynthesis-antenna proteins", "galactose metabolism", "pentose and glucuronate interconversions". In CP and GIP, only one category was enriched, namely "RNA degradation" and "necroptosis", respectively.

3.4.3. DEGs Induced by Inoculation and Wounding

A total of 119 common DEGs (85 upregulated and 34 downregulated) were induced by inoculation and wounding compared with the control. The top 20 categories of significant GO enrichment are shown in Figure 6A. Interestingly, seven DEGs were simultaneously enriched in "amide binding" and "peptide binding". Four of these seven DEGs were enriched in "chloroplast-targeting sequence binding", "intrinsic component of chloroplast outer membrane", and "protein targeting to chloroplast". However, the annotation information of these seven genes is "unknown" in the NR database. Subsequently, the KEGG enrichment analysis showed that the most significantly enriched categories were "nitrogen metabolism", "photosynthesis-antenna proteins", and "glutathione metabolism" (Figure 6B). Notably, most of the categories were related to the chloroplast and photosynthesis.

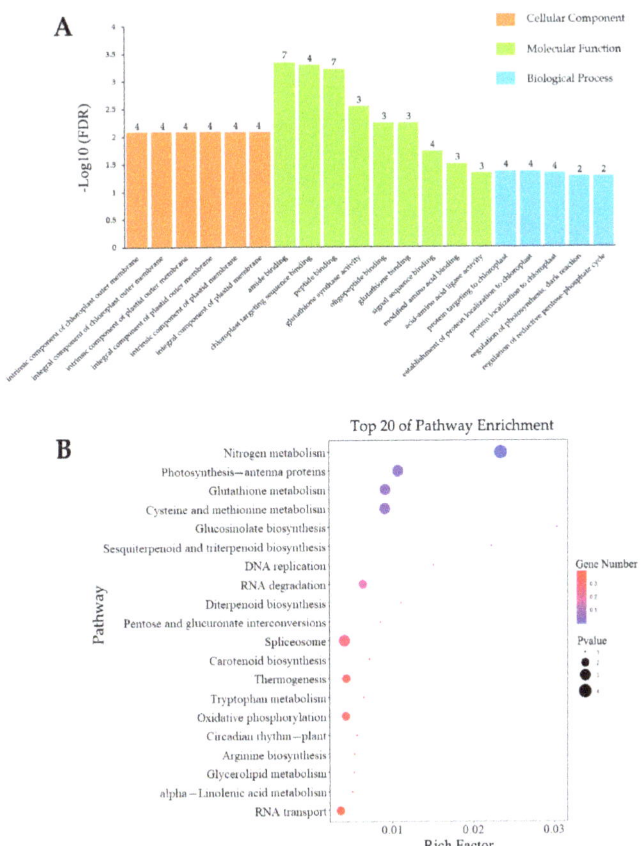

Figure 6. GO (**A**) and KEGG (**B**) analysis of DEGs co-induced by inoculation and wounding.

3.5. Transcription Factors (TFs) in Response to Inoculation and Wounding in P. sylvestris var. mongolica

Of the 706 DEGs specifically induced by *S. noctilio* venom, we identified 7 TF families with 52 DEGs (50 upregulated and 2 downregulated). Among these TFs, the largest number of genes (46) belonged to the ethylene-responsive-factor (AP2) family, with 45 genes being upregulated and 1 gene being downregulated. The number of other TFs, including Zf-CCCH, Zf-C2H2, Myb_DNA_bind, GRAS, DUF573, and DUF260, was small. For wounding-specific DEGs, four TFs were identified with only nine DEGs. The number of genes for AP2, Myb_DNA_bind, and NAM and DUF260 was 4, 3, and 1, respectively. Of the 119 DEGs induced by inoculation and wounding, only 6 DEGs belonged to the three TFs, AP2, Myb_DNA_bind, and DUF260 (Table 2).

Table 2. TF statistics of DEGs under different treatment.

TF Family	Number of TFs		
	Inoculation-Specific	Wounding-Specific	Co-Induced
AP2/ERF	46	4	4
LOB	1	1	1
MYB_superfamily	1	3	1
C2H2	1	N/A	N/A
C3H	1	N/A	N/A
GeBP	1	N/A	N/A
GRAS	1	N/A	N/A
NAC	N/A	1	N/A

4. Discussion

In the past decade, short-read RNA-seq (next-generation sequencing) has been widely used for genetic discovery and research [27]. However, a limitation of the second-generation technologies is that the length of sequenced fragments is usually less than 300 bp [28]. Moreover, obtaining a complete and reliable reference for species without reference genomes is challenging. The third-generation sequencing technique, that is, single-molecule sequencing, which can be used to generate full-length transcripts and near-reference-genome assembly for some species, can reveal the full structure of individual transcripts [29,30]. The present study is the first to conduct transcriptome sequencing of *P. sylvestris* var. *mongolica* needles by using the single-molecule-sequencing technique. A total of 52,963 full-length unigenes (average length, 3625 bp; N50 value, 4581 bp) were generated, which is more than that generated in a previous study (39,231) (N50 value of 1646 bp) [31]. Additionally, the annotated percentage (48,654, 91.86%) of these full-length unigenes was significantly higher than that reported in previous studies [32,33]. The remaining 4309 (8.14%) unannotated unigenes possibly represent a novel gene pool that is specific to the common *P. sylvestris* var. *mongolica*. The potential function of these unannotated genes and their response to *S. noctilio* venom or wounding warrant further investigation.

4.1. Signaling Pathways Involved in the Response of P. sylvestris var. mongolica to Wounding and Inoculation

Plant responses to biological stress are extremely complex because of the presence of various types of interactions between the plant and the pathogen [34]. These interactions serve as the multi-layered immune system that has emerged during plant evolution to prevent or block the colonization of most potential pathogens [35,36]. This in turn promotes the production of specific resistant proteins in plants to recognize pathogen/insect effects [37]. Consistently, in the present study, we found the "plant–pathogen interaction" as the most significant KEGG pathway during the 72 h inoculation (Table S4). Among these genes, those encoding CAM/CML were the most significantly upregulated. During the plant–pathogen interaction, calcium is considered to be an important factor for regulating plant responses to various pathogens and herbivores [38]. Ca2+ signaling by CaM/CML produces NO, which induces defense responses [39]. The CaM/CML protein family regulates cellular responses to various stimuli in grapevine, particularly to biotic stresses [40]. Trujillo-Moya et al. reported a similar conclusion on *P. abies* [41]. In addition, the genes encoding FLAGELLIN-SENSING 2 (FL2), CERK1, and HSP83A were upregulated, all of which have been found to play crucial roles in biotic and abiotic stresses (Table 3) [42–44].

Table 3. The key DEGs involved in plant–pathogen interaction signaling pathway under the *S. noctilio* venom stress.

Gene_ID	Swiss-Prot	Log2FC(PI_72 h/CK_72 h)	*p*-Value
transcript_26258	Probable disease-resistance protein At1g15890 *Arabidopsis thaliana*	1.84	2.31×10^{-6}
transcript_64802	Calmodulin-like protein 3 *Arabidopsis thaliana*	4.40	3.95×10^{-7}
transcript_66692	Putative calmodulin-3 (Fragment) *Solanum tuberosum*	4.78	2.69×10^{-8}
transcript_11085	Calmodulin-like protein 2 *Arabidopsis thaliana*	4.36	1.26×10^{-15}
transcript_52365	LRR receptor-like serine/threonine-protein kinase FLS2 *Arabidopsis thaliana*	4.24	5.83×10^{-5}
transcript_35906	Heat-shock protein 83 *Ipomoea nil*	2.25	2.47×10^{-5}
transcript_25250	Heat-shock protein 83 *Ipomoea nil*	4.02	3.52×10^{-4}

Photosynthesis is the most fundamental and complex physiological process in all green plants, and this process is also severely affected by various biotic and abiotic stresses [45]. The photosynthetic pigments are believed to be damaged by stress factors, which in turn reduces the photosynthetic capacity by reducing the light-absorption efficiency of the photosystem (PSI and PSII) [46,47]. The KEGG-pathway enrichment analysis of the DEGs induced by mechanical specificity and the co-induced DEGs showed that the most significantly enriched pathways were "photosynthesis—antenna proteins" and "nitrogen metabolism". The genes involved in the nitrogen-metabolism pathway were annotated as carbonic anhydrase and glutamine synthetase cytosolic isozyme, both of which were upregulated (Tables S5 and S6). Carbonic anhydrase is an essential enzyme in photosynthesis that exhibits the property of reversibly converting CO_2 to HCO_3^- [48]. In general, wounding mainly induces the differential expression of genes related to photosynthetic metabolic pathways in *P. sylvestris* var. *mongolica*.

4.2. Expression of Typical Venom-Induced Genes in Mongolian Pine

Under stress, plants regulate their homeostatic mechanisms by generating excess ROS, and have evolved with some enzymatic or non-enzymatic detoxification mechanisms to control the excess accumulation of ROS [49–51]. In our study, a total of 706 genes were differentially expressed specifically under venom stress. Of these, we found six DEGs associated with reactive oxygen species (ROS), of which three encode glyceraldehyde-phosphate dehydrogenase (GAPDH), two encode glutathione peroxidase (GPX) and one encodes catalase (CAT). In addition, we also found five DEGs encoding Xyloglucan endo-transglucosylase/hydrolase 2 (XTH2), five putative truncated TIR-NBS-LRR protein, four polyubiquitin, three UDP-glycosyltransferase (UGT), two Hs1pro-1 and one pleiotropic drug-resistance (PDR) protein 1. One of these was a downregulated DEG, and the others were upregulated DEGs. These genes have been shown to be involved in regulating plant immune responses to biological or abiotic stresses, e.g., XTH expression could be rapidly induced in response to various stresses in tobacco [52]. The C-termini of the five putative truncated TIR-NBS-LRR proteins contain a variable LRR domain, which is the most significant structural feature of plant disease-resistance proteins [53]. All of these genes exhibited significantly induced expression 72 h after venom stress, but the expression of these genes was not significant under wounding stress (Table 4). As a related species of *S. noctilio*, *Sirex nitobei* venom proteins have been identified, such as laccase-2, laccase-3, a protein belonging to the Kazal family, chito-oligosaccharidolytic β-N-acetylglucosaminidase, beta-galactosidase, etc. These venom proteins may be the key factors to inducing gene-specific expression in *P. sylvestris* var. *mongolica* [34]. Therefore, the DEGs induced by *S. noctilio* venom were identified as the candidate genes that enhance the self-tolerance of *P. sylvestris* var. *mongolica*, especially for *S. noctilio*. We recommend that researchers study these genes in more detail.

Table 4. DEGs expression under the *S. noctilio* venom stress of Mongolian pine.

Annotation	Gene_ID	Log2FC(PI_72h/CK_72h)
GAPDH	transcript_53182	7.7805
	transcript_44911	4.5701
	transcript_1666	3.1995
GPX	transcript_43923	2.1892
	transcript_54716	1.5788
CAT	transcript_40212	3.5800
	transcript_48021	4.9165
	transcript_59698	3.7113
XTH2	transcript_7449	3.3294
	transcript_52674	2.7133
	transcript_7326	2.5641
	transcript_70636	2.6561
UGT	transcript_40291	1.7841
	transcript_41406	1.6345
	transcript_20771	2.7826
putative truncated TIR-NBS-LRR protein	transcript_19948	2.1045
	transcript_19947	1.9758
	transcript_21018	1.7106
	transcript_20174	1.5549
	transcript_65149	9.5582
polyubiquitin	transcript_55582	2.1239
	transcript_5903	−1.554
	transcript_6253	7.3522
PDR protein 1-like	transcript_37577	6.2179
Hs1pro-1	transcript_23455	2.1044
	transcript_65743	2.1014

4.3. Expression of Typical Wounding-Induced Genes in Mongolian Pine

Under wounding stress, the results showed that 387 DEGs were specifically regulated. Photosynthesis is essential for plant growth and development. It involves the harvesting of light and transfer of solar energy, using light-harvesting chlorophyll-a/b-binding (LHC) proteins. The LHC proteins are among the most abundant in thylakoids, which are encoded by nuclear genes. In high plants, LHC proteins include a large gene family containing 10–12 members, constituting the peripheral light-harvesting antenna of photosystem I (PSI) and photosystem II (PSII). LHCAs are encoded by Lhca1–Lhca6, and LHCBs are encoded by Lhcb1–6 [55–58]. We found 13 DEGs associated with photosynthesis, 6 encoding Lhcb5 protein, 4 encoding LHC protein, partial, 2 encoding oxygen-evolving enhancer protein 1 (OEE) and 1 encoding Lhca4 protein, 5 of which were upregulated and the others were downregulated (Table 5). Many studies have indicated that LHC gene expression is regulated by multiple environmental and developmental cues in higher plants, such as light, oxidative stress, low temperatures, and phytohormone abscisic acid [59–62]. In addition, we also found three casein kinase 1-like protein HD16s, which can change the alteration of photoperiod sensitivity to enhance the adaptability to local environmental conditions in many plants [63]. Therefore, we hypothesize that these genes still function after the plants are subjected to wounding stress.

Table 5. DEGs expression under wounding stress of Mongolian pine.

Annotation	Gene_ID	Log2FC(PW_72 h/CK_72 h)
Lhcb5 protein	transcript_58449	10.3402
	transcript_61176	9.1811
	transcript_71238	5.2818
	transcript_8414	2.5689
	transcript_67898	−7.5324
	transcript_15721	−9.1712
LHC protein, partial	transcript_73156	7.4302
	transcript_67949	−2.3693
	transcript_51999	−2.3040
	transcript_47491	−2.4656
OEE	transcript_7777	−4.6929
	transcript_5485	−2.0221
Lhca4 protein	transcript_49364	−2.3063
	transcript_28611	−7.1177
HD16	transcript_31667	−9.0054
	transcript_17461	−9.0169

4.4. The Role of TFs in Response to Inoculation and Wounding

Families of TFs, as activators and repressors, play an important role in plant resistance to biotic and abiotic stresses [64,65]. In higher plants, approximately 60 TF families have been identified [66], which include AP2/ERF, MYB, and C2H2. The APETALA2/ethylene-responsive element-binding factors (AP2/ERF) represent a large group of factors that are found mainly in plants [67–69]. AP2/ERF are the important regulators involved in plant growth and development [70,71], hormonal regulation, and abiotic stress [72]. They also play a crucial role in plants' defense against biotic stress, including invasion by herbivorous insects and microbial pathogens [73]. In this study, we identified 46 (88.46%) DEGs (45 upregulated and 1 downregulated) associated with the AP2/ERF family, all of which were annotated as ERF in the Swiss-Prot database. However, only four AP2/ERFs were present in wound-specific and co-induced DEGs. The ERF TFs enhance plant resistance to piercing/sucking or chewing insects by stimulating the accumulation of SA, JA, and H2O2 [74]. Thus, *P. sylvestris* var. *mongolica* mainly upregulated the genes encoding ERF in response to *S. noctilio* venom stress. In addition, three upregulated TFs and one downregulated TF, belonging to the GRAS, C3H, C2H2, and GeBP families, respectively, were induced specifically under venom stress. Previous studies have reported that the TF *A. thaliana* GeBP-LIKE 4 was rapidly induced in root tips in response to toxic metals [75]. GRAS-family proteins are present only in plants and not in any other organisms, suggesting that these proteins represent an important and diverse set of regulatory molecules [76]. Moreover, the C3H-family proteins have been reported to participate in lignin synthesis in plants, and the C3H gene downregulation could reduce the lignin content in rice straw. Taken together, most of the TF family members exhibited an inducible expression profile under venom stress in *P. sylvestris* var. *mongolica*, which indicates that these TFs play a vital role in modulating needle signal transduction and molecular regulation.

5. Conclusions

In this study, 4309 (8.14%) new genes of *P. sylvestris* var. *mongolica* were obtained by combining the PacBio ISO-seq and Illumina RNA-seq technology. Under *S. noctilio* and wounding stress, 825 and 506 genes were differentially expressed, respectively, and 119 DEGs were present in both types of stresses. This information provides additional resources to identify and unearth important genes in *P. sylvestris* var. *mongolica*. Inoculation-specific and wounding-specific DEGs were significantly enriched in "plant–pathogen interaction" and "photosynthesis-antenna proteins". In addition, *S. noctilio* venom mainly caused the differential expression of pine resistance genes, such as GAPDH, GPX, CAT, FL2, etc. However, wounding only caused the differential expression of photosynthesis related genes such as Lhcb5 protein, LHC protein and OEE, etc.

Supplementary Materials: The following supporting information can be downloaded at: https://www.mdpi.com/article/10.3390/insects13040338/s1, Figure S1: Length distribution of unigenes. Figure S2: Unigene number distribution of TF families. Table S1: The quality assessment of the 27 samples. Table S2: Summary of RNA-sequencing data from 27 RNA libraries and comparison with full-length unigenes. Table S3: Summary of 706 DEGs specifically induced by *S. noctilio* venom. Table S4: 21 DEGs significantly enriched in "plant–pathogen interaction" KEGG pathway under *S. noctilio* venom-specific induction. Table S5: 14 DEGs significantly enriched in "photosynthesis-antenna proteins" KEGG pathway under wounding-specific induction. Table S6: 4 DEGs significantly enriched in "nitrogen metabolism" KEGG pathway under co-induction of venom and wounding.

Author Contributions: Conceptualization, C.G. and J.S.; methodology, L.R., C.G. and M.W. software, C.G., N.F. and M.W.; resources, C.G., Z.W. and H.W.; writing—original draft preparation, C.G.; writing—review and editing, J.S.; supervision, J.S.; project administration, J.S.; funding acquisition, J.S. All authors have read and agreed to the published version of the manuscript.

Funding: This research was supported by the National Key R&D Program of China (No. 2021YFC2600400), National Natural Science Foundation of China (No. 32171794) and Forestry Science and Technology Innovation Special of Jiangxi Forestry Department (No. 201912).

Institutional Review Board Statement: Not applicable.

Informed Consent Statement: Not applicable.

Data Availability Statement: The raw sequences have been deposited at SRA-NCBI (Accession Number: PRJNA 795839).

Acknowledgments: We greatly appreciated the help from the Eergetu forestry farm in Inner Mongolia Province in the field work. Last but not least, we would like to thank reviewers for their comments and for improving the quality of the article.

Conflicts of Interest: The authors declare no potential conflicts of interest.

References

1. Zhu, J.-J.; Li, F.-Q.; Xu, M.-L.; Kang, H.-Z.; Wu, X.-Y. The role of ectomycorrhizal fungi in alleviating pine decline in semiarid sandy soil of northern China: An experimental approach. *Ann. For. Sci.* **2008**, *65*, 304. [CrossRef]
2. Song, L.; Zhu, J.; Yan, Q.; Li, M.; Yu, G. Comparison of intrinsic water use efficiency between different aged *Pinus sylvestris* var. mongolica wide windbreaks in semiarid sandy land of northern China. *Agrofor. Syst.* **2015**, *89*, 477–489. [CrossRef]
3. Jiao-Jun, Z.; Zhi-Ping, F.; De-Hui, Z.; Feng-Qi, J.; Takeshi, M. Comparison of stand structure and growth between artificial and natural forests of *Pinus sylvestiris* var. mongolica on sandy land. *J. For. Res.* **2003**, *14*, 103–111. [CrossRef]
4. Li, D.; Shi, J.; Lu, M.; Ren, L.; Zhen, C.; Luo, Y. Detection and Identification of the Invasive *Sirex noctilio* (Hymenoptera: Siricidae) Fungal Symbiont, *Amylostereum areolatum* (Russulales: Amylostereacea), in China and the Stimulating Effect of Insect Venom on Laccase Production by *A. areolatum* YQL03. *J. Econ. Entomol.* **2015**, *108*, 1136–1147. [CrossRef]
5. Wang, L.; Ren, L.; Li, C.; Gao, C.; Liu, X.; Wang, M.; Luo, Y. Effects of endophytic fungi diversity in different coniferous species on the colonization of *Sirex noctilio* (Hymenoptera: Siricidae). *Sci. Rep.* **2019**, *9*, 5077. [CrossRef]
6. Sun, X.; Xu, Q.; Luo, Y. A Maximum Entropy Model Predicts the Potential Geographic Distribution of *Sirex noctilio*. *Forests* **2020**, *11*, 175. [CrossRef]
7. Spradbery, J.P.; Kirk, A.A. Aspects of the ecology of siricid woodwasps (Hymenoptera: Siricidae) in Europe, North Africa and Turkey with special reference to the biological control of *Sirex noctilio* F. in Australia. *Bull. Entomol. Res.* **1978**, *68*, 341–359. [CrossRef]
8. Kurt, W. *Proceedings, 17th US Department of Agriculture Interagency Research Forum on Gypsy Moth and Other Invasive Species 2006*; Gen. Tech. Rep. NRS-P-10; US Department of Agriculture, Forest Service, Northern Research Station: Newtown Square, PA, USA, 2007; Volume 10, p. 117.
9. Coutts, M.; Dolezal, J. Some effects of bark cincturing on the physiology of *Pinus radiata*, and on *Sirex* attack. *Aust. For. Res.* **1966**, *2*, 17–28.
10. Borchert, D.; Fowler, G.; Jackson, L. *Organism Pest Risk Analysis: Risks to the Conterminous United States Associated with the Woodwasp, Sirex noctilio Fabricius, and the Symbiotic Fungus, Amylostereum areolatum, (Fries: Fries) Boidin*; USDA-APHIS-PPQ-EDP: Raleigh, NC, USA, 2007; pp. 1–40.

11. Corley, J.C.; Lantschner, M.V.; Martínez, A.S.; Fischbein, D.; Villacide, J.M. Management of *Sirex noctilio* populations in exotic pine plantations: Critical issues explaining invasion success and damage levels in South America. *J. Pest Sci.* **2019**, *92*, 131–142. [CrossRef]

12. Coutts, M.P.; Dolezal, J.E. Polyphenols and Resin in the Resistance Mechanism of *Pinus Radiata* Attacked by the Wood Wasp, *Sirex noctilio*, and Its Associated Fungus, leaflet no. 101. *Forest Res. Inst.* **1966**, 1–19.

13. Coutts, M.P. Rapid physiological change in *Pinus radiata* following attack by *Sirex noctilio* and its associated fungus, *Amylostereum* sp. *Aust. J. Sci.* **1968**, *30*, 275–277.

14. Coutts, M. The Mechanism of Pathogenicity of *Sirex noctilio* on *Pinus Radiata* I. Effects of the Symbiotic Fungus *Amylostereum* sp. (Thelophoraceae). *Aust. J. Biol. Sci.* **1969**, *22*, 915–924. [CrossRef]

15. Coutts, M. The Mechanism of Pathogenicity of *Sirex noctilio* on *Pinus Radiata* II. Effects of *S. noctilio* Mucus. *Aust. J. Biol. Sci.* **1969**, *22*, 1153–1162. [CrossRef]

16. Coutts, M.P. The physiological effects of the mucus secretion of *Sirex noctilio* on *Pinus radiata*. *Aust. For. Res.* **1970**, *4*, 23–26.

17. Spradbery, J.P. A comparative study of the phytotoxic effects of siricid woodwasps on conifers. *Ann. Appl. Biol.* **1973**, *75*, 309–320. [CrossRef]

18. Madden, J.L. Physiological reactions of *Pinus radiata* to attack by woodwasp, *Sirex noctilio* F. (Hymenoptera: Siricidae). *Bull. Èntomol. Res.* **1977**, *67*, 405–426. [CrossRef]

19. Iede, E.T.; Zanetti, R. Ocorrência e recomendações para o manejo de *Sirex noctilio* Fabricius (Hymenoptera, Siricidae) em plantios de *Pinus patula* (Pinaceae) em Minas Gerais, Brasil. *Rev. Bras. Entomol.* **2007**, *51*, 529–531. [CrossRef]

20. Bordeaux, J.M.; Lorenz, W.W.; Dean, J.F. Biomarker genes highlight intraspecific and interspecific variations in the responses of *Pinus taeda* L. and *Pinus radiata* D. Don to *Sirex noctilio* F. acid gland secretions. *Tree Physiol.* **2012**, *32*, 1302–1312. [CrossRef]

21. Fox, H.; Doron-Faigenboim, A.; Kelly, G.; Bourstein, R.; Attia, Z.; Zhou, J.; Moshe, Y.; Moshelion, M.; David-Schwartz, R. Transcriptome analysis of *Pinus halepensis* under drought stress and during recovery. *Tree Physiol.* **2017**, *38*, 423–441. [CrossRef]

22. Visser, E.A.; Wegrzyn, J.L.; Myburg, A.A.; Naidoo, S. Defence transcriptome assembly and pathogenesis related gene family analysis in *Pinus tecunumanii* (low elevation). *BMC Genom.* **2018**, *19*, 632. [CrossRef]

23. Ralph, S.G.; Yueh, H.; Friedmann, M.; Aeschliman, D.; Zeznik, J.A.; Nelson, C.C.; Butterfield, Y.S.N.; Kirkpatrick, R.; Liu, J.; Jones, S.J.M.; et al. Conifer defence against insects: Microarray gene expression profiling of Sitka spruce (*Picea sitchensis*) induced by mechanical wounding or feeding by spruce budworms (*Choristoneura occidentalis*) or white pine weevils (*Pissodes strobi*) reveals large-scale changes of the host transcriptome. *Plant Cell Environ.* **2006**, *29*, 1545–1570. [CrossRef]

24. Li, B.; Ruotti, V.; Stewart, R.M.; Thomson, J.A.; Dewey, C.N. RNA-Seq gene expression estimation with read mapping uncertainty. *Bioinformatics* **2010**, *26*, 493–500. [CrossRef]

25. Li, B.; Dewey, C.N. RSEM: Accurate transcript quantification from RNA-Seq data with or without a reference genome. *BMC Bioinform.* **2011**, *12*, 323. [CrossRef]

26. Wang, L.; Feng, Z.; Wang, X.; Wang, X.; Zhang, X. DEGseq: An R package for identifying differentially expressed genes from RNA-seq data. *Bioinformatics* **2009**, *26*, 136–138. [CrossRef]

27. Yuan, H.; Yu, H.; Huang, T.; Shen, X.; Xia, J.; Pang, F.; Wang, J.; Zhao, M. The complexity of the Fragaria x ananassa (octoploid) transcriptome by single-molecule long-read sequencing. *Hortic. Res.* **2019**, *6*, 46. [CrossRef]

28. Steijger, T.; The RGASP Consortium; Abril, J.F.; Engström, P.G.; Kokocinski, F.; Hubbard, T.J.; Guigó, R.; Harrow, J.; Bertone, P. Assessment of transcript reconstruction methods for RNA-seq. *Nat. Methods* **2013**, *10*, 1177–1184. [CrossRef]

29. Abdel-Ghany, S.E.; Hamilton, M.; Jacobi, J.L.; Ngam, P.; Devitt, N.; Schilkey, F.; Ben-Hur, A.; Reddy, A.S.N. A survey of the sorghum transcriptome using single-molecule long reads. *Nat. Commun.* **2016**, *7*, 11706. [CrossRef]

30. Wang, B.; Tseng, E.; Regulski, M.; Clark, T.A.; Hon, T.; Jiao, Y.; Lu, Z.; Olson, A.; Stein, J.C.; Ware, D. Unveiling the complexity of the maize transcriptome by single-molecule long-read sequencing. *Nat. Commun.* **2016**, *7*, 11708. [CrossRef]

31. Jin, W.-T.; Gernandt, D.S.; Wehenkel, C.; Xia, X.-M.; Wei, X.-X.; Wang, X.-Q. Phylogenomic and ecological analyses reveal the spatiotemporal evolution of global pines. *Proc. Natl. Acad. Sci. USA* **2021**, *118*, e2022302118. [CrossRef]

32. Wachowiak, W.; Trivedi, U.; Perry, A.; Cavers, S. Comparative transcriptomics of a complex of four European pine species. *BMC Genom.* **2015**, *16*, 234. [CrossRef]

33. Vornam, B.; Leinemann, L.; Peters, F.S.; Wolff, A.; Leha, A.; Salinas, G.; Schumacher, J.; Gailing, O. Response of Scots pine (*Pinus sylvestris*) seedlings subjected to artificial infection with the fungus *Sphaeropsis sapinea*. *Plant Mol. Biol. Rep.* **2019**, *37*, 214–223. [CrossRef]

34. Zhang, Y.; Xu, K.; Pei, D.; Yu, D.; Zhang, J.; Li, X.; Chen, G.; Yang, H.; Zhou, W.; Li, C. ShORR-1, a Novel Tomato Gene, Confers Enhanced Host Resistance to *Oidium neolycopersici*. *Front. Plant Sci.* **2019**, *10*, 1400. [CrossRef]

35. Dodds, P.N.; Rathjen, J.P. Plant immunity: Towards an integrated view of plant–pathogen interactions. *Nat. Rev. Genet.* **2010**, *11*, 539–548. [CrossRef]

36. Peng, Y.; Van Wersch, R.; Zhang, Y. Convergent and Divergent Signaling in PAMP-Triggered Immunity and Effector-Triggered Immunity. *Mol. Plant Microbe Interact.* **2018**, *31*, 403–409. [CrossRef]

37. Gassmann, W.; Bhattacharjee, S. Effector-Triggered Immunity Signaling: From Gene-for-Gene Pathways to Protein-Protein Interaction Networks. *Mol. Plant Microbe Interact.* **2012**, *25*, 862–868. [CrossRef]

38. Meena, M.K.; Prajapati, R.; Krishna, D.; Divakaran, K.; Pandey, Y.; Reichelt, M.; Mathew, M.; Boland, W.; Mithöfer, A.; Vadassery, J. The Ca^{2+} Channel CNGC19 Regulates Arabidopsis Defense Against *Spodoptera* Herbivory. *Plant Cell* **2019**, *31*, 1539–1562. [CrossRef]

39. Cheval, C.; Aldon, D.; Galaud, J.-P.; Ranty, B. Calcium/calmodulin-mediated regulation of plant immunity. *Biochim. Biophys. Acta* **2013**, *1833*, 1766–1771. [CrossRef]

40. Vandelle, E.; Vannozzi, A.; Wong, D.; Danzi, D.; Digby, A.-M.; Santo, S.D.; Astegno, A. Identification, characterization, and expression analysis of calmodulin and calmodulin-like genes in grapevine (*Vitis vinifera*) reveal likely roles in stress responses. *Plant Physiol. Biochem.* **2018**, *129*, 221–237. [CrossRef]

41. Trujillo-Moya, C.; Ganthaler, A.; Stöggl, W.; Kranner, I.; Schüler, S.; Ertl, R.; Schlosser, S.; George, J.-P.; Mayr, S. RNA-Seq and secondary metabolite analyses reveal a putative defence-transcriptome in Norway spruce (*Picea abies*) against needle bladder rust (*Chrysomyxa rhododendri*) infection. *BMC Genom.* **2020**, *21*, 336. [CrossRef]

42. Orosa, B.; Yates, G.; Verma, V.; Srivastava, A.K.; Srivastava, M.; Campanaro, A.; De Vega, D.; Fernandes, A.; Zhang, C.; Lee, J.; et al. SUMO conjugation to the pattern recognition receptor FLS2 triggers intracellular signalling in plant innate immunity. *Nat. Commun.* **2018**, *9*, 5185. [CrossRef]

43. Miya, A.; Albert, P.; Shinya, T.; Desaki, Y.; Ichimura, K.; Shirasu, K.; Narusaka, Y.; Kawakami, N.; Kaku, H.; Shibuya, N. CERK1, a LysM receptor kinase, is essential for chitin elicitor signaling in *Arabidopsis*. *Proc. Natl. Acad. Sci. USA* **2007**, *104*, 19613–19618. [CrossRef]

44. Perincherry, L.; Lalak-Kańczugowska, J.; Stępień, Ł. Fusarium-Produced Mycotoxins in Plant-Pathogen Interactions. *Toxins* **2019**, *11*, 664. [CrossRef]

45. Ashraf, M.; Harris, P.J.C. Photosynthesis under stressful environments: An overview. *Photosynthetica* **2013**, *51*, 163–190. [CrossRef]

46. Geissler, N.; Hussin, S.; Koyro, H.-W. Interactive effects of NaCl salinity and elevated atmospheric CO2 concentration on growth, photosynthesis, water relations and chemical composition of the potential cash crop halophyte *Aster tripolium* L. *Environ. Exp. Bot.* **2009**, *65*, 220–231. [CrossRef]

47. Zhang, L.-T.; Zhang, Z.-S.; Gao, H.-Y.; Xue, Z.-C.; Yang, C.; Meng, X.-L.; Meng, Q.-W. Mitochondrial alternative oxidase pathway protects plants against photoinhibition by alleviating inhibition of the repair of photodamaged PSII through preventing formation of reactive oxygen species in *Rumex* K-1 leaves. *Physiol. Plant.* **2011**, *143*, 396–407. [CrossRef]

48. Bhat, F.A.; Ganai, B.A.; Uqab, B. Carbonic anhydrase: Mechanism, structure and importance in higher plants. *Asian J. Plant Sci. Res.* **2017**, *7*, 17–23.

49. Henry, E.; Fung, N.; Liu, J.; Drakakaki, G.; Coaker, G. Beyond Glycolysis: GAPDHs Are Multi-functional Enzymes Involved in Regulation of ROS, Autophagy, and Plant Immune Responses. *PLoS Genet.* **2015**, *11*, e1005199. [CrossRef]

50. Islam, T.; Manna, M.; Kaul, T.; Pandey, S.; Reddy, C.S.; Reddy, M.K. Genome-Wide Dissection of *Arabidopsis* and Rice for the Identification and Expression Analysis of Glutathione Peroxidases Reveals Their Stress-Specific and Overlapping Response Patterns. *Plant Mol. Biol. Rep.* **2015**, *33*, 1413–1427. [CrossRef]

51. Mhamdi, A.; Queval, G.; Chaouch, S.; Vanderauwera, S.; Van Breusegem, F.; Noctor, G. Catalase function in plants: A focus on Arabidopsis mutants as stress-mimic models. *J. Exp. Bot.* **2010**, *61*, 4197–4220. [CrossRef]

52. Wang, M.; Xu, Z.; Ding, A.; Kong, Y. Genome-Wide Identification and Expression Profiling Analysis of the Xyloglucan Endotransglucosylase/Hydrolase Gene Family in Tobacco (*Nicotiana tabacum* L.). *Genes* **2018**, *9*, 273. [CrossRef]

53. Jones, J.D.G.; Dangl, J.L. The plant immune system. *Nature* **2006**, *444*, 323–329. [CrossRef]

54. Gao, C.; Ren, L.; Wang, M.; Wang, Z.; Fu, N.; Wang, H.; Wang, X.; Ao, T.; Du, W.; Zheng, Z.; et al. Proteo-Transcriptomic Characterization of *Sirex nitobei* (Hymenoptera: Siricidae) Venom. *Toxins* **2021**, *13*, 562. [CrossRef]

55. Daum, B.; Nicastro, D.; Austin, J.; McIntosh, J.R.; Kühlbrandt, W. Arrangement of Photosystem II and ATP Synthase in Chloroplast Membranes of Spinach and Pea. *Plant Cell* **2010**, *22*, 1299–1312. [CrossRef]

56. Dekker, J.P.; Boekema, E.J. Supramolecular organization of thylakoid membrane proteins in green plants. *Biochim. Biophys. Acta* **2005**, *1706*, 12–39. [CrossRef]

57. Green, B.R.; Durnford, D.G. The Chlorophyll-Carotenoid Proteins of Oxygenic Photosynthesis. *Annu. Rev. Plant Biol.* **1996**, *47*, 685–714. [CrossRef]

58. Jansson, S. A guide to the Lhc genes and their relatives in Arabidopsis. *Trends Plant Sci.* **1999**, *4*, 236–240. [CrossRef]

59. Caffarri, S.; Frigerio, S.; Olivieri, E.; Righetti, P.G.; Bassi, R. Differential accumulation of Lhcb gene products in thylakoid membranes of *Zea mays* plants grown under contrasting light and temperature conditions. *Proteomics* **2005**, *5*, 758–768. [CrossRef]

60. Ganeteg, U.; Kulheim, C.; Andersson, J.; Jansson, S. Is Each Light-Harvesting Complex Protein Important for Plant Fitness? *Plant Physiol.* **2004**, *134*, 502–509. [CrossRef]

61. Humbeck, K.; Krupinska, K. The abundance of minor chlorophyll a/b-binding proteins CP29 and LHCI of barley (Hordeum vulgare L.) during leaf senescence is controlled by light. *J. Exp. Bot.* **2003**, *54*, 375–383. [CrossRef]

62. Staneloni, R.J.; Rodriguez-Batiller, M.J.; Casal, J.J. Abscisic Acid, High-Light, and Oxidative Stress Down-Regulate a Photosynthetic Gene via a Promoter Motif Not Involved in Phytochrome-Mediated Transcriptional Regulation. *Mol. Plant* **2008**, *1*, 75–83. [CrossRef]

63. Jung, C.; Müller, A.E. Flowering time control and applications in plant breeding. *Trends Plant Sci.* **2009**, *14*, 563–573. [CrossRef]

64. Kiranmai, K.; Gunupuru, L.R.; Nareshkumar, A.; Reddy, V.A.; Lokesh, U.; Pandurangaiah, M.; Venkatesh, B.; Kirankumar, T.V.; Sudhakar, C. Expression Analysis of WRKY Transcription Factor Genes in Response to Abiotic Stresses in Horsegram (*Macrotyloma uniflorum* (Lam.) Verdc.). *Am. J. Mol. Biol.* **2016**, *6*, 125–137. [CrossRef]

65. Zhang, L.; Cheng, J.; Sun, X.; Zhao, T.; Li, M.; Wang, Q.; Li, S.; Xin, H. Overexpression of VaWRKY14 increases drought tolerance in *Arabidopsis* by modulating the expression of stress-related genes. *Plant Cell Rep.* **2018**, *37*, 1159–1172. [CrossRef]

66. Jin, J.; Tian, F.; Yang, D.-C.; Meng, Y.-Q.; Kong, L.; Luo, J.; Gao, G. PlantTFDB 4.0: Toward a central hub for transcription factors and regulatory interactions in plants. *Nucleic Acids Res.* **2016**, *45*, D1040–D1045. [CrossRef]

67. Okamuro, J.K.; Caster, B.; Villarroel, R.; Van Montagu, M.; Jofuku, K.D. The AP2 domain of *APETALA2* defines a large new family of DNA binding proteins in *Arabidopsis*. *Proc. Natl. Acad. Sci. USA* **1997**, *94*, 7076–7081. [CrossRef]

68. Riechmann, J.L.; Meyerowitz, E.M. The AP2/EREBP family of plant transcription factors. *Biol. Chem.* **1998**, *379*, 633–646. [CrossRef]

69. Magnani, E.; Sjölander, K.; Hake, S. From Endonucleases to Transcription Factors: Evolution of the AP2 DNA Binding Domain in Plants[W]. *Plant Cell* **2004**, *16*, 2265–2277. [CrossRef]

70. Chuck, G.; Meeley, R.; Hake, S. Floral meristem initiation and meristem cell fate are regulated by the maize AP2 genes ids1 and sid1. *Development* **2008**, *135*, 3013–3019. [CrossRef]

71. Wang, G.; Wang, H.; Zhu, J.; Zhang, J.; Zhang, X.; Wang, F.; Tang, Y.; Mei, B.; Xu, Z.; Song, R. An expression analysis of 57 transcription factors derived from ESTs of developing seeds in Maize (*Zea mays*). *Plant Cell Rep.* **2010**, *29*, 545–559. [CrossRef]

72. Xie, Z.; Nolan, T.M.; Jiang, H.; Yin, Y. AP2/ERF transcription factor regulatory networks in hormone and abiotic stress responses in Arabidopsis. *Front. Plant Sci.* **2019**, *10*, 228.

73. Hao, D.; Ohme-Takagi, M.; Sarai, A. Unique Mode of GCC Box Recognition by the DNA-binding Domain of Ethylene-responsive Element-binding Factor (ERF Domain) in Plant. *J. Biol. Chem.* **1998**, *273*, 26857–26861. [CrossRef]

74. Lu, J.; Ju, H.; Zhou, G.; Zhu, C.; Erb, M.; Wang, X.; Wang, P.; Lou, Y. An EAR-motif-containing ERF transcription factor affects herbivore-induced signaling, defense and resistance in rice. *Plant J.* **2011**, *68*, 583–596. [CrossRef]

75. Khare, D.; Mitsuda, N.; Lee, S.; Song, W.; Hwang, D.; Ohme-Takagi, M.; Martinoia, E.; Lee, Y.; Hwang, J. Root avoidance of toxic metals requires the GeBP-LIKE 4 transcription factor in *Arabidopsis thaliana*. *New Phytol.* **2016**, *213*, 1257–1273. [CrossRef]

76. Bolle, C. The role of GRAS proteins in plant signal transduction and development. *Planta* **2004**, *218*, 683–692. [CrossRef]

Article

Herbivore-Induced Rice Volatiles Attract and Affect the Predation Ability of the Wolf Spiders, *Pirata subpiraticus* and *Pardosa pseudoannulata*

Jing Liu, Liangyu Sun, Di Fu, Jiayun Zhu, Min Liu, Feng Xiao and Rong Xiao *

Guizhou Provincial Key Laboratory for Agricultural Pest Management of the Mountainous Region, Institute of Entomology, Guizhou University, Guiyang 550025, China; lj1071836108@163.com (J.L.); liangysun@126.com (L.S.); fudi9696@163.com (D.F.); zjy873421380@163.com (J.Z.); liumin8796@163.com (M.L.); xiaofeng1217@163.com (F.X.)
* Correspondence: rong1234xiao@163.com

Simple Summary: The spiders, *Pirata subpiraticus* Bösenberg *et* Strand (Araneae: Lycosidae) and *Pardosa pseudoannulata* Bösenberg *et* Strand (Araneae: Lycosidae) are important natural enemies of many rice pests. Herbivore-induced plant volatiles can attract natural enemies to pest locations and are becoming important in integrated pest management. This study assessed the effects of herbivore-induced rice volatiles on the selection behavior, predation ability and field attraction of two species of spiders. The selection frequency of spiders for methyl salicylate, linalool, and 2-heptanone were significantly greater than the blank group. Methyl salicylate can shorten the predatory latency of male *P. pseudoannulata* and can also trap more *P. pseudoannulata* in the field. Linalool may also shorten the predatory latency of male *P. subpiraticus* and increase the daily predation capacity of female *P. pseudoannulata*. In summary, herbivore-induced rice volatiles attract *P. pseudoannulata* and *P. subpiraticus*, and potentially increase their pest control capability. These results provide support for the practical use of herbivore-induced rice volatiles to attract and retain spiders in rice fields.

Citation: Liu, J.; Sun, L.; Fu, D.; Zhu, J.; Liu, M.; Xiao, F.; Xiao, R. Herbivore-Induced Rice Volatiles Attract and Affect the Predation Ability of the Wolf Spiders, *Pirata subpiraticus* and *Pardosa pseudoannulata*. *Insects* **2022**, *13*, 90. https://doi.org/10.3390/insects13010090

Academic Editors: Gianandrea Salerno, Manuela Rebora and Stanislav N. Gorb

Received: 18 December 2021
Accepted: 9 January 2022
Published: 13 January 2022

Publisher's Note: MDPI stays neutral with regard to jurisdictional claims in published maps and institutional affiliations.

Abstract: Spiders are important natural enemies of rice pests. Studying the effects of herbivore-induced rice volatiles on spider attraction and predation ability may lead to safer methods for pest prevention and control. In this study, four-arm olfactometer, predation ability experiment, and field trapping experiment were used to evaluate the effects of herbivore-induced rice volatiles on *Pirata subpiraticus* Bösenberg *et* Strand (Araneae: Lycosidae) and *Pardosa pseudoannulata* Bösenberg *et* Strand (Araneae: Lycosidae). The 0.5 µg/µL linalool concentration was attractive, and also shortened the predation latency in male *P. subpiraticus* and female *P. pseudoannulata*. The 0.5 µg/µL linalool concentration increased the daily predation capacity of female *P. pseudoannulata*. Male *P. pseudoannulata* were attracted to 1.0 g/L methyl salicylate, which also shortened their predation latency. In field experiments, methyl salicylate and linalool were effective for trapping spiders. Herbivore-induced rice volatiles attract rice field spiders and affect their predatory ability. These results suggest that herbivore-induced rice volatiles can be used to attract spiders and provide improved control of rice pests.

Keywords: rice field spiders; herbivore-induced rice volatiles; selection behavior; predation latency; daily predation capacity

1. Introduction

Rice is an important food crop and rice yield is always a high priority [1]. However, pests such as *Nilaparvata lugens* Stal (Homoptera: Delphacidae), *Sogatella furcifera* Horváth (Homoptera: Delphacidae), *Chilo suppressalis* Walker (Lepidoptera: Pyralidae), and *Cnaphalocrocis medinalis* Guenee (Lepidoptera: Pyralidae) can reduce rice production [2]. Pesticide application is the main approach for managing rice pests. Although chemical pesticides are effective, their widespread use has resulted in an increase in the occurrence of 3R (residue, resistance, and resurgence). This has made the biological management of rice pests more important [3–5]. Herbivore-induced plant volatiles (HIPVs) regulate the interaction of plants, herbivorous insects, and their natural enemies [6,7]. They can be used as attractants for the natural enemies of pests [6,8]. Herbivore-induced rice volatiles (HIRVs) such as (Z)-3-hexen-1-ol, methyl salicylate, 2-heptanone, linalool, and others are produced or increased when rice is damaged by pests [9–13]. HIRVs increase significantly, when rice is eaten by adult females of the brown planthopper *N. lugens* and the white-backed planthopper *S. furcifera* [14]. Some HIRVs are attractive to natural enemies of rice planthoppers, such as *Anagrus nilaparvatae* Pang *et* Wang (Hymenoptera: Mymaridae) [9,15,16], *Haplogonatopus japonicus* Esaki *et* Hashimoto (Hymenoptera: Dryinidae) [17] and *Cyrtorhinus lividipennis* Reuter (Hemiptera: Miridea) [18,19]. Some HIRVs are attractive to *Apanteles chilonis* Munakata (Hymenoptera: Braconidae), which is the natural enemy of *C. suppressalis* [20]. The HIRVs released by *Tibraca limbativentris* Stål (Heteroptera: Pentatomidae) and *Glyphepomis spinosa* Campos *et* Grazia (Heteroptera: Pentatomidae) feeding on rice are attractive to the natural enemy of rice pest, *Telenomus podisi* Ashmead (Hymenoptera: Platygastridae) [21].

Spiders are often referred to as the "Paddy Field Guardian", and are natural enemies of rice pests in the rice fields of China [22,23]. Spiders are carnivorous, with a large food intake, strong predatory ability, high reproductive rate, and high adaptability [24]. Therefore, spiders play a significant role in the biological management of rice pests. In China, the dominant species of rice field spiders include *Tetragnatha maxillosa* Thoren (Araneae: Tetragnathidae), *Pardosa pseudoannulata* Bösenberg *et* Strand (Araneae: Lycosidae), *Pirata subpiraticus* Bösenberg *et* Strand (Araneae: Lycosidae), *Pirala piraticus* Clcrck (Araneae: Lycosidae), *Clubiona japonicola* Bösenberg *et* Strand (Araneae: Clubionidae), and *Oxyopes sertatus* L. Koch (Araneae: Oxyopidae) [25–27]. Netting spiders and wandering spiders are the two main types of dominant spiders. Netting spiders feed on a variety of flying insects while wandering spiders feed on insects in many locations [26]. Wandering wolf spiders are the majority spiders in rice fields. Wolf spiders rely on their sense of smell to discover and locate their prey [28,29]. Cao et al. identified the wolf spider *P. pseudoannulata* having two potential odorant-binding protein genes [30]. This information provides the basis for further research on the olfactory selection behavior of wolf spiders. Spider olfaction is critical in their predatory behavior [31], but it is not known if HIRVs are attractive to spiders. Therefore, we selected two paddy wolf spiders, *P. subpiraticus* and *P. pseudoannulata*, to study the selection behavior of spiders on HIRVs. *P. subpiraticus* often forages on rice, water, and land to prey on rice pests [32]. It has strong predation ability and starvation tolerance. It can consume 6–16 *S. furcifera* every 24 h [33], and survive up to 42.7 d without food under certain humidity conditions [34]. *P. pseudoannulata* is an important predator of rice pests. It is a hunting spider with a wide niche, good running and jumping ability, and strong predatory ability for rice pests [35].

In the study, two rice field spiders and four volatiles were selected to ascertain if HIRVs are attractive to spiders and to determine the optimum concentrations of the attractive volatiles. We used the optimum concentration to verify how HIRVs affect the predatory ability of spiders. HIRVs were also used as attractants to trap spiders in the field.

2. Materials and Methods

2.1. Spiders

P. subpiraticus and *P. pseudoannulata* were collected from a rice field in Yanlou Town, Huaxi District, Guiyang, China (106°6′24″ E, 26°3′19″ N). Spiders of various growth stages were collected and placed in individual plastic test tubes (12 cm × 3.5 cm diam). To maintain humidity, a water-soaked sponge was placed at the bottom of each test tube, and the top of the tube was sealed with a cotton ball. Spiders were raised in a clear artificial climate box at 25 °C ± 1 °C, 75% ± 5% relative humidity, and a 14:10 h (L:D) photoperiod. The spiders were fed *Musca domestica* adults twice a week (2–3 *M. domestica* adults each time) until they were adults.

Table 1. Behavioral responses (observed frequencies (N = 30) compared to expected frequencies assuming random distribution by using a χ^2 test) of spider to volatiles.

Spider	Sex	Treatment	χ^2	p	Significance
P. pseudoannulata	male	CK	2.533	0.469	NS
		MeSA	8.667	0.034	*
		HE	9.200	0.027	*
		CH	6.533	0.088	NS
		LI	10.267	0.016	*
	female	CK	2.267	0.519	NS
		MeSA	1.200	0.753	NS
		HE	0.667	0.881	NS
		CH	3.867	0.276	NS
		LI	13.200	0.004	**
P. subpiraticus	male	CK	5.467	0.141	NS
		MeSA	3.867	0.276	NS
		HE	16.667	0.001	**
		CH	5.467	0.141	NS
		LI	13.467	0.004	**
	female	CK	7.600	0.055	NS
		MeSA	3.333	0.343	NS
		HE	4.400	0.221	NS
		CH	2.267	0.519	NS
		LI	2.533	0.469	NS

Note: CK indicates liquid paraffin, MeSA = methyl salicylate, HE = 2-heptanone, CH = cis-3-hexen-1-ol, LI = linalool; "NS" = no significant difference; "*" denotes a significant difference at the $p < 0.05$ level; "**" denotes a significant difference at the $p < 0.01$.

Spider selection between three concentrations of the same volatile and a negative control was studied using a four-arm olfactometer. The results showed that selection frequency of male *P. subpiraticus* for 2-heptanone (χ^2 = 16.667, p = 0.001) and linalool (χ^2 = 13.467, p = 0.004) were significantly different (Table 1). The male *P. subpiraticus* showed a stronger preference for 0.5 µg/µL linalool and 1.0 µg/µL 2-heptanone than for other concentrations and controls (Figure 2). However, there was no significant difference in selection frequency for cis-3-hexen-1-ol (χ^2 = 5.467, p = 0.141) and methyl salicylate (χ^2 = 3.867, p = 0.276) (Table 1). The selection frequency of female *P. subpiraticus* for linalool (χ^2 = 2.533, p = 0.469), 2-heptanone (χ^2 = 4.400, p = 0.221), methyl salicylate (χ^2 = 3.333, p = 0.343), and cis-3-hexen-1-ol (χ^2 = 2.267, p = 0.519) were not significantly different (Table 1). The selection frequency for methyl salicylate (χ^2 = 8.667, p = 0.034), 2-heptanone (χ^2 = 9.200, p = 0.027), and linalool (χ^2 = 10.267, p = 0.016) in male *P. pseudoannulata* were significantly different (Table 1). Male *P. pseudoannulata* selected 1.0 µg/µL linalool, 1.0 µg/µL methyl salicylate, and 0.5 µg/µL 2-heptanone more frequently than other concentrations and controls (Figure 2). The selection frequency for cis-3-hexen-1-ol (χ^2 = 6.533, p = 0.088) in male *P. pseudoannulata* was not significant (Table 1). Female *P. pseudoannulata* selection frequency for linalool (χ^2 = 13.200, p = 0.004) was significantly different (Table 1). The selection frequency of 0.5 µg/µL linalool by female *P. pseudoannulata* was higher than the other concentrations and controls (Figure 2). Female *P. pseudoannulata* selection frequency for methyl salicylate (χ^2 = 1.200, p = 0.753), 2-heptanone (χ^2 = 0.667, p = 0.881), and cis-3-hexen-1-ol (χ^2 = 3.867, p = 0.276) were not significantly different (Table 1).

We studied the duration that spiders remained in different treatments to see if HIRVs affected their selection behavior. Male *P. pseudoannulata* remained significantly longer in the presence of 1.0 µg/µL methyl salicylate and 1.0 µg/µL linalool than in other concentrations of the same volatiles or the control treatment. Female *P. pseudoannulata* remained significantly longer in the presence of 0.5 µg/µL linalool than in the presence of other concentrations of the same volatile or the control treatment. Male *P. subpiraticus* remained longer with 1.0 μg/μL 2-heptanone and 0.5 μg/μL linalool than with other volatile concentrations or the control treatment (Table 2).

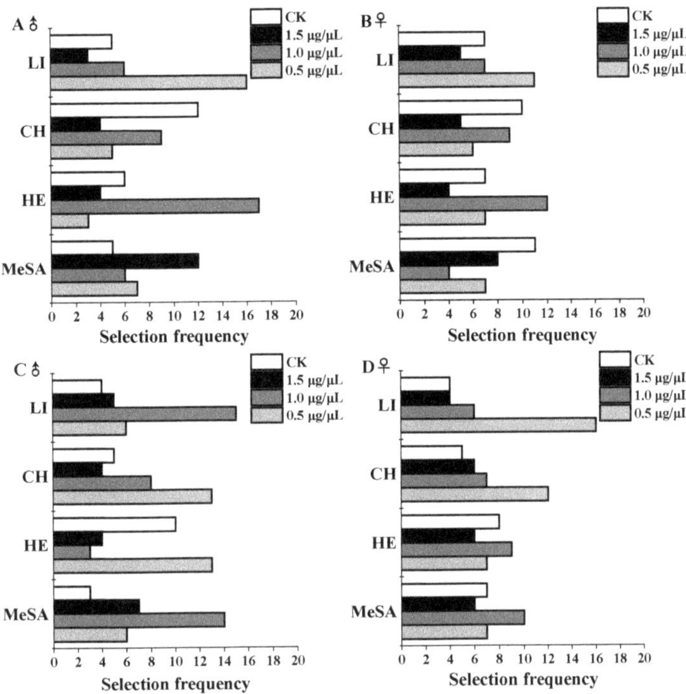

Figure 2. Selection frequency (N = 30), (**A**) male *P. subpiraticus*, (**B**) female *P. subpiraticus*, (**C**) male *P. pseudoannulata*, (**D**) female *P. pseudoannulata*; MeSA indicates methyl salicylate, HE indicates 2-heptanone, CH indicates cis-3-hexen-1-ol, LI indicates linalool.

Table 2. Stay times of spiders in the four test areas.

Spider Name	Sex	Treatment	0.5 μg/μL	1.0 μg/μL	1.5 μg/μL	ck
P. pseudoannulata	male	CH	2.151 ± 0.457 a	1.323 ± 0.407 ab	0.661 ± 0.313 b	0.826 ± 0.343 ab
		MeSA	0.990 ± 0.368 b	2.313 ± 0.459 a	1.157 ± 0.381 b	0.494 ± 0.275 b
		HE	2.110 ± 0.450 a	0.534 ± 0.273 b	0.668 ± 0.312 b	1.652 ± 0.434 a b
		LI	1.171 ± 0.387 b	2.295 ± 0.456 a	0.822 ± 0.335 b	0.661 ± 0.313 b
	female	CH	1.986 ± 0.452 a	1.161 ± 0.391 a	0.993 ± 0.369 a	0.827 ± 0.343 a
		MeSA	1.157 ± 0.390 a	1.651 ± 0.434 a	0.992 ± 0.368 a	1.156 ± 0.389 a
		HE	1.161 ± 0.391 a	1.491 ± 0.423 a	0.990 ± 0.368 a	1.326 ± 0.409 a
		LI	2.640 ± 0.456 a	0.989 ± 0.367 b	0.656 ± 0.311 b	0.655 ± 0.310 b
P. subpiraticus	male	CH	0.827 ± 0.343 a	1.489 ± 0.424 a	0.661 ± 0.313 a	1.987 ± 0.452 a
		MeSA	1.156 ± 0.389 a	1.154 ± 0.390 a	1.821 ± 0.445 a	0.828 ± 0.344 a
		HE	0.825 ± 0.342 b	2.474 ± 0.460 a	0.661 ± 0.313 b	0.991 ± 0.368 b
		LI	2.647 ± 0.460 a	0.991 ± 0.368 b	0.496 ± 0.276 b	0.826 ± 0.343 b
	female	CH	0.993 ± 0.368 a	1.486 ± 0.421 a	0.836 ± 0.343 a	1.656 ± 0.434 a
		MeSA	1.159 ± 0.390 a	0.662 ± 0.313 a	1.324 ± 0.407 a	1.819 ± 0.443 a
		HE	1.167 ± 0.392 a	1.986 ± 0.451 a	0.662 ± 0.313 a	1.157 ± 0.389 a
		LI	1.820 ± 0.444 a	1.158 ± 0.389 a	0.828 ± 0.344 a	1.157 ± 0.386 a

Note: Data are mean ± SE; CK = liquid paraffin, MeSA = methyl salicylate, HE = 2-heptanone, CH = cis-3-hexen-1-ol, LI = linalool; different lowercase letters indicate significant differences in stay time in different areas (ANOVA, Tukey $p < 0.05$).

In summary, comprehensive selection frequency and stay time of *P. pseudoannulata* and *P. subpiraticus*, 1.0 µg/µL methyl salicylate and 1.0 µg/µL linalool were attractive to males of *P. pseudoannulata*; 0.5 µg/µL linalool was attractive to females of *P. pseudoannulata*; 1.0 µg/µL 2-heptanone and 0.5 µg/µL linalool were attractive to males of *P. subpiraticus*.

3.2. Spider's Predatory Ability

3.2.1. Daily Predation Capacity and Predatory Latency of *P. subpiraticus*

There was no difference in the predatory latency of female *P. subpiraticus* between treatments. The predatory latency of male *P. subpiraticus* was not significantly different between control and 2-heptanone (0.5 µg/µL), but linalool (0.5 µg/µL) treatment was significantly shorter than that of control and 2-heptanone (0.5 µg/µL). The daily predation capacity of female and male *P. subpiraticus* was not significantly different between the treatments (Figure 3).

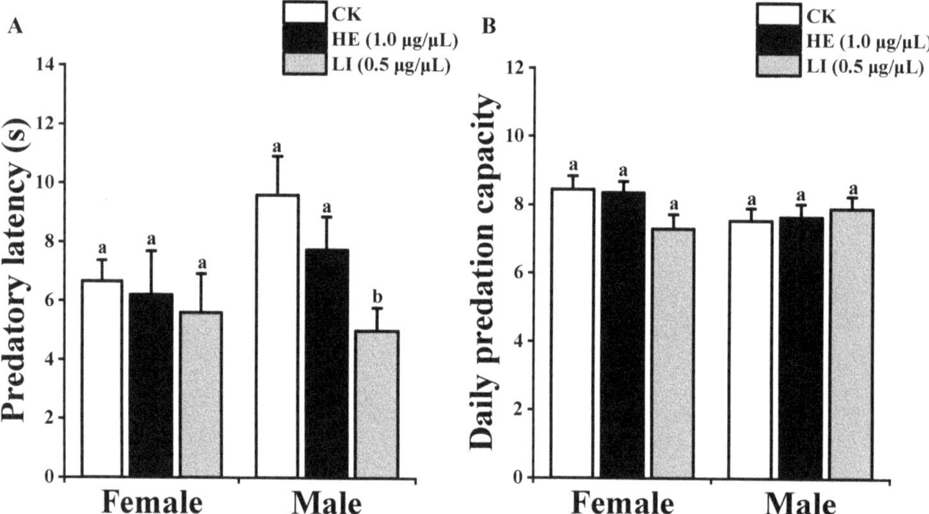

Figure 3. Predatory efficiency of *P. subpiraticus* on *Drosophila melanogaster*: (**A**) indicates the predatory latency of *P. subpiraticus*. (**B**) indicates the daily predation capacity of *P. subpiraticus*. CK indicates liquid paraffin, HE (1.0 µg/µL) indicates 2-heptanone (1.0 µg/µL), LI (0.5 µg/µL) indicates linalool (0.5 µg/µL). The data are expressed as mean ± SE. Vertical bars indicate SE. Columns of the same sex with different lowercase letters are significantly different (ANOVA, Tukey $p < 0.05$).

3.2.2. Daily Predation Capacity and Predatory Latency of *P. pseudoannulata*

The predatory latency of female *P. pseudoannulata* was not significantly different between the treatments. The predatory latency of male *P. pseudoannulata* was not significantly different between the control, linalool (1.0 µg/µL), and linalool (0.5 µg/µL). However, the 1.0 µg/µL methyl salicylate treatment predatory latency was significantly shorter than that of the control, linalool (1.0 µg/µL), and linalool (0.5 µg/µL). The daily predation of male *P. pseudoannulata* was not significantly different between the treatments. The daily predation capacity of female *P. pseudoannulata* was not significantly different between the control, linalool (1.0 µg/µL), and methyl salicylate (1.0 µg/µL), but the linalool (0.5 µg/µL) treatment was significantly greater than the control treatment (Figure 4).

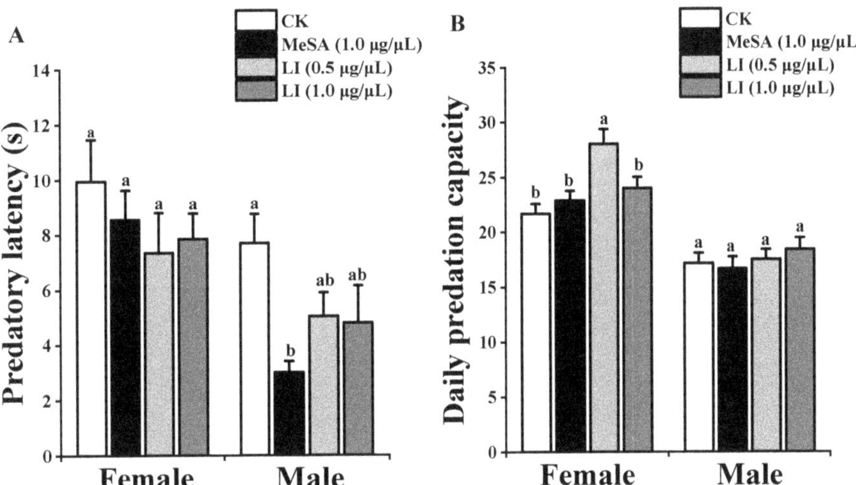

Figure 4. Predatory efficiency of *P. pseudoannulata* on *Drosophila melanogaster*: (**A**) = the predatory latency of *P. pseudoannulata*. (**B**) = the daily predation capacity of *P. pseudoannulata*. CK = liquid paraffin, MeSA (1.0 µg/µL) = methyl salicylate (1.0 µg/µL), LI (1.0 µg/µL) = linalool (1.0 µg/µL), LI (0.5 µg/µL) = linalool (0.5 µg/µL). The data are expressed as mean ± SE. Vertical bars = SE. Columns of the same sex starting with different lowercase letters are significantly different (ANOVA, Tukey $p < 0.05$).

3.3. Field Trapping

The HIRVs were attractive to spiders, and the average catch of traps with HIRVs was higher than that of control traps. In particular, the traps containing methyl salicylate had higher catches than other traps. In the HIRVs traps, the proportions of *P. pseudoannulata* in the total spider number were methyl salicylate (72%), linalool (62.5%), and 2-heptanone (16.67%). The proportions of *P. subpiraticus* in the total spider number were 2-heptanone (16.67%), methyl salicylate (12.5%), and linalool (4%) (Figure 5).

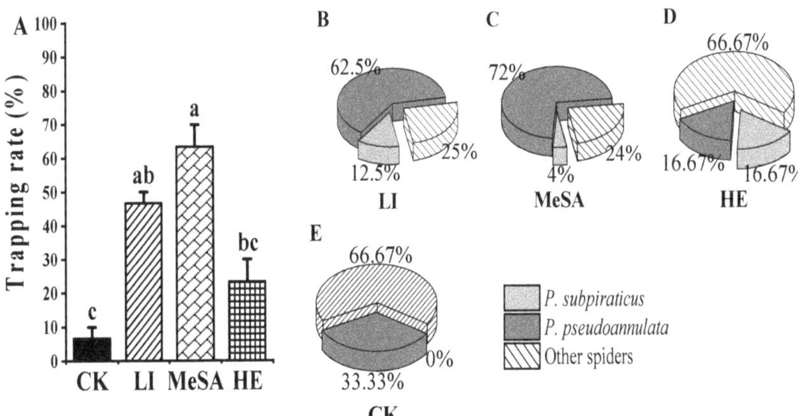

Figure 5. Field trapping rate and spider number: (**A**) = the trapping rate (traps with spiders/total traps), Columns starting with different lowercase letters are significantly different (ANOVA, Tukey $p < 0.05$), N = 3. (**B–E**) = the percentage of *P. subpiraticus*, *P. pseudoannulata*, and other spiders in the total number of spiders. CK = liquid paraffin, MeSA = methyl salicylate, HE = 2-heptanone, CH = cis-3-hexen-1-ol, LI = linalool.

4. Discussion

Spiders are among the most abundant predators in rice fields. *P. subpiraticus* and *P. pseudoannulata* are the predominant species of paddy field spiders. They have large populations and strong predatory ability [22,25]. HIPVs are specific volatile substances that plants release when they are damaged by pest feeding [38,39]. HIPVs are an important component of biological control and attract natural enemies to pest feeding locations [6,8]. The olfactometer is a useful tool that can simulate the emission of volatiles in a field environment and is useful for real-time observation of spider behavior [40]. Olfactometers have been widely used to determine the relationship between insect selection behavior and volatiles such as *Chilo suppressalis* Walker (Lepidoptera: Pyralidae) [41], *Cotesia urabae* Austin *et* Allen (Hymenoptera: Braconidae) [40], *Harmonia axyridis* Pallas (Coleoptera: Coccinellidae) [42], *Coccinella septempunctata* L. (Coleoptera: Coccinellidae) [43], *Diaeretiella rapae* McIntosh (Hymenoptera: Braconidae) [43], *Adoxophyes honmai* Yasuda (Lepidoptera: Tortricidae) [44], and *Diaphania indica* Saunders (Lepidoptera: Crambidae) [45]. We used a four-arm olfactometer to observe the selection behaviors of *P. subpiraticus* and *P. pseudoannulata* on four kinds of HIRVs. We obtained results for *P. subpiraticus* and *P. pseudoannulata* selection frequency and stay time using the four-arm olfactometer. These experiments have practical implications, because the selection behaviors of *P. subpiraticus* and *P. pseudoannulata* were strongly correlated with the types and concentrations of the attractants.

We found that 1.0 µg/µL methyl salicylate was significantly attractive to male *P. pseudoannulata*. Methyl salicylate is also attractive to the natural enemies of other pest insects. For example, Zhu and Park showed that methyl salicylate attracts *Coccinella septempunctata* L. (Coleoptera: Coccinellidae), a predator of *Aphis glycines* Matsumura (Homoptera: Aphididae) [46]. Boer et al. showed that methyl salicylate is attractive to the predatory mite *Phytoseiulus persimilis* Athias-Henriot (Acari: Phytoseiidae) [47]. James et al. demonstrated that methyl salicylate attracted *Chrysopa oculata* Say (Neuroptera: Chrysopidae) [48], *Chrysopa nigricornis* Burmeister (Neuroptera: Chrysopidae) [49], and Syrphidae spp. (Diptera: Syrphidae) [50]. Shimoda showed that methyl salicylate attracts the predatory mite *Neoseiulus californicus* McGregor (Acari: Phytoseidae), which is an important natural enemy of *Tetranychus urticae* Koch. (Acari: Tetranychidae) [51]. Our results showed that 1.0 µg/µL 2-heptanone was significantly attractive to male *P. subpiraticus*. Li et al. also found that 2-heptanone was strongly attractive to *Anagrus nilaparvatae* Pang *et* Wang (Hymenoptera: Mymaridae), the principal parasitoid of rice planthopper eggs [16]. Our results showed that linalool triggered spider olfactory responses. A 1.0 µg/µL linalool concentration was attractive to male *P. pseudoannulata*, while 0.5 µg/µL linalool was attractive to female *P. pseudoannulata* and male of *P. subpiraticus*. Linalool appears to be a very important attractant. Carroll et al. reported that *Spodoptera frugiperda* J. E. Smith (Lepidoptera: Noctuidae) was attracted to linalool [52]. Anderson et al. found that female *Bombyx mori* L. (Lepidoptera: Bombycidae) were attracted to linalool [53]. Reisenman et al. found that linalool could attract and stimulate oviposition in female *Manduca sexta* L. (Lepidoptera: Sphingidae) [54]. Cruz-Lopez et al. studied coffee volatiles and found that *Hypothenemus hampe* Ferrari (Coleoptera: Scolytidae) was attracted to linalool [55]. Mitra et al. found that female *Altica cyanea* Weber (Coleoptera: Chrysomelidae) were attracted to linalool [56].

Our research aimed to determine the influence of HIRVs on spider predation responses. Therefore, the attack latency and daily predation capacity were selected to evaluate the predation ability of *P. subpiraticus* and *P. pseudoannulata*. *Drosophila melanogaster* Meigen (Diptera: Drosophilidae) is a commonly used model organism in the laboratory. *D. melanogaster* is often used as a predator laboratory for alternative prey since it is easy to cultivate, rich in nutrients, and capable of flight [57–60]. Spiders generally only consume living prey and the lively *D. melanogaster* stimulate predation responses in spiders. In this study, *D. melanogaster* was used as an alternative prey for *P. subpiraticus* and *P. pseudoannulata*. The results showed that 1.0 µg/µL methyl salicylate significantly shortened the attack latency of male *P. pseudoannulata* and 0.5 µg/µL linalool significantly shortened the attack latency of *P. subpiraticus* and enhanced the daily predation capacity of female *P.*

pseudoannulata. Lycosid spiders mainly live at the base of rice plants where high densities of brown planthoppers often occur. The brown planthopper is the target prey of many Lycosidae spiders [26]. After the brown planthoppers feed on rice, the rice releases methyl salicylate, linalool, and other volatiles [12,61]. These volatiles stimulate the olfactory selective response in spiders, shortening predation latency and increasing predation. This may be because spiders are attracted to the odor, which may affect their feeding response. Other studies on spider predation have shown that selected chemicals can shorten spider predation latency or enhance their predatory function. Suitable physiologically active plant substances can significantly enhance the predatory function of *P. pseudoannulata* on *N. lugens* [62,63]. Optimal low-dose pesticides, which shorten the subduing and feeding times of spiders on prey, will enhance insect control efficiency and the instant attack rate on the prey of *P. pseudoannulata*, *P. subpiraticus*, *P. astrigera* L. Koch (Araneae: Lycosidae), and *Coleosoma octomaculatum* Bösenberg *et* Strand (Araneae: Theridiidae) [64–67].

Spider density will affect the management of pest populations, and increasing the number of spider predators will aid in insect pest control. There are many examples of using pheromones to attract and retain natural enemies. Simpson et al. used methyl salicylate as an attractant for crops and found that parasitic wasps remain longer in these attractant crops [68]. Jaworski et al. also proved that placing methyl salicylate attractants in orchards increased the effect of predators in controlling pests compared to untreated orchards [69]. Our field experiments showed that the spider trapping rate, using methyl salicylate and linalool as attractants, was greater than that of the control treatment. Zhu and Park also found that traps baited with methyl salicylate were highly attractive to adult *C. septempunctata* in field tests [46]. These results show that natural enemies might use methyl salicylate as an olfactory signal to locate prey. The trapped *P. pseudoannulata* dominated, which showed that HIPVs not only attracts *P. pseudoannulata* in the laboratory but also attracts *P. pseudoannulata* in the field. The *P. subpiraticus* was trapped in smaller numbers, which may be because *P. subpiraticus* mainly wanders on the water surface and the base of rice plants bordering the water surface, while the traps used for this experiment were placed above the water surface. The other spiders trapped were mainly Salticidae and Clubionidae spiders, which are good at jumping, and they may have accidentally fallen into the trap. According to a preliminary field study, *P. subpiraticus* and *P. pseudoannulata* were the dominant species at the experimental site [70], but the number of trapped spiders was lower than expected. There are many uncontrollable factors that affect spider numbers, such as temperature, humidity, wind, the size and location of the trap, and the concentration of attractants.

In conclusion, our results indicate that HIRVs can attract spiders and increase predation. Two dominant species of rice spiders (*P. pseudoannulata* and *P. subpiraticus*) were attracted by one or two concentrations of methyl salicylate, linalool, or 2-heptanone. A predatory ability experiment showed that linalool and methyl salicylate shortened predation latency. Linalool increased the daily predation capacity of the spiders. A field experiment verified that methyl salicylate and linalool are attractive to spiders. In future research, we will determine the reasons why HIRVs enhance the predation ability of spiders and develop practical methods for using HIRVs to attract and retain spiders in rice fields.

Author Contributions: Conceptualization, J.L., R.X. and L.S.; methodology and software, J.L., J.Z. and D.F.; data collection and curation, J.L., F.X. and M.L.; writing—original draft preparation, J.L.; writing—review and editing, J.L. and R.X.; supervision, R.X.; project administration, R.X.; funding acquisition, R.X. All authors have read and agreed to the published version of the manuscript.

Funding: This research work was supported by the Project of Guizhou Science and Technology Department (Qian Ke He Support (2018)2354).

Institutional Review Board Statement: Not applicable.

Data Availability Statement: Data are contained within the article.

Acknowledgments: This work was funded by the project supported by Guizhou Science and Technology Department. We thank all partners associated with the project. We are grateful to Qingshui Wang and Qiu Li for their help with spider rearing.

Conflicts of Interest: All authors declare no conflict of interest.

References

1. Liu, L.L.; Wang, E.L.; Zhu, Y.; Tang, L. Contrasting effects of warming and autonomous breeding on single-rice productivity in China. *Agric. Ecosyst. Environ.* **2012**, *149*, 20–29. [CrossRef]
2. Lou, Y.-G.; Zhang, G.-R.; Zhang, W.-Q.; Hu, Y.; Zhang, J. Biological control of rice insect pests in China. *Biol. Control.* **2013**, *67*, 8–20. [CrossRef]
3. Qin, S.-l.; Lu, X.-y. Do large-scale farmers use more pesticides? Empirical evidence from rice farmers in five Chinese provinces. *J. Integr. Agric.* **2020**, *19*, 590–599. [CrossRef]
4. Tu, L.H.; Boulange, J.; Phong, T.K.; Thuyet, D.Q.; Watanabe, H.; Takagi, K. Predicting rice pesticide fate and transport following foliage application by an updated PCPF-1 model. *J. Environ. Manag.* **2021**, *277*, 111356. [CrossRef] [PubMed]
5. Tuanol, A.P.P.; Xu, Z.; Castillo, M.B.; Mamaril, C.P.; Manaois, R.V.; Romero, M.V.; Juliano, B.O. Content of Tocols, gamma-Oryzanol and Total Phenolics and Grain Quality of Brown Rice and Milled Rice Applied with Pesticides and Organic and Inorganic Nitrogen Fertilizer. *Philipp. Agric. Sci.* **2011**, *94*, 211–216.
6. Dicke, M. Behavioural and community ecology of plants that cry for help. *Plant Cell Environ.* **2009**, *32*, 654–665. [CrossRef]
7. War, A.R.; Sharma, H.C.; Paulraj, M.G.; War, M.Y.; Ignacimuthu, S. Herbivore induced plant volatiles: Their role in plant defense for pest management. *Plant Signal. Behav.* **2011**, *6*, 1973–1978. [CrossRef]
8. Dicke, M.; Baldwin, I.T. The evolutionary context for herbivore-induced plant volatiles: Beyond the 'cry for help'. *Trends Plant Sci.* **2010**, *15*, 167–175. [CrossRef]
9. Mao, G.; Tian, J.; Li, T.; Fouad, H.; Mo, J. Behavioral responses of *Anagrus nilaparvatae* to common terpenoids, aromatic compounds, and fatty acid derivatives from rice plants. *Entomol. Exp. Appl.* **2018**, *166*, 483–490. [CrossRef]
10. Qi, J.; ul Malook, S.; Shen, G.; Gao, L.; Zhang, C.; Li, J.; Zhang, J.; Wang, L.; Wu, J. Current understanding of maize and rice defense against insect herbivores. *Plant Divers.* **2018**, *40*, 189–195. [CrossRef]
11. Zhao, N.; Zhuang, X.; Shrivastava, G.; Chen, F. Analysis of insect-induced volatiles from rice. *Methods Mol. Biol.* **2013**, *956*, 201–208. [CrossRef] [PubMed]
12. Xu, T.; Zhou, Q.; Xia, Q.; Zhang, W.-Q.; Zhang, G.-R.; Gu, D.-X. Effect of insect pest-induced rice volatiles on the host selection behavior of brown planthopper. *Chin. Sci. Bull.* **2002**, 849–853.
13. Yan, F.; Wang, X.; Lu, J.; Pang, B.-P.; Lou, Y.-G. Comparison of rice volatiles induced by the larvae of *Chilo suppressalis* and *Cnaphalocrocis medinalis*. *Chin. J. Appl. Entomol.* **2010**, *47*, 96–101.
14. Liu, X.-L.; Lou, Y.-G. Comparison of the defense responses in rice induced by brown planthopper *Nilaparvata lugens* (Stål) and white-backed planthopper *Sogatella furcifera* (Horváth). *J. Plant Prot.* **2018**, *45*, 971–978.
15. Huang, T.-F.; Ma, Y.; Tang, B.-J.; Wang, B.-Y.; Zeng, R.-X.; Zhou, Q.; Zhang, G.-R. Sexual differences in the response of the parasitoid wasp *Anagrus nilaparvatae* to rice volatiles. *Chin. J. Appl. Entomol.* **2021**, *58*, 876–884.
16. Li, T.; Wang, C.-P.; Jiang, N.-N.; Wei, J.-G.; Mo, J.-C. Attractiveness of rice plant volatiles to *Anagrus nilaparvatae* Pang et Wang. *Chin. J. Appl. Entomol.* **2018**, *55*, 360–367.
17. Li, S.; Chen, W.-L.; Jin, D.-C.; Yang, H. The attraction of diverse rice volatiles to *Haplogonatopus japonicus*. *J. Plant Prot.* **2014**, *41*, 203–209.
18. Jiang, N.-N.; Mao, G.-F.; Li, T.; Mo, J.-C. The olfactory behavior response of *Cytorhinus lividipennis* to single component of rice volatiles. *Ecol. Environ. Sci.* **2018**, *27*, 262–267.
19. Lou, Y.-G.; Chen, J.-A. The role of rice volatiles in the predation behavior of *A. lucorum*. *Entomol. Sin.* **2001**, *8*, 240–250.
20. Zhang, Y.-H.; Li, T.; Mo, J.-C. The attractiveness of rice plant volatiles to *Apanteles chilonis* Munakata and *Anagrus nilaparvatae* Pang et Wang. *Chin. J. Appl. Entomol.* **2016**, *53*, 491–498.
21. Ulhoa, L.A.; Barrigossi, J.A.F.; Borges, M.; Laumann, R.A.; Blassioli-Moraes, M.C. Differential induction of volatiles in rice plants by two stink bug species influence behaviour of conspecifics and their natural enemy *Telenomus podisi*. *Entomol. Exp. Appl.* **2020**, *168*, 76–90. [CrossRef]
22. Sun, L.-Y.; Liu, J.; Li, Q.; Fu, D.; Zhu, J.-Y.; Guo, J.-J.; Xiao, R.; Jin, D.-C. Cloning and differential expression of three heat shock protein genes associated with thermal stress from the wolf spider *Pardosa pseudoannulata* (Araneae: Lycosidae). *J. Asia-Pac. Entomol.* **2021**, *24*, 158–166. [CrossRef]
23. Wang, X.-Q.; Wang, G.-H.; Zhu, Z.-R.; Tang, Q.-Y.; Hu, Y.; Qiao, F.; Heong, K.L.; Cheng, J.-A. Spider (Araneae) predations on white-backed planthopper *Sogatella furcifera* in subtropical rice ecosystems, China. *Pest Manag. Sci.* **2017**, *73*, 1277–1286. [CrossRef] [PubMed]
24. Wang, H.-Q.; Shi, G.-B. Discussion on Dominant Species and Causes of Spiders in Rice Fields in China. *Sci. Agric. Sin.* **2002**, *11*, 85–93.
25. Wang, H.-Q.; Yan, H.-M. Study on the Ecology and Utilization of Spiders in Rice Fields in China. *Sci. Agric. Sin.* **1996**, *29*, 68–75.

26. Wang, Z.; Zeng, B.-P.; Li, W.-J.; Wang, W.-B. Study on the Time Niche of the Paddy field Spiders and Objective Pests. *Hunan Agric. Sci.* **2002**, *3*, 28–29.

27. Yan, H.-M. The Structure and Ecological Distribution of Spider Community in Rice Fields in Southwest China. *J. Nat. Sci. Hunan Norm. Univ.* **1991**, *14*, 80–85.

28. Tan, Z.-J.; Yan, H.-M. Chemical Information between Spiders'Species and Within Species. *Life Sci. Res.* **2015**, *19*, 368–371.

29. Wang, B.; Huang, T.; Han, M.; Wang, X.-L.; Chang, Y.-T.; Huang, X.-Y.; Li, Y.-Y.; Yan, H.-M. The Response of *Pardosa pseudoannulata*'s Chemoreceptor in Locating the Prey. *Sichuan J. Zool.* **2014**, *33*, 86–89.

30. Cao, Y.; Liu, J.; Guo, J.; Wu, H.; Zhang, G. Identification and analysis of odorant-binding protein genes from the wolf spider *Pardosa pseudoannulata* (Araneae: Lycosidae) based on its transcriptome. *Chemoecology* **2018**, *28*, 123–130. [CrossRef]

31. Vizueta, J.; Frias-Lopez, C.; Macias-Hernandez, N.; Arnedo, M.A.; Sanchez-Gracia, A.; Rozas, J. Evolution of Chemosensory Gene Families in Arthropods: Insight from the First Inclusive Comparative Transcriptome Analysis across Spider Appendages. *Genome Biol. Evol.* **2017**, *9*, 178–196. [CrossRef]

32. Yin, C.-M.; Peng, X.-J.; Yan, H.-M. *Zoology of Hunan, Spiders*; Hunan Science and Technology Press: Hunan, China, 2012.

33. Wen, D.-D.; He, Y.-Y.; Lu, Z.-Y.; Yang, H.-M.; Wang, H.-Q. quantitative study of biomass flow in the rice-Sogatella furcifera-Pirata subpiraticus food chain using fluorescent substance tracing. *Acta Entomol. Sin.* **2003**, *46*, 178–183.

34. Xiao, Y.-H.; He, Y.-Y.; Yang, H.-M.; Yan, H.-M. The Study on the Starvation Endurance of Larva *Pirata subpiraticus*. *J. Nat. Sci. Hunan Norm. Univ.* **2004**, *27*, 78–81.

35. Wang, Z. Bionomics and behavior of the wolf spider *Pardosa pseudoannulata* (Araneae: Lycosidae). *Acta Entomol. Sin.* **2007**, *9*, 927–932.

36. Vet, L.E.M.; Lenteren, J.C.V.; Heymans, M.V.; Meelis, E. An airflow olfactometer for measuring olfactory responses of hymenopterous parasitoids and other small insects. *Physiol. Entomol.* **1983**, *8*, 97–106. [CrossRef]

37. Verheggen, F.J.; Fagel, Q.; Heuskin, S.; Lognay, G.; Francis, F.; Haubruge, E. Electrophysiological and behavioral responses of the multicolored Asian lady beetle, Harmonia axyridis pallas, to sesquiterpene semiochemicals. *J. Chem. Ecol.* **2007**, *33*, 2148–2155. [CrossRef] [PubMed]

38. Skoczek, A.; Piesik, D.; Wenda-Piesik, A.; Buszewski, B.; Bocianowski, J.; Wawrzyniak, M. Volatile organic compounds released by maize following herbivory or insect extract application and communication between plants. *J. Appl. Entomol.* **2017**, *141*, 630–643. [CrossRef]

39. Piesik, D.; Rochat, D.; Delaney, K.J.; Marion-Poll, F. Orientation of European corn borer first instar larvae to synthetic green leaf volatiles. *J. Appl. Entomol.* **2013**, *137*, 234–240. [CrossRef]

40. Avila, G.A.; Withers, T.M.; Holwell, G.I. Olfactory cues used in host-habitat location and host location by the parasitoid *Cotesia urabae*. *Entomol. Exp. Appl.* **2016**, *158*, 202–209. [CrossRef]

41. Ghaninia, M.; Amooghli Tabari, M. Olfactory cues explain differential attraction of the striped rice stem borer to different varieties of rice plant. *J. Appl. Entomol.* **2016**, *140*, 376–385. [CrossRef]

42. Adedipe, F.; Park, Y.-L. Visual and olfactory preference of *Harmonia axyridis* (Coleoptera: Coccinellidae) adults to various companion plants. *J. Asia-Pac. Entomol.* **2010**, *13*, 319–323. [CrossRef]

43. Siddique, B.; Tariq, M.; Naeem, M.; Ali, M. Behavioral Responses of *Coccinella septempunctata* and *Diaeretiella rapae* under the Influence of Semiochemicals and Plant Extract in Four Arm Olfactometer. *Pak. J. Zool.* **2019**, *51*, 1403–1411. [CrossRef]

44. Komatsuzaki, S.; Piyasaengthong, N.; Matsuyama, S.; Kainoh, Y. Effect of Leaf Maturity on Host Habitat Location by the Egg-Larval Parasitoid *Ascogaster reticulata*. *J. Chem. Ecol.* **2021**, *47*, 294–302. [CrossRef] [PubMed]

45. Nurkomar, I.; Buchori, D.; Taylor, D.; Kainoh, Y. Innate olfactory responses of female and male parasitoid *Apanteles taragamae* Viereck (Hymenoptera: Braconidae) toward host plant infested by the cucumber moth *Diaphania indica* Saunders (Lepidoptera: Crambidae). *Biocontrol Sci. Technol.* **2017**, *27*, 1373–1382. [CrossRef]

46. Zhu, J.; Park, K.-C. Methyl salicylate, a soybean aphid-induced plant volatile attractive to the predator *Coccinella septempunctata*. *J. Chem. Ecol.* **2005**, *31*, 1733–1746. [CrossRef]

47. De Boer, J.G.; Dicke, M. The role of methyl salicylate in prey searching behavior of the predatory mite *Phytoseiulus persimilis*. *J. Chem. Ecol.* **2004**, *30*, 255–271. [CrossRef]

48. James, D.G. Methyl salicylate is a field attractant for the goldeneyed lacewing, *Chrysopa oculata*. *Biocontrol Sci. Technol.* **2007**, *16*, 107–110. [CrossRef]

49. James, D.G. Field evaluation of herbivore-induced plant volatiles as attractants for beneficial insects: Methyl salicylate and the green lacewing, *Chrysopa nigricornis*. *J. Chem. Ecol.* **2003**, *29*, 1601–1609. [CrossRef]

50. James, D.G.; Price, T.S. Field-Testing of Methyl Salicylate for Recruitment and Retention of Beneficial Insects in Grapes and Hops. *J. Chem. Ecol.* **2004**, *30*, 1613–1628. [CrossRef] [PubMed]

51. Shimoda, T. A key volatile infochemical that elicits a strong olfactory response of the predatory mite *Neoseiulus californicus*, an important natural enemy of the two-spotted spider mite *Tetranychus urticae*. *Exp. Appl. Acarol.* **2010**, *50*, 9–22. [CrossRef] [PubMed]

52. Carroll, M.J.; Schmelz, E.A.; Meagher, R.L.; Teal, P.E. Attraction of *Spodoptera frugiperda* larvae to volatiles from herbivore-damaged maize seedlings. *J. Chem. Ecol.* **2006**, *32*, 1911–1924. [CrossRef] [PubMed]

53. Anderson, A.R.; Wanner, K.W.; Trowell, S.C.; Warr, C.G.; Jaquin-Joly, E.; Zagatti, P.; Robertson, H.; Newcomb, R.D. Molecular basis of female-specific odorant responses in *Bombyx mori*. *Insect Biochem. Mol. Biol.* **2009**, *39*, 189–197. [CrossRef] [PubMed]

54. Reisenman, C.E.; Riffell, J.A.; Bernays, E.A.; Hildebrand, J.G. Antagonistic effects of floral scent in an insect-plant interaction. *Proc. R. Soc. B-Biol. Sci.* **2010**, *277*, 2371–2379. [CrossRef] [PubMed]
55. Cruz-López, L.; Díaz-Díaz, B.; Rojas, J.C. Coffee volatiles induced after mechanical injury and beetle herbivory attract the coffee berry borer and two of its parasitoids. *Arthropod-Plant Interact.* **2016**, *10*, 151–159. [CrossRef]
56. Mitra, S.; Karmakar, A.; Mukherjee, A.; Barik, A. The Role of Leaf Volatiles of *Ludwigia octovalvis* (Jacq.) Raven in the Attraction of *Altica cyanea* (Weber) (Coleoptera: Chrysomelidae). *J. Chem. Ecol.* **2017**, *43*, 679–692. [CrossRef]
57. Oelbermann, K.; Scheu, S. Control of aphids on wheat by generalist predators: Effects of predator density and the presence of alternative prey. *Entomol. Exp. Appl.* **2009**, *132*, 225–231. [CrossRef]
58. Zhang, N.; Xie, L.; Wu, X.; Liu, K.; Liu, C.; Yan, Y. Development, survival and reproduction of a potential biological control agent, *Lasioseius japonicus* Ehara (Acari: Blattisociidae), on eggs of *Drosophila melanogaster* (Diptera: Drosophilidae) and *Sitotroga cerealella* (Lepidoptera: Gelechiidae). *Syst. Appl. Acarol.* **2020**, *25*, 1461–1471. [CrossRef]
59. Cheli, G.; Armendano, A.; Gonzalez, A. Feeding preferences of the spider *Misumenops pallidus* (Araneae: Thomisidae) on potential prey insects from alfalfa crops. *Rev. Biol. Trop.* **2006**, *54*, 505–513. [CrossRef]
60. Hauge, M.S.; Nielsen, F.H.; Toft, S. Weak responses to dietary enrichment in a specialized aphid predator. *Physiol. Entomol.* **2011**, *36*, 360–367. [CrossRef]
61. Chen, C.; Zhang, X.-L.; Liu, B.-L.; Qin, X.-Y.; Guo, H.; Feng, R. Rice volatile components induced by rice planthopper. *J. South. Agric.* **2021**, *52*, 37–44.
62. Sun, Z.-Y. *The Effect of Two Plant Extracts on the Insect Control Potential of Pardosa pseudoannulata*; Hunan Agricultural University: Hunan, China, 2010.
63. Sun, Z.-Y.; Liang, P.-Z.; Hu, R.; Wang, Z. Effect of Plant Physiologically Active Substances on the Predation of *Pardosa pseudoannulata*. *Hunan Agric. Sci.* **2009**, *39*, 85–87. [CrossRef]
64. Li, R.; Li, N.; Liu, J.; Li, S.-C.; Hong, J.-P. The effect of low-dose of pesticide on predation of spider and its preliminary mechanisms. *Acta Ecol. Sin.* **2014**, *34*, 2629–2637.
65. Li, R.; Li, S.-L.; Li, N.; Li, S.-C.; Hong, J.-P. The effects of low doses of insecticides on predation of green peach aphid *Myzus persicae* (Hemiptera: Aphididae) by a wolf spider *Pardosa astrigera* (Araneae: Lycosidae). *J. Plant Prot.* **2014**, *41*, 711–716.
66. Wang, Z.; Song, D.-X.; Zhu, M.-S. Influences of Low-dose Pesticide on the Spatial Niche and Insect-controlling Efficiency of Paddy Field Spiders. *J. Hebei Univ. (Nat. Sci. Ed.)* **2006**, *26*, 278–282.
67. Wang, Z.; Yan, H.-M.; Wang, H.-Q. Effects of low-dose pesticides on insect control ability of rice field spiders. *Acta Ecol. Sin.* **2002**, *22*, 346–351.
68. Simpson, M.; Gurr, G.M.; Simmons, A.T.; Wratten, S.D.; James, D.G.; Leeson, G.; Nicol, H.I.; Orre-Gordon, G.U.S. Attract and reward: Combining chemical ecology and habitat manipulation to enhance biological control in field crops. *J. Appl. Ecol.* **2011**, *48*, 580–590. [CrossRef]
69. Jaworski, C.C.; Xiao, D.; Xu, Q.; Ramirez-Romero, R.; Guo, X.; Wang, S.; Desneux, N. Varying the spatial arrangement of synthetic herbivore-induced plant volatiles and companion plants to improve conservation biological control. *J. Appl. Ecol.* **2019**, *56*, 1176–1188. [CrossRef]
70. Liu, J.; Sun, L.-y.; Fu, D.; Zhu, J.-Y.; Xiao, R. Diversity, and dominant species, of rice field spiders in Guizhou province. *Chin. J. Appl. Entomol.* **2021**, *58*, 142–157.

 *insects*

Article

Combined Effect of Different Flower Stem Features on the Visiting Frequency of the Generalist Ant *Lasius niger*: An Experimental Study

Elena V. Gorb * and Stanislav N. Gorb 

Department of Functional Morphology and Biomechanics, Zoological Institute, Kiel University, Am Botanischen Garten 9, 24098 Kiel, Germany; sgorb@zoologie.uni-kiel.de
* Correspondence: egorb@zoologie.uni-kiel.de; Tel.: +49-431-8804509

Simple Summary: Flowering plants usually attract insect pollinators by offering them nectar, pollen or other energetically valuable sources. To deter ants, which are unreliable pollinators and can act as nectar thieves, plants have developed different systems either inside the flowers or associated with the stems. The latter one, called *greasy pole syndrome*, is based on the combined effect of several stem features hampering the access of ants to the apically located flowers. In this study, we examined the effects of different flower stem features in the round-leaved Alexanders *Smyrnium rotundifolium* on the visiting frequency of the generalist ant species, the black garden ant *Lasius niger*. We conducted the experiments with ants running on dry wooden sticks mimicking four different types of stems. To attract ants, we placed a sweet sugar syrup droplet on a stick tip. Ants visited different types of stem-mimicking sticks with significantly different frequencies. The highest number of insects were registered on untreated stick samples, whereas the lowest visiting frequency was observed on sticks bearing cuff-like structures (serving as macroscopic physical barriers) covered with a nano/microparticle film, which caused the slipperiness of the surface. Thus, by combining macroscopic obstacles and slippery surfaces, plants can protect their flowers from undesirable crawling visitors such as ants.

Citation: Gorb, E.V.; Gorb, S.N. Combined Effect of Different Flower Stem Features on the Visiting Frequency of the Generalist Ant *Lasius niger*: An Experimental Study. *Insects* **2021**, *12*, 1026. https://doi.org/10.3390/insects12111026

Academic Editor: Donato Antonio Grasso

Received: 13 October 2021
Accepted: 12 November 2021
Published: 14 November 2021

Publisher's Note: MDPI stays neutral with regard to jurisdictional claims in published maps and institutional affiliations.

Abstract: In order to understand the effects of the morphology and surface texture of flower stems in *Smyrnium rotundifolium* on the visiting frequency of generalist ants, we conducted experiments with *Lasius niger* ants running on dry wooden sticks mimicking different types of stems: (1) intact (grooved) sticks; (2) sticks painted with slaked (hydrated) lime (calcium carbonate coverage) imitating plant epicuticular wax coverage; (3) intact sticks with smooth polyester plate-shaped cuffs imitating upper leaves; and (4) intact sticks bearing cuffs painted with slaked lime. Ants were attracted by the sweet sugar syrup droplets placed on a stick tip, and the number of ants visiting the drops was counted. Our data showed significant differences in the visiting frequencies between the different types of stem-mimicking samples. The number of recorded ants progressively decreased in the following order of samples: intact sticks—painted sticks—sticks with intact cuffs—sticks with painted cuffs. These results clearly demonstrated that micro/nanoscopic surface coverages and macroscopic physical barriers, especially if combined, have a negative impact on the attractiveness of stems to ants. This study provides further evidence for the hypothesis that having a diversity of plant stems in the field, generalist ants prefer substrates where their locomotion is less hindered by obstacles and/or surface slipperiness.

Keywords: calcium carbonate coverage; cuffs; cuticular folds; epicuticular wax projections; greasy pole syndrome; slaked (hydrated) lime; nectar robbing; *Smyrnium rotundifolium*

1. Introduction

To attract insect pollinators, mostly from the orders Hymenoptera, Diptera, Lepidoptera, and Coleoptera, the angiosperms (flowering plants) usually produce and offer them nectar, pollen or other energetically valuable resources that are in the main associated with flowers [1]. On the other hand, plants have also developed different systems, primarily inside the flowers, such as toxic nectar or chemical/morphological modifications of the floral tissues, in order to deter ants [2–9] that generally belong to the floral antagonists, which are unreliable pollinators and can act as nectar thieves [10,11]. In addition to consuming nectar, ants sometimes attack and deter legitimate pollinators, in that way decreasing pollination success [7,12–16]. The *greasy pole syndrome* is another defense mechanism preventing ants from visiting flowers and robbing nectar [17–20]. It was described in plants from the genera *Salix* (Salicaceae*), Hypenia* and *Eriope* (both Lamiaceae) and is based on the combined effect of several non-floral stem features hampering the access of ants to the apically located reproductive organs of plants. These features include slender elongate erect stems, rigid spreading trichomes (i.e., hair-like protuberances extending from the epidermis of aerial plant tissues) on the lower internodes, three-dimensional epicuticular wax coverage (composed of projections of hydrophobic cuticular lipids deposited onto the aerial primary surfaces of higher plants) and often swellings in the upper internodes [19]. Admittedly, the trichomes deter ground foraging ants, whereas the waxes prevent ants from climbing the upper stem portions. The stem shape is presumably responsible for its extreme motility in windy conditions, and the fistulose swellings provide additional physical barriers hampering ant locomotion.

Until now, the main attention was given to the contribution of the wax coverage on stems to ant deterrence. First, Kerner von Marilaun [17] detected the presence of wax on *S. daphnoides* stems and suggested its hampering effect on ants trying to reach the nectar-bearing flowers. Later field observations on *Eriope* plants showed that ants of all sizes failed to climb up these stems covered with microscopic wax projections [18]. Simple experiments with *H. vitifolia* and *Lasius niger* (Hymenoptera: Formicidae) demonstrated that ants were not able to gain a foothold on the waxy stems, perhaps because they dislodged wax projections and therefore fell to the ground [19]. Our previous experiments, where we recorded the traversed distances of *L. niger* ants on the waxy flower stems of *Anethum graveolens* (Apiaceae*), Dahlia pinnata* and *Tagetes patula* (both Asteraceae) clearly showed that ants were able to walk significantly lower distances on the wax-covered stems compared to the reference wax-free ones [21]. Ants moved much more slowly and carefully on wax-covered stems; however, in all the tests performed, not a single ant fell down from either surface. The later study on the frequency of plant visits by *L. niger* ants was performed with five plant species bearing different surface structures on their stems: *Alchemilla mollis* (Rosaceae), with wax projections and long thread-shaped trichomes; *Lilium lancifolium* (Liliaceae), without wax, but with ribbon-shaped trichomes and cuticular folds (i.e., surface microstructures usually caused by the folding of the cuticle over the outer cell wall of epidermal cells); *Salvia nemorosa* (Lamiaceae), without wax, but with trichomes of various lengths and cuticular folds; *Tulipa gesneriana* (Liliaceae), with wax projections and lacking trichomes; *Paeonia lactiflora* (Paeoniaceae), having neither of the above surface features [22]. It was found that, on the one hand, ants avoided climbing the wax-covered stems, especially if trichomes were lacking. On the other hand, some trichome-bearing stems having specific trichome micromorphologies were also ignored by ants. The last result is in line with that of the only previous study on the contribution of the effect of trichomes to the greasy pole syndrome [19], showing that *L. niger* ants, which had been introduced at the ground level, became entangled in the trichomes near the base of the *H. vitifolia* stem.

Whereas the impact of the waxes and trichomes covering the stems of plants showing the greasy pole syndrome on ant deterrence has been previously examined, the effect of macroscopic characteristics (e.g., the swellings in the upper internodes mentioned above) has not been studied so far. In the present study, besides the microscopic surface features, we are also focusing on the macroscopic stem structures, which can potentially serve as a physical barrier for ants—the sessile upper leaves clasping the stem like cuffs in the round-leaved Alexanders *Smyrnium rotundifolium* (Apiaceae) (Figure 1c). In this plant, the small flowers that are assembled in branched umbels (Figure 1b) are exposed and openly offer, like a cafeteria, nectar and pollen for pollinators. However, ants are also readily attracted by the nectar (Figure 1b). To ensure visitation from only winged insects, purposely so to be more easily cross-pollinated by them and to avoid the visits of nectar-robbing ants, this plant has developed a system of different characteristics, both macroscopic (cuffs formed by upper leaves) and microscopic (cuticular folds and three-dimensional epicuticular wax coverage), on their flower stems, hindering the attachment and locomotion of ants.

Figure 1. The plant *Smyrnium rotundifolium*. (**a**) General view of the plant in the natural habitat. (**b**) Branched umbel with an ant foraging on flowers. (**c**) Upper leaves forming a kind of cuff around flower stems. Note a water pool visible in the cuff located in the center of the picture.

In order to understand the effects of the flower stem morphology and surface texture in *S. rotundifolium* on the visiting frequency of generalist ants, we conducted the experiment with *L. niger* ants and dry wooden sticks mimicking different types of plant stems, among them that of *S. rotundifolium*. Ants were attracted by the sweet sugar syrup droplets placed on a stick tip, and the number of ants visiting the droplets on the different stick samples was counted. Our zero hypothesis was that the frequency of ant visits should be similar for all the stick sample types tested. We have found that both the microscopic and macroscopic stem characteristics negatively impact ant visits, and if combined, have a much more pronounced effect. The obtained results support the previously proposed hypothesis [22] that having a diversity of plant stems in the field, generalist ants prefer surfaces where their attachment and locomotion are less hindered by the stem features.

2. Materials and Methods

2.1. Plants and Insects

The round-leaved Alexanders *S. rotundifolium* Mill., also treated at the rank of the subspecies under *S. perfoliatum* (*S. perfoliatum* subsp. *rotundifolium* (Mill.) Hartvig) [23] or even as a variety (*S. perfoliatum* var. rotundifolium (Mill.) Fiori) [24], is an erect, branching perennial [25] of 30–60 cm height, with a ridged, but wingless, not hollow stem and broadly oblong, divided (biternate), green lower (basal and in the lower stem portion) leaves (Figure 1a). Bright, rounded upper leaves, being actually the bracts, have more or less entire margins, are sessile and clasp the stem (Figure 1c). Tiny yellow flowers assembled in branched umbels, which have up to 12 rays (Figure 1b), are produced in late spring and early summer (description after [26,27]). The plants are common in olive groves, thickets and in dry locations [26] and occur in the central and eastern Mediterranean regions at an elevation of up to 800 m [28]. No specialized ant species associated with *S rotundifolium* has been reported so far.

The black garden ant *L. niger* L. is an omnivorous species that originated from South America and Africa [29] but is recently spread over Europe, where it became widely distributed at sites of human disturbance (e.g., roadsides, gardens, etc.). Workers of this ant species regularly forage on diverse plants [30], can rob nectar from flowering plants and collect honeydew from aphids living on plants. We selected *L. niger* as a model generalist ant species for experiments because (1) it is associated with a variety of plant species (personal observations) and (2) it was available in a great number at the study site. Attachment system of *L. niger* is composed of paired claws and a pad-like, smooth arolium on each foot (for details see [21]).

2.2. Microscopy

Plant material for cryo-scanning electron microscopy (SEM) examination was collected from the plants growing in the olive grove near the Gardiki Castle (surroundings of Agios Matthaios, Corfu, Greece, 39.476500° N, 19.884983° E). Upper leaves and upper portions of the stem were cut off from living plants and kept for 24 h inside small plastic vials containing wet paper in order to prevent desiccation of the plant material. Then, small samples (1 cm × 1 cm) from the middle region of the leaves and from the stem were cut out, attached mechanically to a small vice on a metal holder and frozen in a cryo-stage preparation chamber at −140 °C (Gatan ALTO 2500 Cryo Preparation System, Gatan Inc., Abingdon, UK). Frozen samples of the upper (adaxial) and lower (abaxial) leaf sides and of the stem surface were sputter coated with gold–palladium (thickness 6 nm) and examined in a frozen condition in a cryo-SEM Hitachi S-4800 (Hitachi High-Technologies Corporation, Tokyo, Japan) at 3 kV accelerating voltage and −120 °C temperature. Types of wax projections were identified according to Barthlott et al. [31].

To visualize the surface textures of dry wooden sticks, transparent polyester film and calcium carbonate coverage used in the experiment with ants (see Section 2.3), small pieces of the intact stick and polyester foil as well as painted stick and cuff (ca. 1 cm^2) were sputter coated with gold–palladium (6 nm thickness) and studied in the SEM at 3 kV accelerating voltage and room temperature (ca. 20 °C).

Morphometrical variables of surface features were measured from digital images using SigmaScan Pro 5 software (SPSS Inc., Chicago, IL, USA). These data are presented in the text as mean $\pm$ SD for $n = 10$.

2.3. Experiment

For the experiment with ants, instead of native plant stems of *S. rotundifolium*, we used dry wooden sticks (50 cm long, 0.3 cm in diameter, with grooved surface; Gardol Splittstab; BAHAG AG, Mannheim, Germany) dug to a 5 cm depth into the soil. We prepared four types of sticks mimicking different types (i.e., having diverse morphologies and surface textures) of plant stems: (1) intact, untreated sticks; (2) sticks painted with slaked, or hydrated, lime, thereby imitating plant epicuticular wax coverage; (3) untreated sticks with smooth plate-shaped cuffs imitating upper leaves (Figure 2b); and (4) untreated sticks bearing cuffs painted with slaked lime (Figure 2c). The cuffs were prepared in the following way: discs (ca. 5 cm in diameter) with a small hole in the center were cut out from the transparent polyester film (color laser transparency film 5752341, 125 μm thick; Office Depot, La Venio, the Netherlands) and firmly attached to sticks at the heights of 15 and 30 cm from the ground level (two cuffs per stick) (Figure 2d). To get wax-mimicking coverage on the surface, we painted sticks/cuffs with slaked lime (calcium hydroxide Ca(OH)$_2$, Art-Nr. KK03.1; Carl Roth GmbH, Karlsruhe, Germany) dissolved in double-distilled water in the ratio 1:1 (lime to water). After drying under air conditions (in the presence of carbon dioxide CO_2), the limewater resulted in the formation of a micro/nanostructured layer of calcium carbonate ($CaCO_3$) on the surface.

The experiment was carried out in a private garden (Kronshagen, Schleswig-Holstein, Germany; 54°20′11.354″ N, 10°5′15.547″ E). Five sticks of each type (altogether 20 sticks) were placed at nearly similar distance (ca. 10 cm) from each other; the order of different stick types was randomized (Figure 2e). This approach allowed us a direct comparison of samples located at a very narrow space (ca. 1 m^2), and therefore exposed to approximately the same number of foraging ants. The position of each stick was changed every day during the experiment. A small droplet (ca. 15 μL) of the customly prepared sugar syrup (double-distilled water to sugar weight ratio was 1:1) was deposited on the stick tip. After 12 h, each single stick sample was individually checked, and the number of ants feeding on the syrup droplet (Figure 2a) was registered. After counting ants, the droplets were renewed. The experiment was performed during 14 dry days (336 h) in June (from 10 to 23 of June) at a temperature of 20–25 °C and 40–65% relative humidity. We counted ants twice a day, in the morning (at 8:00) and in the evening (at 18:00–22:00); on some weekend days (15.06, 16.06, and 22.06), and additionally during the daytime (at 11:00–16:00). In all, 31 recordings on each single stick sample were made and the visits of 470 ants were registered.

To determine whether there is a significant difference between the expected frequencies and the observed frequencies of ant visits in different stick types, the number of ant visits per stick type was tested using the Chi-square test (software Good Calculator, free online calculators). Pairwise multiple comparisons between the stick types were performed using the Chi-square test with Bonferroni correction ($\alpha = 0.05/6 = 0.0083$). For each test, the expected frequencies in each stick type were calculated as expected = total number of registered ants/number of stick types.

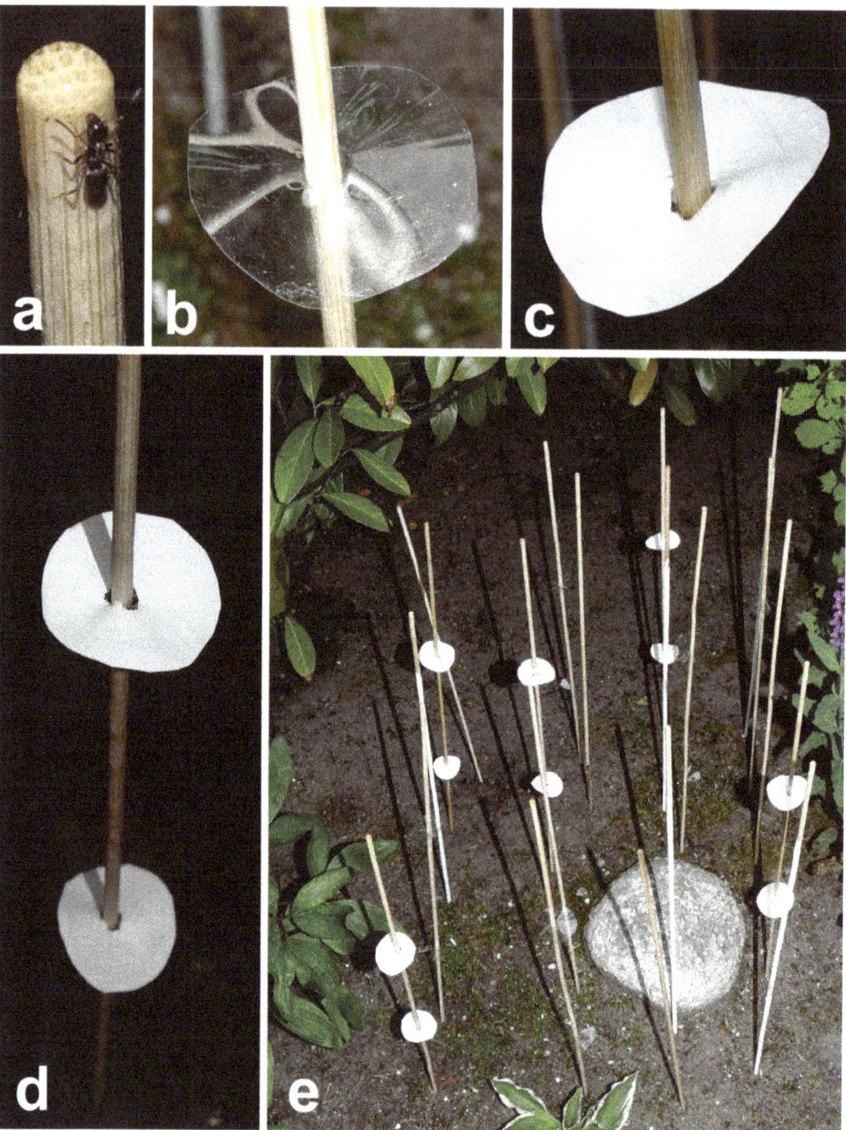

Figure 2. Set-up for the experiment with ants. (**a**) An ant feeding on a sweet droplet deposited on the stick tip. (**b**) Transparent cuff made out of transparent polyester film. (**c**) The cuff bearing calcium carbonate coverage that mimics an epicuticular plant wax. (**d**) Two cuffs places on a stick sample. (**e**) Arrangement of test sticks at the study site.

3. Results

3.1. Micromorphology of Plant Surfaces

The flower stem surface of *S. rotundifolium* is noticeably structured with rather uniform, straight cuticular folds (width: 1.68 ± 0.34 µm; height: 0.68 ± 0.16 µm) running parallel along the longitudinal stem axis (Figure 3). Usually, each 5–10 (7.2 ± 1.3) folds build clusters separated through somewhat wider and deeper grooves (Figure 3a). On top, cuticular folds bear numerous, nearly rounded nanostructures (diameter: 235 ± 73 nm), apparently epicuticular wax granules (Figure 3c).

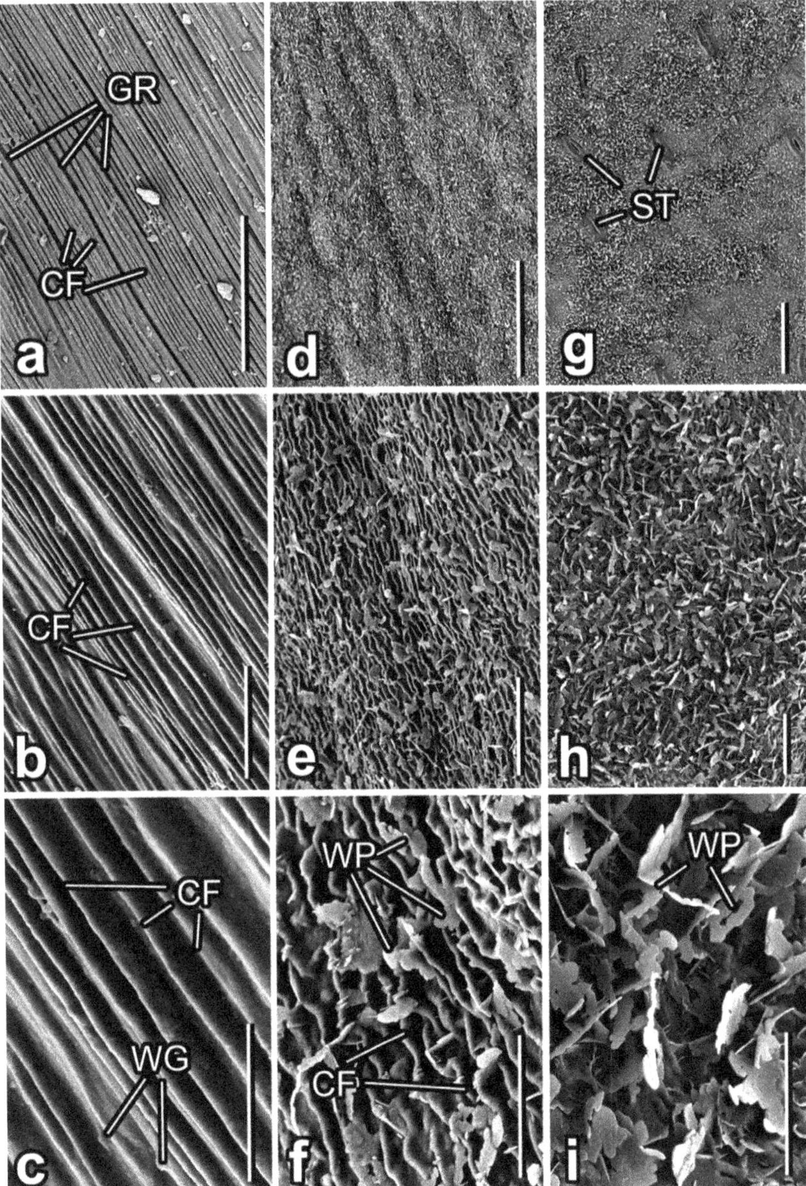

Figure 3. Surfaces of different organs in the *Smyrnium rotundifolium* plant (cryo-SEM). (**a–c**) Flower stem. (**d–f**) Upper (adaxial) leaf side. (**g–i**) Lower (abaxial) leaf side. Abbreviations: CF—cuticular folds; GR—grooves; ST—stomata; WG—wax granules; WP—wax platelets. Scale bars: 50 µm (**a,d,g**), 10 µm (**b,e,h**), 5 µm (**c,f,i**).

The adaxial leaf side is slightly uneven because of the convex shape of the epidermal cells (Figure 3d), and has a prominent hierarchical microstructure composed of a dense network of winding cuticular folds and scattered (occurrence: ca. 0.5 µm^{-2}) epicuticular wax projections (Figure 3e). The folds, being relatively short, narrow and shallow (length: 3.80 ± 1.13 µm; width: 0.53 ± 0.07 µm; height: 0.39 ± 0.06 µm), are responsible for the irregular corrugate surface appearance (Figure 3f). Small (length: 1.17 ± 0.26 μm; width: 0.57 ± 0.19 μm; thickness: 0.05 ± 0.01 μm) flat wax projections (irregular platelets) having highly variable shapes and non-entire margins protrude from the cuticle at various, often acute, angles and do not show any characteristic orientation (Figure 3f).

The abaxial leaf side bears stomata having smooth guard cells (Figure 3g) and a multilayered, extremely dense (occurrence in the external layer: ca. 2 μm^{-2}) coverage of flat, often interconnected epicuticular wax projections (membranous platelets) in the regions between the stomata (Figure 3h). These platelets with irregular margins or even filiform extensions vary greatly in shape and size (length: 2.00 ± 0.36 μm; width: 0.79 ± 0.16 μm; thickness: 0.04 ± 0.01 μm) (Figure 3i). They have neither specific orientation nor distinct arrangement on the surface.

3.2. Microtexture of Samples Used in the Experiment

The surface of the wooden sticks has straight grooves running parallel along the longitudinal stick axis (Figure 4a). Both the width and the height/depth of both the grooves and especially the elevations (groove width: 5.64 ± 4.14 μm; elevation width: 23.74 ± 14.69 μm) vary greatly at different stick portions. Moreover, non-uniform, often flake-like microscopic (1–4 μm) irregularities having various dimensions (2.31 ± 1.27 μm in length/diameter) protrude from the surface (Figure 4b). On the contrary, the polyester film surface is rather smooth at the microscopic scale (Figure 4c).

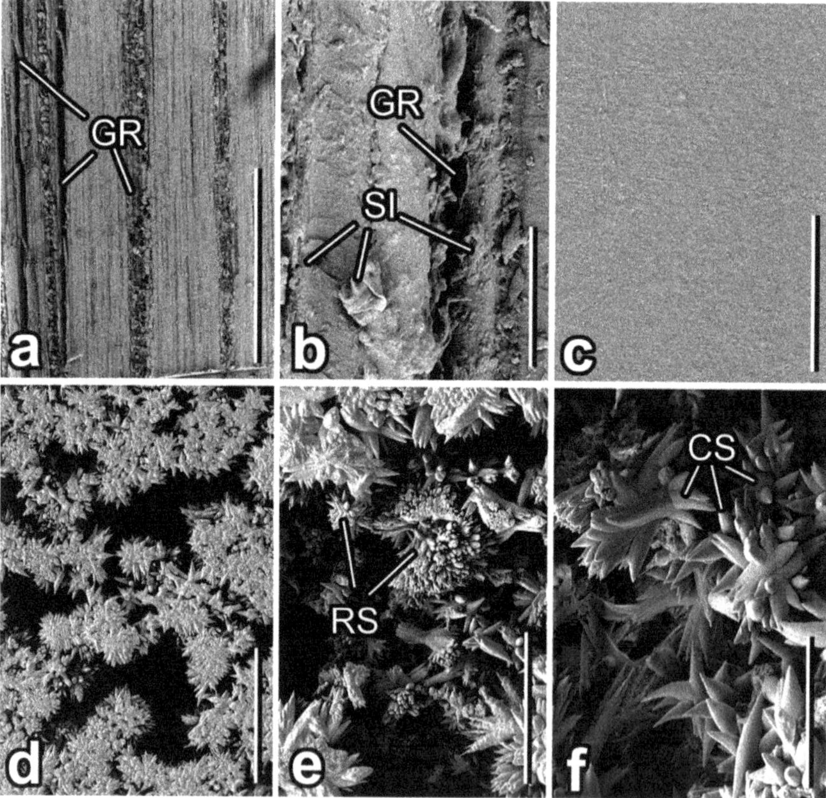

Figure 4. Surfaces of different experimental samples (SEM). (**a**,**b**) Intact stick. (**c**) Transparent polyester film. (**d**–**f**) Calcium carbonate (CaCO₃) coverage. Abbreviations: CS—cone-shaped structures; GR—grooves; RS—double rosettes; SI—surface irregularities. Scale bars: 1 mm (**a**), 20 μm (**d**), 10 μm (**b**,**e**), 5 μm (**c**,**f**).

Calcium carbonate coverage obtained from an aqueous solution of calcium hydroxide after air drying represents a rather loose layer (Figure 4d) composed of double rosettes of numerous (more than ten in each rosette) accrete, solid, cone-shaped microscopic (length: 1.82 ± 0.46 µm) structures (Figure 4e,f). The cones have either a sharp or rounded tip (however, the same within one individual rosette) and the varying ratio of cross section diameters between the basal, middle, and upper parts (basal: 0.53 ± 0.18 µm; middle: 0.53 ± 0.18 µm; upper: 0.23 ± 0.05 µm). This variability gives different tapered shapes of structures ranging from the relatively slender to almost barrel-like. The voids between the rosettes and between the cones within a rosette are responsible for the high porosity of the material that the coverage is made of.

3.3. Ant Visits to Different Samples

Data on the number of ant individuals visiting each sample during the experimental time (31 recordings per sample) are presented in Supplementary Materials (Table S1). All the sticks were visited at least twice (out of 31 recordings) by ants. The maximum number (12) of ant individuals seen at the same time was detected on the intact stick, followed by 10 ants recorded on both the intact and the painted stick samples. On the stick type with transparent cuffs, a maximum four ants at once were detected, whereas on sticks with painted cuffs, not more than two individuals at the same time were registered.

In Figure 5, the number of registered ant individuals (counted during all 31 recordings) for each stick and each sample type is shown. The number of recorded ants progressively decreased in the following order of samples: intact sticks—painted sticks—sticks with intact cuffs—sticks with painted cuffs (Figure 5b). Ants demonstrated clear, significantly different preferences ($\chi^2 = 129.5084$, d.f. $= 3$, $p < 0.0001$). Most preferable samples were intact sticks that differed significantly from all other (modified) types of sticks (intact sticks vs. painted sticks: $\chi^2 = 102.9845$, $p < 0.0001$; intact sticks vs. intact cuffs: $\chi^2 = 59.1793$, $p < 0.0001$; intact sticks vs. painted cuffs: $\chi^2 = 16.7906$, $p < 0.001$, Chi-square test with Bonferroni correction ($\alpha = 0.05/6 = 0.0083$), d.f. $= 1$). Moreover, the modified stick types differed significantly from each other (painted sticks vs. intact cuffs: $\chi^2 = 8.1270$, $p < 0.005$; painted sticks vs. painted cuffs: $\chi^2 = 41.5934$, $p < 0.0001$; intact cuffs vs. painted cuffs; $\chi^2 = 14.1402$, $p < 0.001$, Chi-square test with Bonferroni correction ($\alpha = 0.05/6 = 0.0083$), d.f. $= 1$).

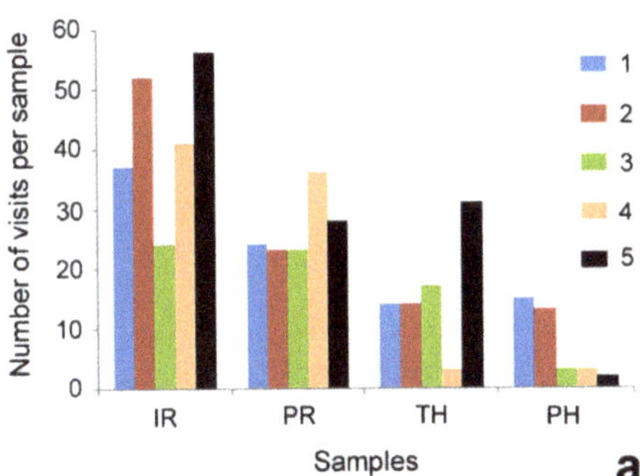

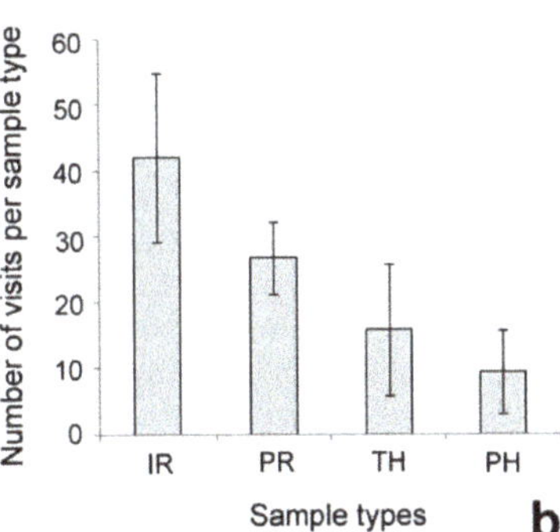

Figure 5. Number of ant visits per sample (**a**) and per sample type (mean and SD) (**b**). Abbreviations: IR—intact sticks; PH—sticks with painted (i.e., bearing calcium carbonate coverage) cuffs; PR—painted sticks bearing calcium carbonate coverage; TH—sticks with unpainted, transparent cuffs; 1, 2, 3, 4, 5—different individual sticks in each sample type.

4. Discussion

Despite the world abundance of ants, very few plant species (ten so far) are described as ant-pollinated, and ants are regarded as undesirable flower visitors that are only capable of nectar thieving [10]. Moreover, ants can scare pollinators away and even assail them [7,12–16]. To avoid such negative effects caused by ants on pollination, many plants have evolved different chemical/morphological floral adaptations [2–9] or a complex of morphological stem features preventing the access of ants to the flowers (i.e., greasy pole syndrome) [17–22], especially if flowers are openly placed and bear nonhidden (i.e., freely accessible) nectaries, like in the round-leaved Alexanders *S. rotundifolium* studied here.

Smyrnium rotundifolium possesses several stem- and leaf-related macroscopic and micro/nanoscopic characteristics, such as clusters of microscopic cuticular folds decorated with nanoscopic epicuticular wax projections on the flower stems and the cuffs, formed by the upper leaves bearing either wax projections on the adaxial side or a combination of cuticular folds and wax projections on the abaxial side. In our experiments, in order to evaluate the effect of different features on the frequency of ant visits, we used different types of wooden sticks mimicking stems with different characteristics: intact sticks with a grooved surface imitating stems with grooves between the clusters of cuticular folds; painted sticks imitating stems with wax projections; sticks with transparent polyester cuffs mimicking upper leaves; and sticks with painted cuffs imitating upper leaves bearing wax projections.

Our experimental results clearly demonstrated that both the macroscopic structures and the micro/nanoscopic surface coverages significantly reduced the visiting frequency of ants. In the first case, the upper leaves, completely wrapping the stems and making the cuffs, represent a physical barrier where ants have to overcome three transitions: (1) from the vertical stem to the adaxial side of the cuff (ceiling situation) [32,33]; (2) from the adaxial to abaxial side of the cuff; and (3) from the abaxial side of the cuff to the vertical stem. The first two transitions are particularly complicated, since insects can easily lose the grip there and fall down. Moreover, the cuffs usually collect rain and dew water and form a kind of pool (see such a pool in Figure 1c), creating an additional physical barrier for ants to overcome. In our study, we did not test the effect of water pools experimentally, but observed their presence in nature. However, it has been also previously shown for one ant species, *Camponotus schmitzi* (Hymenoptera: Formicidae), living in symbiosis with the Bornean pitcher plant, *Nepenthes bicalcarata* (Nepenthaceae), that it can swim and dive in the pitcher's fluid to forage for food, but compared to running, the swimming involved lower stepping frequencies and larger phase delays within the legs of each tripod [34]. Even though this kind of behavior is known for this particular ant species, it seems that these ants are strongly specialized in this symbiosis for their diving behavior in order to get some reward in the form of protein-rich food from the plant. That is why we do not expect that *L. niger* and a majority of other ant species associated with the habitats where *S. rotundifolium* occurs, can overcome the water filled out pools in order to get access to the flower nectar.

Additionally, our experiments (see Figure 5b) showed even lower numbers of ants on the unpainted cuffs (solely the effect of the cuffs) than on the painted sticks (effect of wax). This fact demonstrates the rather strong effectiveness of the cuff-like morphology against nectar-robbing ants.

As for the effect of the cuticular folds, we did not mimic these microscopic surface structures in our experimental samples. In the literature, it has been hypothesized that the folds support the attachment and locomotion of pollinators on flowers [35,36], but recent experimental studies performed with non-pollinating beetles on various leaf and petal surfaces showed a decrease of insect attachment forces on the cuticular folds compared to flat substrates due to the reduction of the real contact area [37–40]. However, in our previous study on the frequency of plant visits by *L. niger* ants that was performed with five plant species having different surface structures on their stems, *Lilium lancifolium* flower stems covered with both cuticular folds and long soft trichomes showed a rather high number of ant visits [22]. We suggested that in this plant species, the soft twisted trichomes probably "compensate" for an unsuitable quality of the stem surface caused by the cuticular folds.

The wax projections were simulated by calcium carbonate crystals in our experimental samples (painted sticks and painted cuffs). Hydrated lime is one of the first minerals used in "ancient particle film technology" [41], which alone or in mixture represents the prevailing insect-repelling material used in agriculture in the early 1900s [42] and is still presently applied against insect pest and plant pathogens. Whereas in livestock farming it is mostly used for disinfectant purposes, producing a dry and alkaline environment in which bacteria do not readily multiply, in horticultural farming it serves as an insect repellent causing no harm to either pests or plants [43]. In particular, lime-based products have been reported to function as oviposition deterrents for the *Rhagoletis indifferens* fly (Diptera: Tephritidae) [44]. Hydrated lime treatments may also repel insects, and these effects have been confirmed for different insect species [45–47]. Our recent experimental study revealed that calcium carbonate coverage greatly diminished the attachment of the bug *Nezara viridula* (Heteroptera: Pentatomidae) due to its distinct microrough surface topology and, to a lesser extent, due to the contaminating effect on the insect adhesive organs [41]. Moreover, the high absorption ability, in particular the water absorption ability, of the calcium carbonate film described in [41] may contribute to insect attachment reduction, as has been previously found for both *Coccinella septempunctata* and *Harmonia axyridis* beetles (both Coleoptera: Coccinellidae) on nanoporous substrates [48,49].

Both stick sample types bearing the calcium carbonate coverage (painted sticks and painted cuffs) showed significantly lower visiting frequencies of ants compared to the corresponding untreated samples (intact sticks and transparent cuffs, respectively). These data are in line with the results of numerous previous experimental studies performed with many insect and plant species (reviewed in [50]), showing that prominent epicuticular wax coverage in plants usually reduces insect attachment using different mechanisms: (1) the reduction of the real contact area between the substrate and the tips of insect attachment organs (the roughness hypothesis); (2) the contamination of insect adhesive organs by the wax projections (the contamination hypothesis); (3) the adsorption of fluid secretion from the insect adhesive pads due to the high capillarity of the wax coverage (the fluid absorption hypothesis); (4) hydroplaning caused by the appearance of a thick layer of fluid caused by the dissolving of the wax material in insect adhesive fluid (the wax dissolving hypothesis); and (5) the formation of a separation layer between the insect attachment organs and the substrate [51,52]. Moreover, our previous studies with *L. niger* ants and the wax-bearing stems of *Anethum graveolens, Dahlia pinnata, Tagetes patula* and *Tulipa gesneriana* showed that ants avoided these stems but were still able to walk on such antiadhesive vertical substrates when they had no other choice [21,22]. It was concluded that the reason why nonspecialized ants usually do not climb wax-covered stems is that the additional locomotory efforts are needed to master climbing on "greasy" stems. In the present study, ants still climbed up the painted sticks and visited, although much more rarely, the painted cuffs. The fact that we still observed ants on these sample types means that ants can hold and walk on a waxy surface. However, very few ant visits detected on the sticks with painted cuffs indicated that if macroscopic obstacles like cuffs are combined with micro/nanoscopic coverages (i.e., waxes in the case of plants and calcium carbonate film in our experimental samples), such substrates become extremely challenging for ant locomotion. Thus, by having flower stems with both macroscopic barriers and prominent micro/nanoscopic coverages, plants protect their flowers bearing openly placed nectaries from undesirable crawling visitors such as ants.

Thus, in contrast to the previous experimental studies on the greasy pole syndrome that were performed either with plants or plant samples [19,21,22], we here for the first time have used artificial samples mimicking different types of plant flower samples. For this, we applied dry wooden sticks instead of the original plant stems and imitated the microscopic surface characteristics and macroscopic obstacles by using calcium carbonate particle coverage and transparent polyester film, respectively. The role of the macroscopic stem structures in plant defense against the generalist ant species was first tested here. We confirmed once more the deterring effect of microscopic surface features on insects [50–52] and showed experimentally for the first time a negative influence of the macroscopic barriers on the visiting frequency of ants. We discovered that a combination of microscopic (three-dimensional epicuticular wax coverage) and macroscopic (cutts formed by upper leaves) stem related characteristics has a great impact on ant visits: in this way, plants can protect their generative organs from these unwanted creeping visitors. The stem-related characteristics are plant adaptations to avoid the visits of nectar-robbing ants, probably by hindering their attachment and locomotion, in order to ensure the visitation of pollinators.

5. Conclusions

Our data show significant differences in ants' visiting frequencies among the different types of stem-mimicking samples. The results clearly demonstrate that micro/nanoscopic surface structures and macroscopic physical barriers, especially if combined, have a negative impact on the attractiveness of the flower stems to ants. This study provides further evidence for the hypothesis that having a diversity of plant stems in the field, generalist ants prefer substrates where their locomotion is less hindered by obstacles or surface slipperiness [22].

Supplementary Materials: The following is available online at https://www.mdpi.com/article/10.3390/insects12111026/s1, Table S1: Number of ant individuals counted on stick samples mimicking different features of the *Smyrnium rotundifolium* flower stem at the time of 31 visits of experimenters during 14 days.

Author Contributions: E.V.G. and S.N.G. have equally contributed to the work reported. All authors have read and agreed to the published version of the manuscript.

Funding: This research received no external funding.

Institutional Review Board Statement: Not applicable.

Data Availability Statement: The complete experimental data set supporting reported results is provided in the supplementary Table S1.

Conflicts of Interest: The authors declare no conflict of interest.

References

1. Waser, N.M. Specialization and generalization in plant-pollinator interactions: A historical perspective. In *Plant-Pollinator Interactions: From Specialization to Generalization*; Waser, N.M., Ollerton, J., Eds.; The University of Chicago Press: Chicago, IL, USA, 2006; pp. 3–17.
2. Van der Pijl, L. Some remarks on myrmecophytes. *Phytomorphology* **1955**, *5*, 190–200.
3. Janzen, D.H. Why don't ants visit flowers? *Biotropica* **1977**, *9*, 252. [CrossRef]
4. Guerrant, E.O.; Fiedler, P.L. Flower defenses against nectar-pilferage by ants. *Biotropica* **1981**, *13*, 25–33. [CrossRef]
5. Junker, R.R.; Blüthgen, N. Floral scents repel potentially nectar-thieving ants. *Evol. Ecol. Res.* **2008**, *10*, 295–308.
6. Junker, R.R.; Gershenzon, J.; Unsicker, S.B. Floral odor bouquet loses its ant repellent properties after inhibition of terpene biosynthesis. *J. Chem. Ecol.* **2011**, *37*, 1323–1331. [CrossRef]
7. Galen, C.; Cuba, J. Down the tube: Pollinators, predators, and the evolution of flower shape in the alpine skypilot, *Polemonium viscosum*. *Evolution* **2001**, *55*, 1963–1971. [CrossRef]
8. Tagawa, K. Repellence of nectar-thieving ants by a physical barrier: Adaptive role of petal hairs on *Menyanthes trifoliata* (Menyanthaceae). *J. Asia Pac. Entomol.* **2018**, *21*, 1211–1214. [CrossRef]
9. Villamil, N.; Boege, K.; Stone, G.N. Testing the distraction hypothesis: Do extrafloral nectaries reduce ant-pollinator conflict? *J. Ecol.* **2019**, *107*, 1377–1391. [CrossRef] [PubMed]
10. Peakall, R.; Handel, S.N.; Beattie, A. The evidence for, and importance of, ant pollination. In *Ant-Plant Interactions*; Huxley, C.R., Cutler, D.E., Eds.; Oxford University Press: Oxford, UK, 1991; pp. 421–429.
11. Willmer, P.G.; Nuttman, C.V.; Raine, N.E.; Stone, G.N.; Pattrick, J.G.; Henson, K.; Stillman, P.; McIlroy, L.; Potts, S.G.; Knudsen, J.T. Floral volatiles controlling ant behaviour. *Func. Ecol.* **2009**, *23*, 888–900. [CrossRef]
12. Tsuji, K.; Hasyim, A.; Harlion; Nakamura, K. Asian weaver ants, *Oecophylla smaragdina*, and their repelling of pollinators. *Ecol. Res.* **2004**, *19*, 669–673. [CrossRef]
13. Ness, J.H. A mutualism's indirect costs: The most aggressive plant bodyguards also deter pollinators. *Oikos* **2006**, *113*, 506–514. [CrossRef]
14. Lach, L. Argentine ants displace floral arthropods in a biodiversity hotspot. *Divers. Distrib.* **2008**, *14*, 281–290. [CrossRef]

15. Hansen, D.M.; Müller, C.B. Invasive ants disrupt gecko pollination and seed dispersal of the endangered plant *Roussea simplex* in Mauritius. *Biotropica* **2009**, *41*, 202–208. [CrossRef]
16. Cembrowski, A.R.; Tan, M.G.; Thomson, J.D.; Frederickson, M.E. Ants and ant scent reduce bumblebee pollination of artificial flowers. *Am. Nat.* **2014**, *183*, 133–139. [CrossRef] [PubMed]
17. Kerner von Marilaun, A. *Flowers and their Unbidden Guests*; C. Kegan Paul & Co.: London, UK, 1878.
18. Harley, R.M. Evolution and distribution of Eriope (Labiatae) and its relatives in Brazil. In *Proceedings of a Workshop on Neotropical Distributions*; Academia Brasileira de Ciencias: Rio de Janeiro, Brazil, 1988; pp. 71–120.
19. Harley, R. The greasy pole syndrome. In *Ant-Plant Interactions*; Huxley, C.R., Cutler, D.E., Eds.; Oxford University Press: Oxford, UK, 1991; pp. 430–433.
20. Juniper, B.E. Waxes on plant surfaces and their interactions with insects. In *Waxes: Chemistry, Molecular Biology and Functions*; Hamilton, R.J., Ed.; Oily Press West Ferry: Dundee, UK, 1995; pp. 157–174.
21. Gorb, E.; Gorb, S. How a lack of choice can force ants to climb up waxy plant stems. *Arthropod-Plant Interact.* **2011**, *5*, 297–306. [CrossRef]
22. Gorb, S.N.; Gorb, E.V. Frequency of plant visits by the generalist ant *Lasius niger* depends on the surface microstructure of plant stems. *Arthropod-Plant Interact.* **2019**, *13*, 311–320. [CrossRef]
23. Strid, A. *Mountain Flora of Greece*; Cambridge University Press: Cambridge, UK, 1986; 852p.
24. Fiori, A. *Nuova Flora Analitica d'Italia*; Tipografia di M. Ricci: Florence, Italy, 1925; p. 95.
25. Kiehn, M. Pflanzen mit invasivem Potenzial in Botanischen Gärten X: *Smyrnium perfoliatum* L. (Apiaceae). *Carinthia II (Klagenfurt, Austria)* **2015**, *295/125*, 73–82.
26. Papiomioglou, V. *Wild Flowers of Greece*; Mediterraneo Editions: Rethymno, Greece, 2006; p. 181.
27. Unterarten von Smyrnium perfoliatum. Available online: http://www.mittelmeerflora.de/Zweikeim/Apiaceae/smyr_perfoliatum.htm (accessed on 10 March 2020).
28. Jäger, E.J.; Reckardt, K. Beiträge zur Wuchsform und Biologie der Gefäßpflanzen des herzynischen Raumes. 2: *Smyrnium perfoliatum* L. (Apiaceae). *Hercynia N. F.* **1998**, *31*, 103–116.
29. Klotz, J.H.; Hansen, L.; Pospischil, R.; Rust, M. *Urban Ants of North America and Europe: Identification, Biology, and Management*; Cornell University Press: Ithaca, NY, USA, 2008; 201p.
30. Hölldobler, B.; Wilson, E.O. *The Ants*; Harvard University Press: Cambridge, MA, USA, 1990; 732p.
31. Barthlott, W.; Neinhuis, C.; Cutler, D.; Ditsch, F.; Meusel, I.; Theisen, I.; Wilhelmi, H. Classification and terminology of plant epicuticular waxes. *Bot. J. Linn. Soc.* **1998**, *126*, 237–260. [CrossRef]
32. Ritzmann, R.E.; Quinn, R.D.; Fischer, M.S. Convergent evolution and locomotion through complex terrain by insects, vertebrates and robots. *Arthropod Struct. Dev.* **2004**, *33*, 361–379. [CrossRef]
33. Bußhardt, P.; Gorb, S.N.; Wolf, H. Activity of the claw retractor muscle in stick insects in wall and ceiling situations. *J. Exp. Biol.* **2011**, *214*, 1676–1684. [CrossRef]
34. Bohn, H.F.; Thornham, D.G.; Federle, W. Ants swimming in pitcher plants: Kinematics of aquatic and terrestrial locomotion in *Camponotus schmitzi*. *J. Comp. Physiol.* **2012**, *198*, 465–476. [CrossRef]
35. Stork, N.E. Experimental analysis of adhesion of *Chrysolina polita* (Chrysomelidae, Coleoptera) on a variety of surfaces. *J. Exp. Biol.* **1980**, *88*, 91–107. [CrossRef]
36. Barthlott, W.; Ehler, N. Raster-Elektronenmikroskopie der Epidermisoberflächen von Spermatophyten. *Trop. Subtrop. Pflanzenwelt* **1977**, *19*, 367–467.
37. Prüm, B.; Seidel, R.; Bohn, H.F.; Speck, T. Plant surfaces with cuticular folds are slippery for beetles. *J. R. Soc. Interface* **2011**, *9*, 127–135. [CrossRef] [PubMed]
38. Prüm, B.; Bohn, H.F.; Seidel, R.; Rubach, S.; Speck, T. Plant surfaces with cuticular folds and their replicas: Influence of microstructuring and surface chemistry on the attachment of a leaf beetle. *Acta Biomater.* **2013**, *9*, 6360–6368. [CrossRef] [PubMed]
39. Voigt, D.; Schweikart, A.; Fery, A.; Gorb, S. Leaf beetle attachment on wrinkled surfaces: Isotropic friction on anisotropic surfaces. *J. Exp. Biol.* **2012**, *215*, 1975–1982. [CrossRef] [PubMed]
40. Bräuer, P.; Neinhuis, C.; Voigt, D. Attachment of honeybees and greenbottle flies to petal surfaces. *Arthropod-Plant Interact.* **2017**, *11*, 171–192. [CrossRef]
41. Salerno, G.; Rebora, M.; Piersanti, S.; Saitta, V.; Kovalev, A.; Gorb, E.; Gorb, S. Reduction in insect attachment caused by different nanomaterials used as particle films (kaolin, zeolite, calcium carbonate). *Sustainability* **2021**, *13*, 8250. [CrossRef]
42. Glenn, D.M.; Puterka, G.J. Particle films: A new technology for agriculture. *Hort. Rev.* **2005**, *31*, 1–44.
43. Agricultural Lime. Available online: https://en.wikipedia.org/wiki/Agricultural_lime (accessed on 31 July 2020).
44. Yee, W.L. Behavioural responses by *Rhagoletis indifferens* (Dipt., Tephritidae) to sweet cherry treated with kaolin- and limestone-based products. *J. Appl. Entomol.* **2012**, *136*, 124–132. [CrossRef]
45. Barata, J.M.S.; Santos, J.L.F.; Da Rosa, J.A.; Gomes, R.D. Evaluation of triatomine behavior under the effect of contact with calcium hydroxide CaOH$_2$: Mortality rates of *Triotoma infestans* and *Rhodnius neglectus* (Hemiptera, Reduviidae). *An. Soc. Entomol. Bras.* **1992**, *21*, 169–177. [CrossRef]
46. Boucher, J.; Adams, R.; Johnson, F.; Pachauskas, R. Eggplant: Hydrated lime as an insect repellent. *Insectic. Acaricidae Tests* **1993**, *18*, 131. [CrossRef]

47. Strack, T.; Cahenzli, F.; Daniel, C. Kaolin, lime and rock dusts to control *Drosophila suzukii*. In *Ökologschen Landbau Weiterdenken: Verantwortung Übernehmen, Vertrauen Stärken*; Wolfrum, S., Heuwinkel, H., Wiesinger, K., Reents, H.J., Hülsbergen, K.-J., Eds.; Verlag Dr. Köster: Berlin, Germany, 2017; pp. 262–263.
48. Gorb, E.; Hosoda, N.; Miksch, C.; Gorb, S. Slippery pores: Anti-adhesive effect of nanoporous substrates on the beetle attachment system. *J. R. Soc. Interface* **2010**, *7*, 1571–1579. [CrossRef]
49. Gorb, E.V.; Lemke, W.; Gorb, S.N. Porous substrate affects a subsequent attachment ability of the beetle *Harmonia axyridis* (Coleoptera, Coccinellidae). *J. R. Soc. Interface* **2019**, *16*, 20180696. [CrossRef]
50. Gorb, E.V.; Gorb, S.N. Anti-adhesive effects of plant wax coverage on insect attachment. *J. Exp. Bot.* **2017**, *68*, 5323–5337. [CrossRef] [PubMed]
51. Gorb, E.V.; Gorb, S.N. Attachment ability of the beetle *Chrysolina fastuosa* on various plant surfaces. *Entomol. Exp. Appl.* **2002**, *105*, 13–28. [CrossRef]
52. Gorb, E.V.; Purtov, J.; Gorb, S.N. Adhesion force measurements on the two wax layers of the waxy zone in *Nepenthes alata* pitchers. *Sci. Rep.* **2014**, *4*, 1–7. [CrossRef]

Article

Defensive Traits during White Spruce (*Picea glauca*) Leaf Ontogeny

Antoine-Olivier Lirette and Emma Despland *

Biology Department, Concordia University, 7141 Sherbrooke West, Montreal, QC H4B 1R6, Canada;
a-o.lirette@hotmail.com
* Correspondence: Emma.Despland@concodia.ca

Simple Summary: Leaves can only toughen after they have finished growing and, as a result, many herbivorous insects specialize in newly developing leaves because softer leaves are easier to chew. The foliage of conifer trees is particularly tough, and so one would expect conifers to invest more defensive chemicals into soft vulnerable growing needles than into tough mature ones. We summarize the literature describing how chemical defenses of foliage change during the growing season in white spruce, an economically important conifer tree. We next report measurements of the toughness of white spruce buds as they swell, burst, and grow into young needles. As expected, buds soften as they swell in spring, but after budburst, needles become tougher until they are similar to previous-year foliage in mid-summer. Leaves grown in the sun are slightly tougher than leaves grown in the shade. However, there was no indication that trees invest more in chemical defense of growing leaves than of mature leaves.

Abstract: Changes during leaf ontogeny affect palatability to herbivores, such that many insects, including the eastern spruce budworm (*Choristoneura fumiferana* (Clem.)), are specialist feeders on growing conifer leaves and buds. Developmental constraints imply lower toughness in developing foliage, and optimal defense theory predicts higher investment in chemical defense in these vulnerable yet valuable developing leaves. We summarize the literature on the time course of defensive compounds in developing white spruce (*Picea glauca* (Moench) Voss) needles and report original research findings on the ontogeny of white spruce needle toughness. Our results show the predicted pattern of buds decreasing in toughness followed by leaves increasing in toughness during expansion, accompanied by opposite trends in water content. Toughness of mature foliage decreased slightly during the growing season, with no significant relationship with water content. Toughness of sun-grown leaves was slightly higher than that of shade-grown leaves. However, the literature review did not support the expected pattern of higher defensive compounds in expanding leaves than in mature leaves, suggesting that white spruce might instead exhibit a fast-growth low-defense strategy.

Keywords: plant-insect interactions; leaf defensive traits; herbivory; spruce budworm; toughness; leaf physical traits; phenology

Citation: Lirette, A.-O.; Despland, E. Defensive Traits during White Spruce (*Picea glauca*) Leaf Ontogeny. *Insects* **2021**, *12*, 644. https://doi.org/10.3390/insects12070644

Academic Editor: Brian T. Forschler

Received: 12 June 2021
Accepted: 11 July 2021
Published: 15 July 2021

1. Introduction

Physical and chemical changes occur during leaf ontogeny that change palatability to herbivores. Expanding leaves are generally less tough and more nutritious than mature foliage [1–3]. This is directly linked to the process of leaf growth: young leaves cannot accumulate lignin, cellulose, or other cell wall components until leaves have ceased growing [4], and as a result, they are both low in toughness and high in nitrogen content. Similarly, sugars and amino acids are translocated to expanding leaves to build tissues. At the end of expansion, leaves change from being a photosynthate sink to a source, and their cells sclerify [5]. The accompanying changes in chemical defense traits remain poorly understood, yet are essential to understanding host tissue choices by insects and the distribution of herbivory damage on plants [1]. Since expanding foliage is both less well mechanically defended and more nutritious, it is historically predicted that plant secondary compounds are maintained at higher concentrations in expanding foliage, to protect this vulnerable stage from herbivores [6]. Furthermore, optimal defense theory predicts high investment in defense of expanding foliage because the same amount of herbivory has more impact on a growing structure than on a fully grown one [1].

Nonetheless, many herbivores still preferentially feed on, and perform better on, expanding foliage. Examples among evergreen plants that retain their leaves over many years include boreal conifers [7–10], tropical angiosperms [3], and cycads [11]. Ontogenetic changes in foliage quality thus limit the period when the foliage is suitable for herbivorous insects, promoting tight phenological host–herbivore relationships that define a window of opportunity [12–14]. The magnitude of these differences between young and old foliage defines the strictness of the window of opportunity for insect herbivores on evergreen trees [15].

White spruce (*Picea glauca* (Moench) Voss) is an economically important resource and a foundation species in North American boreal forests. Its genome has recently been sequenced [16,17], and considerable research is underway to examine the genetic architecture of white spruce traits, including investigating the biosynthesis pathways of defensive compounds [17–19] and attempting to select traits for resistance against herbivores [20]. However, the phenology and ontogeny of defensive compound expression in foliage has received little attention, despite the fact that many defoliators on white spruce, including the notorious spruce budworm (*Choristoneura fumiferana* (Clem.)), preferentially feed on expanding foliage [21–23]. The window of opportunity for these herbivores has been linked to nutrient content, allelochemicals and toughness [22,24], but contributions of these variables are not clear. In particular, the hypothesis that secondary chemicals should be more concentrated in expanding foliage has not received much attention in spruce.

This paper combines a literature review on the ontogeny of white spruce foliar defense with original research addressing the knowledge gap on seasonal patterns of leaf toughness. We first summarize past work on ontogenetic progression of defense traits in expanding white spruce foliage and highlight knowledge gaps for future research. Review of the literature highlights that toughness is one trait that has received relatively little attention, despite evidence for an important role in defense against the spruce budworm [15]. Therefore, we measured white spruce leaf toughness during the growing season in both expanding and mature foliage in order to improve understanding of the ontogenetic trajectory of mechanical defense during the window of opportunity for herbivores. Leaf physical trait data from this boreal conifer will also contribute to generalize understanding of ontogenetic shifts in plant defense syndromes during leaf development [1].

2. Literature Review on Phenology of Spruce Defensive Traits

2.1. Leaf Nutritional Traits

Conifer growth in the spring begins before Nitrogen uptake by roots, and therefore depends on remobilization of both Carbon and Nitrogen from reserves stored in one-year-old foliage as they are translocated to developing buds [25]. Indeed, most of the N in new conifer shoots is derived from storage in old needles. Soluble N increases in year-old conifer foliage prior to budburst in spring then quickly drops as N is translocated to developing buds [26,27]. Primary meristems are the source of signal perception for the resumption of growth, responding to both growth regulators and sugars [28].

In spring, before budburst, carbon supply is greater than carbon demand [29], and the carbon needed for primary growth accumulates, mostly in old needles, mainly in the form of starch. Reserves are depleted in the period between budburst and the carbon autonomy of new leaves [28], as sugars are translocated to expanding shoots. In conifers, newly assimilated carbon by older needles is mainly allocated to the canopy during primary growth, with only a minor fraction translocated to the lower stem [30].

Measurement of the elemental composition of white spruce leaves showed that the greatest quantitative changes occurred during the periods of active shoot growth [22]: Total N and P levels were highest in the swelling buds, declined rapidly during initial shoot elongation, and then remained uniformly low and stable through the summer. Levels of K also declined during shoot elongation but did not reach their lowest points until after cessation of shoot elongation. Levels of Ca, Mg, Cu, and Zn in current-year foliage generally decreased during shoot elongation, but then increased afterwards. Total foliar sugar (fructose + glucose + sucrose) in new foliage peaked near the middle of shoot elongation and thereafter generally decreased [22]. To our knowledge, these trends have not been linked to the physiological processes described in the preceding paragraphs, and therefore more research is needed to trace movement of nutrients in expanding shoots with modern methods, and in particular to distinguish between different forms of organic Carbon and Nitrogen only some of which are bioavailable to herbivorous insects [12,31].

2.2. Phenolics

Early phenological studies showed that total phenolics in current-year foliage decreased during shoot elongation but increased thereafter: they fell to their lowest levels of 4–5% dw at ca. 80–90% shoot elongation in early June [22]. Mattson et al. [32] similarly reported that total phenolic levels peaked at the time of budbreak in white spruce and declined consistently during shoot growth.

However, total phenolic content is no longer considered a good measure of chemical defense against insects [33]. In more recent years, the structural and functional diversity of spruce phenolics have been much better characterized, and their synthesis from the shikimate pathway is being unravelled [17]. Complex phenolics like lignins clearly play a structural role in spruce leaves [5], but other molecules are thought to defend against fungal pathogens or insect attack. This defensive role is supported by the fact that phenolic synthesis pathways are upregulated in several spruce tissues, including foliage following fungal infection, insect attack, or treatment with the defense hormone methyl jasmonate [17]. Nonetheless, few individual phenolic compounds have convincingly been tested for anti-herbivore activity and the evidence for a defensive role of conifer phenolics remains poorly understood [34]. Some phenolic compounds do accumulate in foliage [17], but to our knowledge, their phenology is unknown except for the two acetophenones discussed below.

Two phenolic compounds have been suggested to play an important role in the defense of coniferous trees against defoliation by the spruce budworm [35,36]. Two sets of acetophenones have been identified from white spruce trees resistant to budworm attack which suffered only light defoliation when other trees around them were heavily damaged: piceol and pungenol (aglycones) and picein and pungenin (their glycosides). The aglycones increase mortality and slow growth in bioassays, but the glycosylated forms appear to have no effect on the budworm [35]. These acetophenones have been shown to be broadly distributed across coniferous trees; in general, the glycosides were found alone or at higher concentrations than the aglycones [36]. Genetic analysis of white spruce trees showed that a glucosyl hydrolase gene, *PgBgluc-1*, was constitutively highly expressed in resistant trees, catalyzing formation of the aglycones from the glycosylated compounds [37]. Levels of both the gene transcripts and the aglycones are highly heritable [36]. In white spruce, the aglycones begin to be expressed in current-year foliage near the end of shoot elongation, when the budworm reach the final instars [36]. These compounds thus do not follow the predicted trend of higher concentration in expanding foliage.

2.3. Terpenoids

Terpenoids are considered important defensive compounds of conifers, and their role has been well-studied in defense against stem-boring insects. Oleoresin is comprised of a diverse array of terpenoid compounds mobilized to the site of wounding on stems. Oleoresin can physically 'pitch out' or entomb attacking insects as well as clean and seal the wound from microorganisms. In white spruce, induction of a local terpenoid response can be triggered by stem damage and lead to the formation of specialized traumatic resin ducts [18,38].

Terpene Synthase genes exhibit high functional diversity, and differential expression leads to a wide range of different blends [17]. Herbivore damage, pathogen infection or MeJA treatment upregulate terpenoids in different tissues, but it is difficult to disentangle stem vs. leaf responses [39]. Indeed, the ecological role of foliar terpenoids pools is not clear [40]. Several examples show a correlation between conifer foliar terpenoid concentrations and insect resistance [8,41,42]: in particular, spruce terpenoids have been suggested to be implicated in natural resistance against both the eastern and western spruce budworm [43]. However, manipulative experiments with bioassays included varying monoterpene concentrations give ambiguous results on the performance of folivores [32,44].

Emission of terpenoid volatiles can be induced in conifer foliage following insect damage [40,45,46]; these volatile compounds may attract natural enemies, but their direct effect on folivores is not clear. However, increases in synthesis of volatile monoterpenes do not necessarily lead to increases in foliar pools because of increased emissions [40,46]. Foliar monoterpene pools are at a metabolic crossroads, between synthesis, emission, storage and translocation to other tissues and may not be under selective pressure from folivores [18]. More research is therefore needed to characterize the patterns and phenology of terpenoid synthesis, emission, and translocation both into and out of leaves.

A few studies have characterized the time course of monoterpene foliar pools: white spruce foliar monoterpenes increased over the growing season in a warm, dry year but stayed low in a cooler, wetter year [47]. A similar seasonal increase has previously been observed in white spruce [48] and other conifers [45]. In Norway spruce (*Picea abies* (L.) H. Karst, 1881), expression of a terpene synthase gene was shown to be low in buds and to increase during bud swelling and shoot elongation [17].

Examination of individual monoterpenes showed a spring peak in d-3-carene, followed by a gradual increase in bornyl acetate over the summer [47]. D-3-carene is of particular interest as it is similarly expressed only during leaf expansion in Scots pine (*Pinus sylvestris* L. [49]) and defines two chemotypes of this species. Moreover, high and low levels of d-3-carene in growing shoots define Sitka spruce (*Picea sitchensis* (Bong.) Carrière) genotypes resistant and susceptible to white pine weevil [50,51]. This compound seems the best candidate among monoterpenes for a defensive effect in conifer foliage.

2.4. Alkaloids

Several 2,6 disubstituted piperidine alkaloids have been detected in the wood, bark, roots and foliage of conifers; in *Picea*, the two main ones are *cis*-pinidinol and epidihydropinidine [52,53]. The function of these is still uncertain; they might be involved in winter hardiness or defense against pathogens, but they are thought to deter herbivores. Some evidence suggests that *cis*-pinidinol and epidihydropinidine deter feeding by spruce budworm larvae and oviposition by female moths [54,55].

In Norway spruce, total alkaloid concentrations peak at budburst and decrease during shoot elongation. However, the peak is composed mostly of accumulation of precursor compounds used in biosynthesis of *cis*-pinidinol and epidihydropinidine, and the concentrations of these two main compounds remain stable in mature foliage, including multi-year old foliage [56].

Concentrations of these alkaloids are low compared to those of phenolics and terpenoids and their role in spruce metabolism or defense has received little attention [57,58].

2.5. Toughness

Toughness is a highly effective defense against insect herbivores that reduces the nutritional value of the leaf and presents mechanical problems for chewing insects [59]. In evergreen tropical forest plants, toughness is a strong predictor of herbivory rates, and most leaf damage occurs during leaf expansion before maximal toughness is achieved [60,61]. Foliar toughness is particularly problematic for young larvae that do not have the mandibular morphology necessary to pierce tough tissues [62–64].

Mature needle toughness has been suggested to determines the window of opportunity for insect herbivores on boreal conifers [15], yet has seldom been measured empirically. Foliar toughness was the most likely explanation for the inability of young larvae of hemlock looper (*Lambdina fiscellaria fiscellaria* (Guenée)) to initiate feeding on old foliage of eastern hemlock (*Tsuga canadensis* (L.) Carrière) [7]. Similarly, young larvae of the palewinged gray moth (*Iridopsis ephyraria* (Walker)) exhibited high mortality when feeding on old foliage of balsam fir (*Abies balsamea* (L.) Miller) and their inability to feed was attributed to leaf toughness [65]. Toughness of mature foliage is an important mechanism inhibiting initiation of feeding by emerging eastern spruce budworm on black spruce (*Picea mariana* (Mill.) B.S.P.) [15,24]. Although it is well understood in tropical angiosperms, the role of leaf toughness in concentrating herbivore damage on young expanding conifer foliage deserves further attention. Therefore, we next present results from original research on the seasonal pattern of toughness of both mature and expanding white spruce foliage in order to test the hypothesis that expanding foliage is less tough than mature leaves.

3. Materials and Methods

Leaf water content and toughness were measured at weekly intervals from 24 May to 26 July 2018 (N = 10 measurement dates) on 12 white spruce trees at the Valcartier Forestry Research Station Arboretum Serge Légaré (latitude, 46.95° N; longitude, 71.48° W; elevation, 152 m). These trees represented 3 clones each of 4 genotypes. Branches were selected based on the presence of buds and year-old needles and were always taken from the bottom half of the crown ca 1.5–2 m above the ground. The trees used in this experiment were from seed lot C9612893, which was the result of an intra-provenance cross performed in the white spruce hybridization parc "Parc Algonquin" at Cap Tourmente, Québec. The female clone (ALG-10) in the cross was found at a latitude of 45.50° N and a longitude of 78.30° W at an elevation of 400 m and the male clone (ALG-8) at a latitude of 45.50° N and longitude of 78.30° W and an elevation of 400 m. The selected trees (6M, 1A, 6O, 12F, 6S, 2F, 12A, 1C) were the result of somatic embryogenesis as in [66], from the seed lot C9612893.

Toughness was measured with a penetrometer [15,24] on 10 individual needles and 10 buds per tree and these were pooled to create one value per tree per date for both needle and bud toughness. The penetrometer consists of a modified microscope with removed lenses to which a Medio-Line Spring Scale (Pesola AG, Baar, Switzerland) is affixed, reversed in a way that pushing instead of pulling will yield a measurement. A syringe head is attached to the end of the measuring device facing towards the needle or bud. Toughness is then measured by slowly raising the scale until the syringe head pierces the needle or bud and the measurement is read from the scale and recorded. Single needles or buds were plucked from freshly cut branches with the cut end kept in water before being placed on the penetrometer. To ensure consistency between the readings, the syringe needle used to pierce through the spruce needles and measure their toughness was changed every 5 readings; any more was determined to dull the edges of the needle and affect the toughness reading. Groups of 10 needles were weighed on a Sartorius analytical balance (0.01 mg accuracy), then air-dried at 60 °C for 48 h, and water content was measured as the difference between wet and dry weight. Needles were weighed in groups due to their low mass.

Measurements were taken on previous-year foliage and expanding buds or foliage. At each sampling date, buds were sampled for toughness measurement and bud stage was recorded according to the scale developed by [67] and illustrated in this field guide [68]. The scale is numerical from 0 to 6 each representing a stage of development of the bud, which can be assessed visually as follows: (0) closed and dormant buds, (1) Bud scales are opening, and a white spot is visible at the top of the bud, (2) buds are elongating, (3) buds are swelling, (4) buds scales are translucent, and the needles are partly visible,

(5) bud scales are ripped at the base of the bud and needles are tightly bundled, (6) needles are elongating and expanding laterally. After buds reached stage 6 (buds fully open), expanding needles were sampled instead. Current-year measurements were considered as buds (stages 0–6) from 24 May to 4 June, and as expanding needles from 28 June to 26 July. Data from 7 June to 21 June (N = 3 measurements) were omitted from analyses including current-year foliage because the opening buds were too soft for meaningful measurement.

A mixed model was run in R including individual tree as a random factor, and foliage type (previous year foliage, expanding bud or current-year foliage), genotype, measurement date (as a Julian date) and water content as fixed factors. Model simplification was based on AIC. Relationship between toughness and water content and between toughness and Julian date was estimated with Pearson correlations. Genotype was included as a variable in our model as a potential explanation for the difference in leaf toughness but did not contribute significantly to the model.

To test for potential environmental effects on leaf toughness, toughness was also compared between 10 previous-year sun-exposed and shade leaves on the same 12 trees on one date in mid-July. Shade leaves were defined as leaves on branches in the lower crown that were continuously in the shade, whereas sun leaves were leaves in the lower crown that received at least 4 h direct sunlight daily. Leaf sampling was done during those hours of sun exposure on a sunny day. The effect of sun exposure on leaf toughness was estimated with a mixed linear model including individual tree as a random factor and genotype and sun exposure as fixed factors.

4. Results

The model including foliage type, Julian date, the interaction between the two, and water content (AIC = 1426) was retained by model simplification over the full model including also genotype (AIC = 1489). The significant interactions between foliage type and Julian date (current year foliage * Julian date coefficient: 5.02, $p < 0.0001$; previous year foliage * Julian date: 3.42, $p < 0.001$) imply that the trajectory of toughness over the growing season differs between swelling buds, current year foliage and previous year foliage and thus that main effects cannot be directly interpreted. The effect of water content on leaf toughness was not quite significant (0.5573, $p = 0.06$), but its removal did not improve the model.

Figure 1 shows how bud toughness decreases as the buds swell (stage 2 to 6), the gap between the bud measurements and the expanding foliage measurements is due to the first few weeks after the buds open and the foliage was too soft for meaningful measurement. After budburst, the toughness of expanding foliage gradually increases until it is similar to that of previous-year foliage at the end of expansion.

Pearson correlations showed that toughness of mature foliage decreased slightly during the season (Pearson correlation coefficient: -0.251; $p = 0.006$) but was not related to water content (Pearson correlation coefficient: -0.081; $p = 0.37$)—see Figure 2. The toughness of expanding buds also showed a negative relationship with Julian date (Pearson correlation coefficient: -0.64; $p < 0.0001$), and decreased with water content (Pearson correlation coefficient: -0.506; $p = 0.0016$) as well. Current-year leaves toughness increased as the season progressed (Pearson correlation coefficient: 0.96; $p < 0.0001$), but decreased with water content (Pearson correlation coefficient: -0.61; $p < 0.0001$).

The mixed model comparing toughness between sun leaves and shade leaves showed a significant interaction between sun exposure and genotype, showing that sun leaves were tougher than shade leaves in 2 of the 4 genotypes (coefficient: 12.733; $p < 0.0001$; coefficient: 9.27; $p < 0.0001$)—see Figure 3. Nonetheless, an overall paired t-test showed that sun leaves are significantly tougher than shade leaves (t = -3.55, df = 231.08, p-value = 0.0004). Pearson correlation showed no correlation between leaf toughness and water content (Pearson correlation coefficient: 0.089; $p = 0.68$).

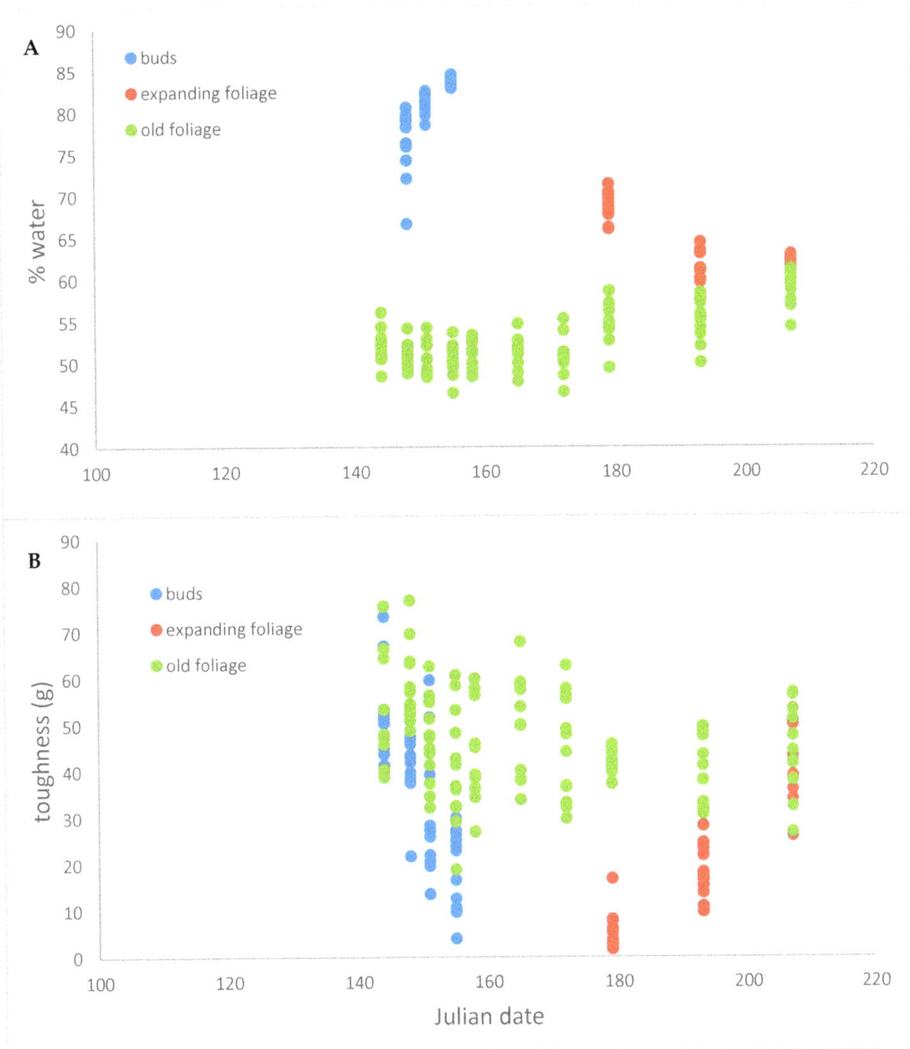

Figure 1. Time course of white spruce (**A**) water content and (**B**) leaf toughness. Green represents previous-year needles, blue represents swelling buds and red the foliage that emerges from those buds. There is a two-week gap between measurement of buds and expanding foliage in which the opening bud was too soft to permit toughness measurement.

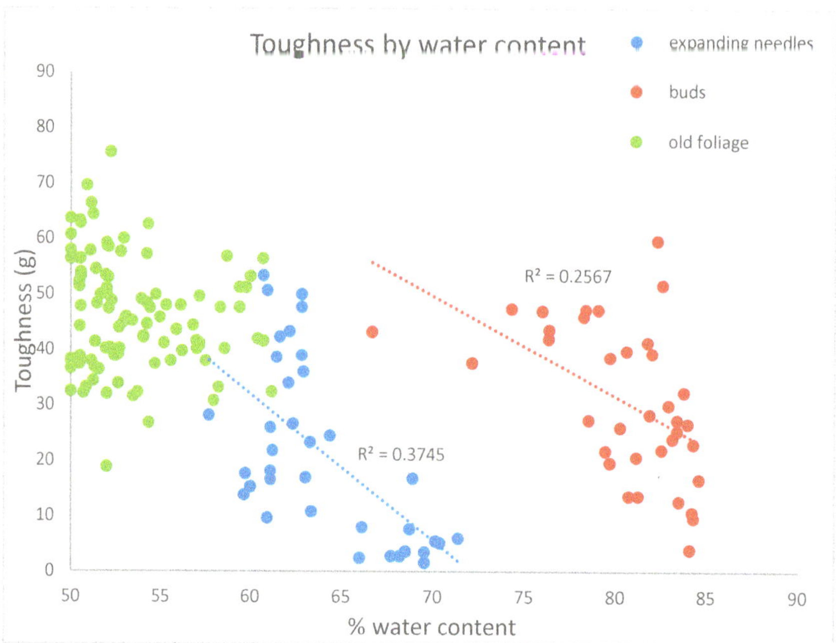

Figure 2. Regressions of toughness against water content. Regression lines and R^2 values are shown for expanding foliage and for buds, but not for mature foliage because the relationship was not statistically significant.

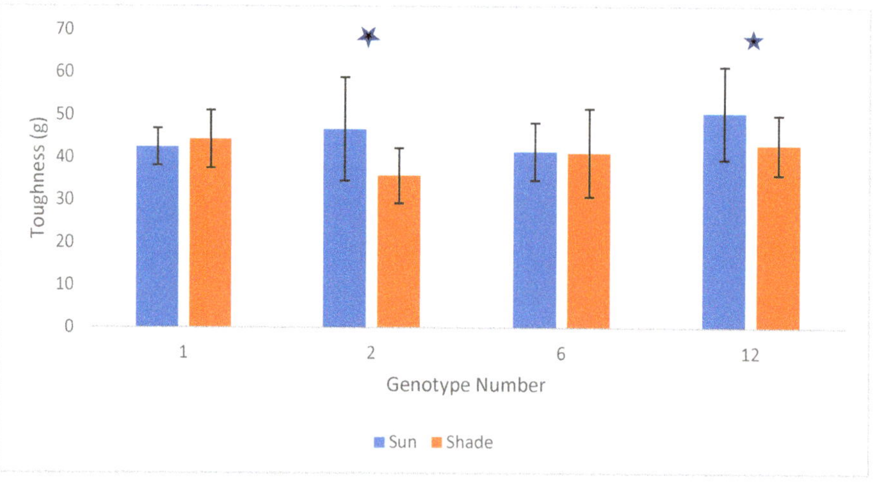

Figure 3. Average ($+/-$ standard deviation) toughness of sun vs. shade leaves in four different white spruce genotypes. The numbers 1, 2, 6, and 12 represent the different genotypes of clones chosen at the Arboretum for this study. Stars indicate the two genotypes for which significant differences between sun and shade foliage were observed.

5. Discussion

Our results show that leaf expansion in white spruce thus generates the predicted window of opportunity in terms of providing less tough foliage for a short period before growth is completed. Review of the literature suggests that nutritional content, mostly

Nitrogen content, also decreases with leaf expansion, but that patterns of defensive chemistry are less clear. Indeed, some compounds decrease as expected with leaf expansion (e.g., total phenolics, alkaloids, d-3-carene), while others, by contrast, increase (e.g., total monoterpenes, acetophenones). However, the defensive role of these compounds remains poorly defined, especially given that many specialist herbivores, including the spruce budworm, possess effective detoxification enzymes [69]. It must be noted that many of these compounds could also be implicated in plant metabolism in roles other than anti-herbivore defense and these other roles might determine their seasonal time course.

As expected, toughness of expanding foliage was inversely proportional to water content. However, water content was a poor predictor of toughness of mature foliage, and this is consistent with a previous study that links conifer toughness to fiber and particularly to hemicellulose content in balsam fir and black spruce [15]. Previous work [22] showed that water content of current-year white spruce foliage peaks soon after budbreak in mid-May as water flows in for cell expansion [26], and then steadily decreases during the shoot growth period until mid-July. Toughness of the new needles increased rapidly during the period of shoot elongation as water content dropped [22].

Sun leaves were found to be slightly tougher than shade leaves, similar to tropical angiosperms where the high leaf specific area of sun leaves contributes to higher toughness [70]. However, this effect only attained statistical significance in two of four genotypes, and effect size was smaller than that of leaf phenology. It nonetheless suggests that sun exposure should be controlled for when estimating foliage toughness in conifers.

During leaf ontogeny, cells continue to grow in size after cell division, mostly via expansion of a water-filled intracellular vacuole. This growth is accompanied by the development of a primary cell wall, but the sclerification and thickening of the secondary cell wall that increases toughness and strengthens the leaf only begins after the cell has reached its final size [4]. Plant cell walls are made of long cellulose microfibrils whose spatial organization in a matrix of polysaccharides (e.g., hemicellulose), proteins, and phenolics (e.g., lignin) promotes toughness [71]. The thickness and chemical composition of cell walls is variable, and thick cell walls rich in hemicellulose and lignin generally imply higher biomechanical strength and toughness [71]. Spruce needles comprise three tissue layers: the protective dermal tissue includes the epidermis and hypodermis and is overlain by a heavy waxy cuticle, the mesophyll is the bulk of the needle where photosynthesis occurs and contains two resin canals, and the vascular tissue in the center is surrounded by an endodermis [72]. In a mature spruce needle, sclerified thick-walled cells are found in the single-layer epidermis, in the hypodermis, near the resin canals and in the endodermis [5]. The mechanism of toughening during white spruce leaf ontogeny is not clear, but likely includes thickening and strengthening of secondary cell walls in the hypodermis and possible in the endodermis [4].

Despite the increase in toughness in developing white spruce leaves that we demonstrate here, the literature does not provide clear evidence for the compensatory higher investment in chemical defense of young leaves. Total phenolics appear to follow the expected trends, but the focus on total phenolics is out-of-date, as these compounds are very diverse and their effects on the alkaline Lepidopteran midgut depends on their chemical structure [33]. The best characterized phenolic compounds, the acetophenones picein and pungenin show the opposite pattern and only accumulate in foliage once it matures [19]. The time course of terpenoids has not received much attention with modern methods; what little evidence exists suggests that overall monoterpene concentration increases in maturing foliage (against the expected trend), but that one compound, d-3-carene, shows the predicted trend of greater concentration in expanding foliage and hence might warrant further attention [47].

An alternative strategy for leaf defense during expansion involves investing in fast growth to escape the vulnerable stage, and investing in defense later [73]. Indeed, growth and defense functions are often considered to trade-off in plants [6,74], and one expression of this trade-off occurs in leaf ontogeny where plants can invest either in rapid leaf growth to attain the defended mature stage quickly or in defense of slower growing expanding leaves. In general, caterpillars grow faster on species with fast-expanding than slow-expanding leaves, but have less time to complete their development before the leaf matures and defenses set in [73]. White spruce has been shown to exhibit faster leaf expansion than related species [32,75], suggesting that it exhibits the fast growth strategy and might invest less in defenses of expanding foliage. White spruce is considered a good host for several defoliators, the best-studied of which is the spruce budworm, but escapes too heavy foliage loss because of fast foliage growth [76]; these are both traits consistent with a fast-growth low-defense strategy.

A recent meta-analysis of globally important insect pests suggests that many show responses to climate change that will lead to increased damage [77]. Phenological synchrony is increasingly recognized as important in plant-insect interactions, including in the population dynamics of outbreaking forest pests, as both host plant characteristics and insect requirements change over their respective ontogenies [12,31,78,79]. Understanding ecology and trophic relationships under a changing climate requires a focus on the ontogenetic and physiological underpinnings of phenology of the interacting partners [80]. Plant phenology is usually considered in terms of visible changes, like budburst, but phenological traits that are fundamental to plant relationships with herbivores are often 'cryptic', or not visible to the human observer, and hence much less well understood. These cryptic traits are increasingly under molecular and genomic investigation [17,18], and studies need to consider the phenology of these traits. Investigation of cryptic phenologies, like the toughness, chemical and nutritional content traits covered here, is essential to understand plant-mediated effects on herbivores, including the spruce budworm [81].

Author Contributions: The experiment was designed by E.D. and carried out by A.-O.L. Data were analyzed by E.D. First draft was written by E.D. and both authors edited the final manuscript. Both authors have read and agreed to the published version of the manuscript. All authors have read and agreed to the published version of the manuscript.

Funding: This research was funded by a Discovery Grant from the Natural Sciences and Engineering Research Council (Canada) to E.D.

Institutional Review Board Statement: Not applicable.

Data Availability Statement: Data are available at Concordia University's online data repository, Spectrum.

Acknowledgments: Thanks to Éric Dussault and Marie-Claude Gros-Louis for assistance in the field. Thanks to the Laurentian Forestry Centre of the Canadian Forest Service for use of laboratory space and access to the arboretum.

Conflicts of Interest: The authors declare no conflict of interest.

References

1.   Barton, K.E.; Edwards, K.F.; Koricheva, J. Shifts in Woody Plant Defence Syndromes during Leaf Development. *Funct. Ecol.* **2019**, *33*, 2095–2104. [CrossRef]
2.   Coley, P.D. Effects of Leaf Age and Plant Life History Patterns on Herbivory. *Nature* **1980**, *284*, 545–546. [CrossRef]
3.   Kursar, T.A.; Coley, P.D. Convergence in Defense Syndromes of Young Leaves in Tropical Rainforests. *Biochem. Syst. Ecol.* **2003**, *31*, 929–949. [CrossRef]
4.   Albersheim, P.; Darvill, A.; Roberts, K.; Sederoff, R.; Staehelin, A. *Plant Cell Walls*; Garland Science: New York, NY, USA, 2011.
5.   Esau, K. *Anatomy of Seed Plants*, 2nd ed.; Wiley: New York, NY, USA, 1977; ISBN 978-0-471-24520-9.
6.   Boege, K.; Marquis, R.J. Facing Herbivory as You Grow up: The Ontogeny of Resistance in Plants. *Trends Ecol. Evol.* **2005**, *20*, 441–448. [CrossRef]
7.   Carroll, A.L. Physiological Adaptation to Temporal Variation in Conifer Foliage by a Caterpillar. *Can. Entomol.* **1999**, *131*, 659–669. [CrossRef]
8.   Chen, Z.; Kolb, T.E.; Clancy, K.M. The Role of Monoterpenes in Resistance of Douglas Fir to Western Spruce Budworm Defoliation. *J. Chem. Ecol.* **2002**, *28*, 897–920. [CrossRef]
9.   Nealis, V. The Phenological Window for Western Spruce Budworm: Seasonal Decline in Resources Quality. *Agric. For. Entomol.* **2012**, *14*, 4. [CrossRef]

10. Watt, A.D. The Effect of Shoot Growth Stage of *Pinus contorta* and *Pinus sylvestris* on the Growth and Survival of Panolis Flammea Larvae. *Oecologia* **1987**, *72*, 429–433. [CrossRef]

11. Prado, A.; Sierra, A.; Windsor, D.M.; Bede, J.C. Leaf Traits and Herbivory Levels in a Tropical Gymnosperm, *Zamia stevensonii* (Zamiaceae). *Am. J. Bot.* **2014**, *10*, 437–447. [CrossRef]

12. van Asch, M.; Visser, M.E. Phenology of Forest Caterpillars and Their Host Trees: The Importance of Synchrony. *Annu. Rev. Entomol.* **2007**, *52*, 37–55. [CrossRef]

13. Feeny, P. Plant Apparency and Chemical Defense. In *Biochemical Interaction Between Plants and Insects*; Recent Advances in Phytochemistry; Wallace, J.W., Mansell, R.L., Eds.; Springer: Boston, MA, USA, 1976; pp. 1–40, ISBN 978-1-4684-2646-5.

14. Hunter, A.F.; Lechowicz, M.J. Foliage Quality Changes during Canopy Development of Some Northern Hardwood Trees. *Oecologia* **1992**, *89*, 316–323. [CrossRef]

15. Fuentealba, A.; Sagne, S.; Legendre, G.; Pureswaran, D.; Bauce, E.; Despland, E. Leaf Toughness as a Mechanism of Defense against Spruce Budworm. *Arthropod Plant Interact.* **2020**, *14*, 481–489. [CrossRef]

16. Birol, I.; Raymond, A.; Jackman, S.; Pleasance, S.; Coope, R.; Taylor, G.; Yuen, M.M.S.; Keeling, C.; Brand, D.; Vandervalk, B.; et al. Assembling the 20 Gb White Spruce (*Picea glauca*) Genome from Whole-Genome Shotgun Sequencing Data. *Bioinformatics* **2013**, *29*, 1492–1497. [CrossRef]

17. Hammerbacher, A.; Wright, L.P.; Gershenzon, J. Spruce Phenolics: Biosynthesis and Ecological Functions. In *The Spruce Genome*; Compendium of Plant Genomes; Porth, I.M., De la Torre, A.R., Eds.; Springer International Publishing: Cham, Switzerland, 2020; pp. 193–214, ISBN 978-3-030-21001-4.

18. Keeling, C.I.; Bohlmann, J. Genes, Enzymes and Chemicals of Terpenoid Diversity in the Constitutive and Induced Defence of Conifers against Insects and Pathogens. *New Phytol.* **2006**, *170*, 657–675. [CrossRef]

19. Parent, G.J.; Méndez-Espinoza, C.; Giguère, I.; Mageroy, M.H.; Charest, M.; Bauce, É.; Bohlmann, J.; MacKay, J.J. Hydroxyace-tophenone Defenses in White Spruce against Spruce Budworm. *Evol. Appl.* **2020**, *13*, 62–75. [CrossRef]

20. Beaulieu, J.; Nadeau, S.; Ding, C.; Celedon, J.M.; Azaiez, A.; Ritland, C.; Laverdière, J.-P.; Deslauriers, M.; Adams, G.; Fullarton, M.; et al. Genomic Selection for Resistance to Spruce Budworm in White Spruce and Relationships with Growth and Wood Quality Traits. *Evol. Appl.* **2020**, *13*, 2704–2722. [CrossRef] [PubMed]

21. Fuentealba, A.; Pureswaran, D.S.; Bauce, E.; Despland, E. How Does Synchrony with Host Plant Affect Spruce Budworm Performance? *Oecologia* **2017**, *184*, 847–857. [CrossRef]

22. Lawrence, R.K.; Mattson, W.J.; Haack, R.A. White Spruce and the Spruce Budworm: Defining the Phenological Window of Susceptibility. *Can. Entomol.* **1997**, *129*, 291–318. [CrossRef]

23. Quiring, D.T. Rapid Change in Suitability of White Spruce for a Specialist Herbivore, *Zeiraphera canadensis*, as a Function of Leaf Age. *Can. J. Zool.* **1992**, *70*, 2132–2138. [CrossRef]

24. Fuentealba, A.; Sagne, S.; Pureswaran, D.; Bauce, É.; Despland, E. Defining the Window of Opportunity for Feeding Initiation by Second-Instar Spruce Budworm Larvae. *Can. J. For. Res.* **2018**, *48*, 285–291. [CrossRef]

25. Millard, P.; Grelet, G. Nitrogen Storage and Remobilization by Trees: Ecophysiological Relevance in a Changing World. *Tree Physiol.* **2010**, *30*, 1083–1095. [CrossRef] [PubMed]

26. Dhuli, P.; Rohloff, J.; Strimbeck, G.R. Metabolite Changes in Conifer Buds and Needles during Forced Bud Break in Norway Spruce (*Picea abies*) and European Silver Fir (*Abies alba*). *Front. Plant Sci.* **2014**, *5*. [CrossRef] [PubMed]

27. Durzan, D.J. Nitrogen Metabolism of Picea Glauca. I. Seasonal Changes of Free Amino Acids in Buds, Shoot Apices, and Leaves, and the Metabolism of Uniformly Labelled 14C-L-Arginine by Buds during the Onset of Dormancy. *Can. J. Bot.* **1968**, *46*, 909–919. [CrossRef]

28. Deslauriers, A.; Fournier, M.-P.; Cartenì, F.; Mackay, J. Phenological Shifts in Conifer Species Stressed by Spruce Budworm Defoliation. *Tree Physiol.* **2019**, *39*, 590–605. [CrossRef]

29. Wiley, R.G.; Helliker, A. Re-evaluation of Carbon Storage in Trees Lends Greater Support for Carbon Limitation to Growth. *New Phytol.* **2012**, *195*, 285–289. [CrossRef]

30. Heinrich, S.; Dippold, M.A.; Werner, C.; Wiesenberg, G.L.; Kuzyakov, Y.; Glaser, B. Allocation of Freshly Assimilated Carbon into Primary and Secondary Metabolites after in Situ 13C Pulse Labelling of Norway Spruce (*Picea abies*). *Tree Physiol.* **2015**, *35*, 1176–1191. [PubMed]

31. Despland, E. Effects of Phenological Synchronization on Caterpillar Early-Instar Survival under a Changing Climate. *Can. J. For. Res.* **2018**, *58*, 247–254. [CrossRef]

32. Mattson, W.J.; Haack, R.A.; Lawrence, R.K.; Slocum, S.S. Considering the Nutritional Ecology of the Spruce Budworm and Its Management. *For. Ecol. Manag.* **1991**, *39*, 83–210. [CrossRef]

33. Salminen, J.P.; Karonen, M. Chemical Ecology of Tannins and Other Phenolics: We Need a Change in Approach. *Funct. Ecol.* **2011**, *25*, 325–338. [CrossRef]

34. Mumm, R.; Hilker, M. Direct and Indirect Chemical Defence of Pine against Folivorous Insects. *Trends Plant Sci.* **2006**, *11*, 351–358. [CrossRef]

35. Delvas, N.; Bauce, E.; Labbé, C.; Ollevier, T.; Bélanger, R. Phenolic Compounds That Confer Resistance to Spruce Budworm. *Entomol. Exp. Appl.* **2011**, *141*, 35–44. [CrossRef]

36. Parent, G.; Giguere, I.; Germanos, G.; Lamara, M.; Bauce, E.; MacKay, J.J. Insect Herbivory (*Choristoneura fumiferana*, Tortricidea) Underlies Tree Population Structure (*Picea glauca*, Pinaceae). *Sci. Rep.* **2017**, *7*, 1–11. [CrossRef] [PubMed]

37. Mageroy, M.H.; Parent, G.; Germanos, G.; Giguere, I.; Delvas, N.; Maaroufi, H.; Bauce, E.; Bohlmann, J.; Mackay, J. Expression of the Beta-Glucosidase Gene Pg-Glu-1 Underpins Natural Resistance of White Spruce against Spruce Budworm. *Plant J.* **2014**, *81*, 68–80. [CrossRef] [PubMed]
38. Alfaro, R.I. An Induced Defense Reaction in White Spruce to Attack by the White Pine Weevil, *Pissodes strobi*. *Can. J. For. Res.* **1995**, *25*, 1725–1730. [CrossRef]
39. Huber, D.P.; Philippe, R.N.; Madilao, L.L.; Sturrock, R.N.; Bohlmann, J. Changes in Anatomy and Terpene Chemistry in Roots of Douglas-Fir Seedlings Following Treatment with Methyl Jasmonate. *Tree Physiol.* **2005**, *25*, 1075–1083. [CrossRef]
40. Martin, D.M.; Gershenzon, J.; Bohlmann, J. Induction of Volatile Terpene Biosynthesis and Diurnal Emission by Methyl Jasmonate in Foliage of Norway Spruce. *Plant Physiol.* **2003**, *132*, 1586. [CrossRef] [PubMed]
41. Cates, R.G.; Henderson, C.B.; Redak, R.A. Responses of the Western Spruce Budworm to Varying Levels of Nitrogen and Terpenes. *Oeocologia* **1987**, *73*, 312–316. [CrossRef]
42. Ikeda, T.; Matsumura, F.; Benjamin, D.M. Chemical Basis for Feeding Adaptation of Pine Sawflies *Neodiprion rugions* and *Neodiprion swainei*. *Science* **1977**, *197*, 497–498. [CrossRef]
43. Nealis, V.G.; Nault, J.R. Seasonal Changes in Foliar Terpenes Indicate Suitability of Douglas-Fir Buds for Western Spruce Budworm. *J. Chem. Ecol.* **2005**, *31*, 683–696. [CrossRef]
44. Gershenzon, J.; Dudareva, N. The Function of Terpene Natural Products in the Natural World. *Nat. Chem. Biol.* **2007**, *3*, 408–414. [CrossRef]
45. Litvak, M.A.; Monson, R.K. Patterns of Induced and Constitutive Monoterpene Production in Conifer Needles in Relation to Insect Herbivory. *Oecologia* **1998**, *114*, 531–540. [CrossRef]
46. Miller, B.; Madilao, L.L.; Ralph, S.; Bohlmann, J. Insect-Induced Conifer Defense. White Pine Weevil and Methyl Jasmonate Induce Traumatic Resinosis, de Novo Formed Volatile Emissions, and Accumulation of Terpenoid Synthase and Putative Octadecanoid Pathway Transcripts in Sitka Spruce. *Plant Physiol.* **2005**, *137*, 369–382. [CrossRef] [PubMed]
47. Despland, E.; Bourdier, T.; Dion, E.; Bauce, E. Do White Spruce Epicuticular Was Monoterpenes Follow Foliar Patterns? *Can. J. For. Res.* **2016**, *46*, 1051–1058. [CrossRef]
48. Rudloff, E.V. Seasonal Variation in the Composition of the Volatile Oil of the Leaves, Buds, and Twigs of White Spruce (*Picea glauca*). *Can. J. Bot.* **1972**, *50*, 1595–1603. [CrossRef]
49. Thoss, V.; O'Reilly-Wapstra, J.; Iason, G.R. Assessment and Implications of Intraspecific and Phenological Variability in Monoterpenes of Scots Pine (*Pinus sylvestris*) Foliage. *J. Chem. Ecol.* **2007**, *33*, 477–491. [CrossRef]
50. Hall, D.E.; Robert, J.A.; Keeling, C.I.; Domanski, D.; Quesada, A.L.; Jancsik, S.; Kuzyk, M.A.; Hamberger, B.; Borchers, C.H.; Bohlmann, J. An Integrated Genomic, Proteomic and Biochemical Analysis of (+)-3-Carene Biosynthesis in Sitka Spruce (*Picea sitchensis*) Genotypes That Are Resistant or Susceptible to White Pine Weevil. *Plant J.* **2011**, *65*, 939–948. [CrossRef]
51. Roach, C.R.; Hall, D.E.; Zerbe, P.; Bohlmann, J. Plasticity and Evolution of (+)-3-Carene Synthase and (-)-Sabinene Synthase Functions of a Sitka Spruce Monoterpene Synthase Gene Family Associated with Weevil Resistance. *J. Biol. Chem.* **2014**, *289*, 23859–23869. [CrossRef]
52. Schneider, M.J.; Montali, J.A.; Hazen, D.; Stanton, C.E. Alkaloids of Picea. *J. Nat. Prod.* **1991**, *54*, 905–909. [CrossRef]
53. Stermitz, F.R.; Kamm, C.D.; Tawara, J.N. Piperidine Alkaloids of Spruce (Picea) and Fir (Abies) Species. *Biochem. Syst. Ecol.* **2000**, *28*, 177–181. [CrossRef]
54. Kamm, C.D.; Tawara, J.N.; Stermitz, F.R. Spruce Budworm Larval Processing of Piperidine Alkaloids from Spruce Needles. *J. Chem. Ecol.* **1998**, *24*, 1153–1160. [CrossRef]
55. Zhao, B.; Grant, G.G.; Langevin, D.; MacDonald, L. Deterring and Inhibiting Effects of Quinolizidine Alkaloids on Spruce Budworm (Lepidoptera: Tortricidae) Oviposition. *Environ. Entomol.* **1998**, *27*, 984–992. [CrossRef]
56. Virjamo, V.; Julkunen-Tiitto, R. Shoot Development of Norway Spruce (*Picea abies*) Involves Changes in Piperidine Alkaloids and Condensed Tannins. *Trees* **2014**, *28*, 427–437. [CrossRef]
57. Arimura, G.; Maffei, M. *Plant Specialized Metabolism: Genomics, Biochemistry, and Biological Functions*; CRC Press: Boca Raton, FL, USA, 2016; ISBN 978-1-315-35337-1.
58. Zidorn, C. Seasonal Variation of Natural Products in European Trees. *Phytochem. Rev.* **2018**, *17*, 923–935. [CrossRef]
59. Juniper, B.; Southwood, R. *Insects and the Plant Surface*; Arnold, E., Ed.; Edward Arnold: Baltimore, MD, USA, 1986; p. 360, ISBN 0-7131-2909-3.
60. Coley, P.D. Herbivory and Defensive Characteristics of Tree Species in a Lowland Tropical Forest. *Ecol. Monogr.* **1983**, *53*, 209–234. [CrossRef]
61. Lowman, M.; Box, J.D. Variation in Leaf Toughness and Phenolic Content among Five Species of Australian Rain Forest Trees. *Aust. J. Ecol.* **1983**, *8*, 17–25. [CrossRef]
62. Clissold, F. The biomechanics of chewing and plant fracture: Mechanisms and implications. In *Advances in Insect Physiology: Insect Mechanics and Control*; Casas, J., Simpson, S.J., Eds.; Academic Press: London, UK, 2008; Volume 34, pp. 317–372.
63. Hochuli, D.F. Insect Herbivory and Ontogeny: How Do Growth and Development Influence Feeding Behaviour, Morphology and Host Use? *Austral Ecol.* **2001**, *26*, 563–570. [CrossRef]
64. Zalucki, M.P.; Clarke, A.R.; Malcolm, S.B. Ecology and Behavior of First Instar Larval Lepidoptera. *Annu. Rev. Entomol.* **2002**, *47*, 361–393. [CrossRef] [PubMed]

65. Pinault, L.; Thurston, G.; Quiring, D. Interaction of Foliage and Larval Age Influences Preference and Performance of a Geometrid Moth. *Can. Entomol.* **2009**, *141*, 136–144. [CrossRef]

66. Klimaszewska, K.; Overton, C.; Stewart, D.; Rutledge, R.G. Initiation of Somatic Embryos and Regeneration of Plants from Primordial Shoots of 10-Year-Old Somatic White Spruce and Expression Profiles of 11 Genes Followed during the Tissue Culture Process. *Planta* **2011**, 635–647. [CrossRef] [PubMed]

67. Dorais, L.G.; Kettala, E. *Revue, Par Région, Des Techniques d'inventaire Entomologique et d'évaluation Des Programmes de Pulvérisation à Grande Échelle Contre La Tordeuse Des Bourgeons de l'épinette Choristoneura Fumiferana (Clem.)*; Ministère de l'Énergie et des Ressources du Québec: Québec, QC, Canada, 1982; p. 51.

68. Dhont, C.; Sylvestre, P.; Gros-Louis, M.C.; Isabel, N. *Guide-Terrain Pour l'identifcation Des Stades de Débourrement et de Formation Du Bourgeon Apical Chez l'épinette Blanche*; Natural Resources Canada, Laurentian Forestry Centre: Ste-Foy, QC, Canada, 2010.

69. Donkor, D.; Mirzahosseini, Z.; Bede, J.; Bauce, E.; Despland, E. Detoxification of Host Plant Phenolic Aglycones by the Spruce Budworm. *PLoS ONE* **2019**, *14*, e0208288. [CrossRef]

70. Kitajima, K.; Llorens, A.M.; Stefanescu, C.; Timchenko, M.V.; Lucas, P.W.; Wright, S.J. How Cellulose-Based Leaf Toughness and Lamina Density Contribute to Long Leaf Lifespans of Shade-Tolerant Species. *New Phytol.* **2012**, *195*, 640–652. [CrossRef]

71. Lucas, P.W.; Turner, I.M.; Dominy, N.J.; Yamashita, N. Mechanical Defences to Herbivory. *Ann. Bot.* **2000**, *86*, 913–920. [CrossRef]

72. Marco, H.F. The Anatomy of Spruce Needles. *J. Agric. Res.* **1939**, *58*, 357–368.

73. Coley, P.D.; Lokvam, J.; Rudolph, K.; Bromberg, K.; Sackett, T.E.; Wright, L.; Brenes-Arguedas, T.; Dvorett, D.; Ring, S.; Clark, A.; et al. Divergent Defensive Strategies of Young Leaves in Two Species of Inga. *Ecology* **2005**, *86*, 2633–2643. [CrossRef]

74. Donaldson, J.R.; Lindroth, R.L. Genetics, environment, and their interaction determine efficacy of chemical defense in trembling aspen. *Ecology* **2007**, *88*, 729–739. [CrossRef]

75. Nealis, V.G.; Régnière, J. Insect-Host Relationships Influencing Disturbance by the Spruce Budworm in a Boreal Mixedwood Forest. *Can. J. For. Res.* **2004**, *34*, 1870–1882. [CrossRef]

76. Blais, J.R. Trends in the Frequency, Extent, and Severity of Spruce Budworm Outbreaks in Eastern Canada. *Can. J. Zool.* **1983**, *13*, 539–547. [CrossRef]

77. Lehmann, P.; Ammunét, T.; Barton, M.; Battisti, A.; Eigenbrode, S.D.; Jepsen, J.U.; Kalinkat, G.; Neuvonen, S.; Niemelä, P.; Terblanche, J.S.; et al. Complex Responses of Global Insect Pests to Climate Warming. *Front. Ecol. Environ.* **2020**, *18*, 141–150. [CrossRef]

78. Bellemin-Noel, B.; Bourassa, S.; Despland, E.; De Grandpré, L.; Pureswaran, D. Improved Performance of the Eastern Spruce Budworm on Black Spruce as Warming Temperatures Disrupt Phenological Defenses. *Glob. Chang. Biol.* **2021**. [CrossRef]

79. Ekholm, A.; Tack, A.J.M.; Pulkkinen, P.; Roslin, T. Host Plant Phenology, Insect Outbreaks and Herbivore Communities—The Importance of Timing. *J. Anim. Ecol.* **2019**, *89*, 829–841. [CrossRef] [PubMed]

80. Forrest, J.; Miller-Rushing, A.J. Toward a Synthetic Understanding of the Role of Phenology in Ecology and Evolution. *Philos. Trans. R. Soc. B Biol. Sci.* **2010**, *365*, 3101–3112. [CrossRef] [PubMed]

81. Albert, L.P.; Restrepo-Coupe, N.; Smith, M.N.; Wu, J.; Chavana-Bryant, C.; Prohaska, N.; Taylor, T.C.; Martins, G.A.; Ciais, P.; Mao, J.; et al. Cryptic Phenology in Plants: Case Studies, Implications, and Recommendations. *Glob. Chang. Biol.* **2019**, *25*, 3591–3608. [CrossRef] [PubMed]

Article

Role of Fruit Epicuticular Waxes in Preventing *Bactrocera oleae* (Diptera: Tephritidae) Attachment in Different Cultivars of *Olea europaea*

Manuela Rebora [1], Gianandrea Salerno [2,*], Silvana Piersanti [1], Elena Gorb [3] and Stanislav Gorb [3]

[1] Department of Chemistry, Biology and Biotechnology, University of Perugia, Via Elce di Sotto 8, 06121 Perugia, Italy; manuela.rebora@unipg.it (M.R.); silvana.piersanti@unipg.it (S.P.)
[2] Department of Agricultural, Food and Environmental Sciences, University of Perugia, Borgo XX Giugno, 06121 Perugia, Italy
[3] Department of Functional Morphology and Biomechanics, Zoological Institute, Kiel University, Am Botanischen Garten 9, 24098 Kiel, Germany; egorb@zoologie.uni-kiel.de (E.G.); sgorb@zoologie.uni-kiel.de (S.G.)
* Correspondence: gianandrea.salerno@unipg.it; Tel.: +39-075-585-6034

Received: 18 February 2020; Accepted: 15 March 2020; Published: 17 March 2020

Abstract: The olive fruit fly *Bactrocera oleae* (Diptera: Tephritidae) is the major pest of cultivated olives (*Olea europaea* L.), and a serious threat in all of the Mediterranean Region. In the present investigation, we demonstrated with traction force experiments that *B. oleae* female adhesion is reduced by epicuticular waxes (EWs) fruit surface, and that the olive fruit fly shows a different ability to attach to the ripe olive surface of different cultivars of *O. europaea* (Arbequina, Carolea, Dolce Agogia, Frantoio, Kalamata, Leccino, Manzanilla, Picholine, Nostrale di Rigali, Pendolino and San Felice) in terms of friction force and adhesion, in relation with different mean values of olive surface wettability. Cryo-scanning morphological investigation revealed that the EW present on the olive surface of the different analyzed cultivars are represented by irregular platelets varying in the orientation, thus contributing to affect the surface microroughness and wettability in the different cultivars, and consequently the olive fruit fly attachment. Further investigations to elucidate the role of EW in olive varietal resistance to the olive fruit fly in relation to the olive developmental stage and environmental conditions could be relevant to develop control methods alternative to the use of harmful pesticides.

Keywords: friction; adhesion; olive fruit fly; pulvilli; claws; wax crystals; biomechanics

1. Introduction

The interaction between plants and their environment is mediated by a series of complex chemical and physical factors, among which epicuticular waxes (EWs), covering the surface of most plant organs, have a fundamental functional role [1,2]. EWs are a complex mixture of cyclic (triterpenoids) or long chain aliphatic substances, such as primary and secondary alcohols, primary aldehydes, fatty acids and alkenes [1,3], and constitute two-dimensional films/layers or three-dimensional micro- or nanoscale projections covering the plant cuticle [4]. EWs represent the primary barrier against biotic and abiotic stress, exhibiting a multitude of functions, such as being a barrier against water loss [5], offering protection against incident radiation by favoring light reflection [6], protection from surface contamination by dust particles [7,8] or from pathogenic microorganisms [9].

Because of the long coevolution between plants and insects, EWs have an important role also in mediating the insect–plant interaction by protecting the plant from herbivores, or helping it in the capture of pollinators, and preventing the escape of prey insects from carnivorous plants (see review in [10]). In this regard, wax projections can decrease the ability of insects to attach to the plant cuticle surface (see review in [11,12]).

The fly family Tephritidae contains nearly 4500 known species, including some of the world's most significant agricultural insect pests, among which the olive fruit fly *Bactrocera oleae* (Rossi) (Diptera: Tephritidae). This fly is the major pest of commercial olives worldwide, and represents a major pest in the Mediterranean basin (see reviews in [13,14]). The larvae are monophagous on olive fruit in the genus *Olea*, including *Olea europaea* L. (cultivated and wild). On both cultivated and wild olives, females of *B. oleae* lay their eggs in ripening and ripe fruit, right beneath the olive epicarp, providing direct access to food for larvae just after their egg emergence. Larvae feed upon the olive pulp, thus causing losses of up to 80% of the oil value [15]. Although *B. oleae* is a key pest in the olive crop, some *O. europaea* cultivars are less susceptible to *B. oleae* adult females [14–17]. Based upon the investigations on the susceptibility of different olive cultivars to *B. oleae* [18–20], it was concluded that female attraction toward the different olive cultivars is due to the interaction of several physical (e.g., fruit size, weight, volume, color) and chemical factors. Different studies reported that olives lacking the usual waxy coverage were more susceptible to olive fruit fly attack than normal olives [18,19]. In particular, Neuenschwander et al. [18] attributed this reduction of susceptibility to the thickness of the waxy coverage, able to shield the attractant chemicals in the olive cultivars. Vichi et al. [21] observed differences in the chemical composition of the EWs of nine olive varieties grown in the same geographical area, and hypothesized a possible relationship between EW composition and varietal resistance to several biotic and abiotic factors, highlighting the need of further investigations to elucidate the role of EWs in olive varietal resistance to insect pests and environmental conditions.

The EW role in reducing insect attachment to the olive surface has not been previously investigated in detail. In this study, we tested the hypothesis that the attachment of the female of *B. oleae* to the olive fruit surface can vary on the ripe fruits of different cultivars of *O. europaea*, and that this could be related to the EW features, such as wettability and nanostructure in different cultivars. In particular, we used traction force experiments to test *B. oleae* female friction and adhesion to the ripe olive surface of different cultivars of *O. europaea* (Arbequina, Carolea, Dolce Agogia, Frantoio, Kalamata, Leccino, Manzanilla, Picholine, Nostrale di Rigali, Pendolino and San Felice), and we characterized the olive surface of the different cultivars with cryo-scanning morphological investigation and contact angle measurements.

2. Materials and Methods

2.1. Insects

B. oleae adults emerged from pupae obtained from olives collected in the field around Perugia (Umbria, Italy) during October 2018. Olives were kept in the laboratory in a controlled condition chamber (14 h photophase, temperature of 25 ± 1 °C; RH of $60 \pm 10\%$), on a net in order to collect pupae falling down. Pupae were kept inside net cages (300 mm x 300 mm x 300 mm) in a controlled condition chamber (14 h photophase, temperature of 25 ± 1 °C; RH of $60 \pm 10\%$) until the adult emergence. Females and males after emergence were maintained in the same cages and provided with water and crystallized sucrose. Only mated females 10–15 days old were used in the experiments.

2.2. Olive Fruits

Olive fruits belonging to eleven cultivars of *O. europaea* were collected in November 2018 from the germoplasm collection of the Department of Agricultural, Food and Environmental Science of the University of Perugia, located nearby Perugia town (43°04′54.58″ N, 12°22′53.41″ E). Healthy, ripe olive fruit samples were collected from the selected Mediterranean cultivars Arbequina, Carolea, Dolce Agogia, Frantoio, Kalamata, Leccino, Manzanilla, Picholine, Nostrale di Rigali, Pendolino and San Felice. All of the cultivars were subjected to the same cultivation practices, and kept at the same environmental conditions. Only fruits with intact EWs were used in the experiments. The investigations were carried out on ripe olives to have the possibility to compare simultaneously the attachment of *B. oleae* on different cultivars at the same level of ripeness.

2.3. Cryo Scanning Electron Microscopy (Cryo-SEM)

The shock-frozen samples of olive fruit surfaces of the selected cultivars and the tarsi of *B. oleae* insects (females) were studied in a scanning electron microscope (SEM) Hitachi S-4800 (Hitachi High-Technologies Corp., Tokyo, Japan) equipped with a Gatan ALTO 2500 cryo-preparation system (Gatan Inc., Abingdon, UK). For details of sample preparation and mounting for cryo-SEM, see Gorb and Gorb [22]. Whole mounts of olive fruit surface pieces and insect tarsi were sputter-coated in frozen conditions with gold-palladium (thickness 10 nm), and examined at 3 kV acceleration voltage and temperature of $-120\,^{\circ}\mathrm{C}$ at the cryo-stage within the microscope.

2.4. Evaluation of Pulvilli Area

To evaluate the area of pulvilli in both sexes of *B. oleae*, the adults' tarsi were dissected from 20 anesthetized insects (10 males and 10 females), mounted on glass slides with a drop of glycerine, and observed with reflection interference contrast microscopy (RICM) using an inverted bright-field microscope ZEISS Axio Observer A1 (Carl Zeiss Microscopy GmbH, Jena, Germany). Areas of the pulvilli were measured from pad digital images taken with a Sony 3CCD video camera, and using the open source image processing program ImageJ [23]. Measurements were made individually for all 20 insects (240 pulvilli, 12 pulvilli per insect), and recorded separately for the fore-, mid- and hindlegs.

2.5. Characterization of Wettability of Natural (Olive Fruits) and Artificial (Hydrophilic and Hydrophobic Glass) Surfaces

The wettability of ripe, olive fruit surfaces in different cultivars and of hydrophilic and hydrophobic glass was characterized by determining the contact angles of water (aqua millipore, droplet size = 1 µl, sessile drop method) using a high-speed optical contact angle measuring instrument OCAH 200 (Dataphysics Instruments GmbH, Filderstadt, Germany). The contact angle between a liquid and a solid is the angle within the body of the liquid formed at the gas–liquid–solid interface. If the water contact angle on a surface is < 90°, the surface is considered hydrophilic, if the contact angle is ≥ 90° the surface is hydrophobic. Ten measurements ($n = 10$) were performed for each substrate. To increase confidence in the collected data for the olive fruits, wettability measurements were repeated after one year on fruits of the same cultivars grown in the same field. To confirm that the hydrophobic coating on the glass remained over repeated use/time, water contact angle measurements were repeated at the end of the force experiments. There was no significant difference between the contact angles of water on hydrophobic glass measured before and after the experiments ($t = 1.11$; $d.f. = 13$; $p = 0.2886$).

2.6. Force Measurements

The experiments were performed using force measuring experimental set ups for testing the insect attachment to natural surfaces (olive fruits) and artificial surfaces (hydrophilic and hydrophobic glass). Forces were measured using the load cell force sensor FORT-10 (10 g capacity; World Precision Instruments Inc., Sarasota, FL, USA) connected to a force transducer MP 100 (Biopac Systems Ltd., Goleta, CA, USA) [24]. Data were recorded using AcqKnowledge 3.7.0 software (Biopac Systems Ltd., Goleta, CA, USA). Force values were estimated using the force–time curves as the maximum of the detected forces. Only females were tested, owing to their need to firmly attach to the fruit surface during oviposition.

Prior to the force measurements, females of *B. oleae* were weighed upon a micro-balance (Mettler Toledo AG 204 Delta Range, Greifensee, Switzerland). Experimental insects were anesthetized with carbon dioxide for 60 s, and were made incapable of flying by carefully gluing their wings together with a small droplet of melted wax. For force tests, one end of about 15 cm long human hair was fixed on the insect thorax with a droplet of melted wax. Before starting experiments, insects were left to recover for 30 min. All of the experiments were performed during the daytime at 25 ± 1 °C temperature and 60 ± 10% relative humidity.

To test insects on natural fruit surfaces, each olive fruit was firmly attached to a microscope slide, which was fixed to a motorized micromanipulator DC3314R and a controller MS314ZU (World Precision Instruments Inc., Sarasota, FL, USA). At the beginning of each experiment, the insect, attached to the force sensor by means of the hair glued to its thorax, was placed on the olive fruit surface to be tested (see Figure 1 for detail).

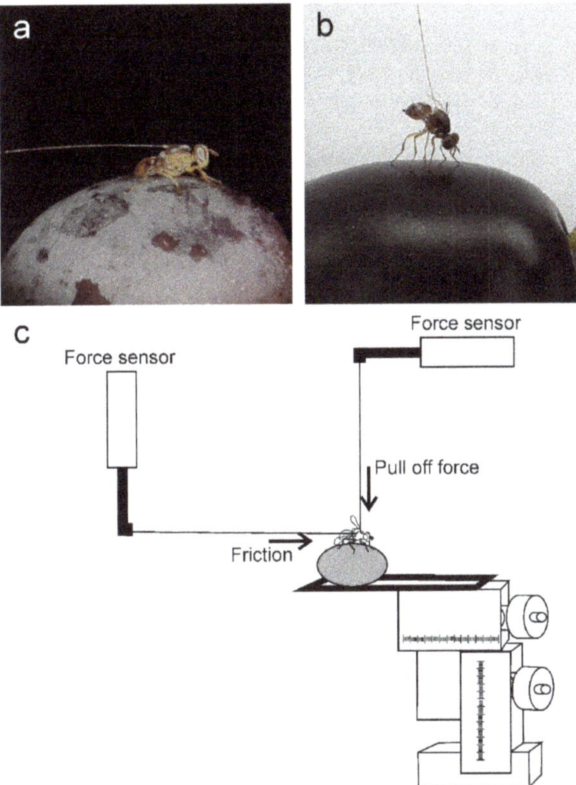

Figure 1. Experimental set ups for testing *Bactrocera oleae* attachment to olive fruits belonging to different cultivars. The insect is attached to the force sensor by means of a hair glued to its thorax. (**a**) Friction force measurements; (**b**) Adhesion (pull off) force measurements; (**c**) The olive fruit is attached to a microscope slide fixed to a motorized micromanipulator and a controller. To measure the friction force, the sensor (on the left side) is kept vertically and the olive is moved in a direction parallel to the olive surface and opposite to the pulling direction of an insect (horizontal arrow). To measure the pull off force, the sensor (on the right side) is kept horizontally and the olive is moved down in a direction perpendicular to the olive surface (vertical arrow).

Two sets of experiments have been performed on natural fruit surfaces. In the first set carried out to measure the insect friction force, the sensor was kept in the vertical position, and the olive was moved in a direction parallel to the olive surface and opposite to the pulling insect until the insect detachment from the olive surface (Figure 1a,c, horizontal arrow).

In the second set of experiments carried out to measure the insect adhesion (pull off) force, the sensor was kept in the horizontal position, and the olive was moved in a direction perpendicular to the olive surface until the insect detached from the olive surface (Figure 1b,c, vertical arrow). In both experiments, the olive was moved at a continuous speed of 200 μm s^{-1} using the motorized micromanipulator. The olive fruits were changed every three tested insects to avoid dehydration of fruits, and a possible effect of changes in EW. Each insect was tested on the olive fruit surface of each cultivar presented in a random order. Additionally, at the beginning and at the end of the sequence of olive fruits, each insect individual was tested on glass to evaluate the effect of possible reduction in insect attachment ability due to possible pulvilli contamination by eroded wax. One olive cultivar (Kalamata) was tested intact and with mechanically removed waxes (dewaxed). The waxes were mechanically removed by gently cleaning the olive fruit surface with a soft paper towel. Between the two sets of experiments, the flies were left to recover for about 30 min. In total, 15 females were tested.

In another set of force experiments for testing insect attachment to artificial surfaces (hydrophilic and hydrophobic glass), the insect attached to the force sensor by means of the hair glued to its thorax was allowed to move on the substrate to be tested in a direction perpendicular to the force sensor. The force generated by the insect walking horizontally on the test substrates (traction force) was measured. In total, 10 females were tested.

To prepare hydrophobic glass, a glass disk was rinsed and then sonicated in an ultrasound bath (Bandelin electronic, Berlin, Germany) for 5 min, first with ethanol (70%), and then with distilled water. This washing procedure was repeated several times and the remaining liquid on the substrate was blown by compressed air. Cleaned glass was subjected to air plasma treatment (Diener electronic, Ebhausen, Germany), and then vacuum pumped (BÜCHI Labortechnik, Flawil, Switzerland) in a desiccator together with a glass vial containing 200 μl of dichlorodimethylsilane (Merck Schuchardt, Hohenbrunn, Germany). The vacuum pump was disconnected from the desiccator once the silane started to boil, and the desiccator was left closed for 5 h to achieve sufficient deposition of the silane on the glass. The surface-modified glass was rinsed thoroughly with isopropanol and distilled water, and blown dry by compressed air.

2.7. Statistical Analysis

To compare the area of pulvilli between the sexes and among the fore-, mid- and hindlegs, two-way repeated measures analysis of variance (ANOVA), considering the insect sex and the leg type as main factors, was used. For significant factors, the Tukey's HSD test was used as post hoc test.

The water contact angles on olive fruit surfaces were compared among the different cultivars using one-way ANOVA and the Tukey's HSD post-hoc test for multiple comparisons between means. The water contact angles on dewaxed and intact olives of the Kalamata cultivar were compared using the *t*-test for independent samples.

In the force experiments, the safety factors (force divided by the weight of the insect individual) obtained for the friction and pull off forces of *B. oleae* females on the fruit surfaces were analyzed with one-way repeated measures ANOVA and the Tukey's HSD post-hoc test for multiple comparisons between means to verify differences between the tested cultivars of *O. europaea*.

For all the performed ANOVA, F-tests were used to assess the significance of the effects and their interactions (Statistica 6.0, StatSoft Inc., Tulsa, OK, USA.).

The *t*-test for dependent samples was used to compare the safety factor values obtained on glass at the beginning and at the end of experiments with each set of olives, to compare the safety factors obtained on the fruit surface in intact fruit and fruit with mechanically removed waxes (dewaxed), and to compare the safety factors obtained on hydrophilic glass and hydrophobic glass.

The relationship between the water contact angle and the normalized safety factor (safety factor normalized on that obtained on glass, to reduce the variability caused by individual insects), both for friction and pull off forces, was tested using linear regression analysis [25].

Before all the analysis, the data were subjected to Box–Cox transformations, in order to reduce data heteroscedasticity [26].

3. Results

3.1. Morphology of Bactrocera oleae Attachment Organs

The attachment devices of the adult *B. oleae* are located at the pretarsus, and are composed of two dorsally situated and ventrally curved claws, and two ventrally situated hairy pulvilli (Figure 2a,b). An unguitractor plate with a distal empodium consisting of a long-tapered hair is located in the basal region of the pretarsus (Figure 2a). Each pulvillus has an oval shape, and on its ventral side consists of numerous distally-oriented tenent setae (Figure 2). The dorsal side of each pulvillus is constituted of a central core, from which numerous cuticular digitations depart, giving rise to the tenent setae (Figure 2b). Each tenent seta consists of a setal shaft and a terminal plate (endplate), whose shape changes from the proximal to the distal portion of the pulvillus along its ventral side (Figure 2c,d). The distal setae have a circular terminal plate (Figure 2c), while the proximal setae have a narrower terminal plate (Figure 2d).

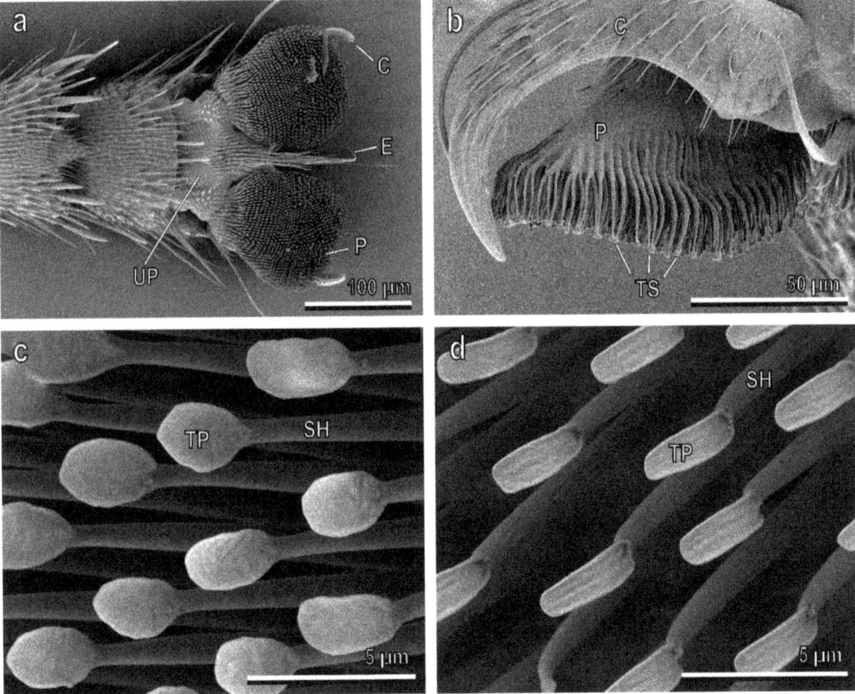

Figure 2. Pretarsal attachment devices of the female of *Bacrocera oleae* in the cryo-SEM. (**a**) Ventral view of hairy pulvilli (P) and curved claws (C). E, empodium, UP, unguitractor plate; (**b**) Lateral view of a pulvillus with its ventral tenent setae (TS). C, claw; (**c**) Detail of the distal tenent setae constituted of a setal shaft (SH) and a circular terminal plate (TP); (**d**) Detail of the proximal tenent setae with a narrow terminal plate (TP). SH, setal shaft.

The female pulvilli are wider than the male pulvilli. Moreover, in both sexes, the area of each pulvillus increases from the forelegs to the hindlegs (sex: $F = 66.9$, *d.f.* $=1, 18$, $p < 0.0001$; legs: $F = 33.1$, *d.f.* $= 2, 36$, $p < 0.0001$; sex x legs: $F = 0.4$, *d.f.* $= 2, 36$, $p = 0.7036$) (Table 1).

Table 1. Area of single pulvillus in the different legs of female and male *Bactrocera oleae*.

Legs	Pulvillus Area (μm^2)	n
♀*		
foreleg	10440 ± 234.0 [b]	10
midleg	12644 ± 353.4 [a]	10
hindleg	14118 ± 340.5 [a]	10
♂		
foreleg	7437 ± 236.9 [A]	10
midleg	8879 ± 292.8 [B]	10
hindleg	9512 ± 284.5 [B]	10

Data are presented as mean ± the standard error of the mean (s.e.m.). The asterisk and different letters show statistical differences at $p < 0.05$, two-way repeated measures ANOVA.

3.2. Surface Morphology and Wettability of Olive Fruits

In all studied cultivars of *O. europaea*, the ripe fruit cuticle is densely covered by EW, with a complex three-dimensional structure (Figure 3a,c). According to the classification proposed by Barthlott et al. [1], flat wax projections composing this coverage are in the form of irregular platelets, with irregular sinuate margins (Figure 3b,d). The platelets are attached to the olive fruit surface by their narrow side, and protrude from the surface at different angles. In most cases, they are oriented nearly perpendicular to the fruit surface, such as in the Picholine cultivar (Figure 3c,d), creating by this prominent fine surface roughness, but in some cultivars (e.g., Manzanilla), they have rather shallow angles with the fruit surface (Figure 3a,b) that leads to a more flattened surface profile.

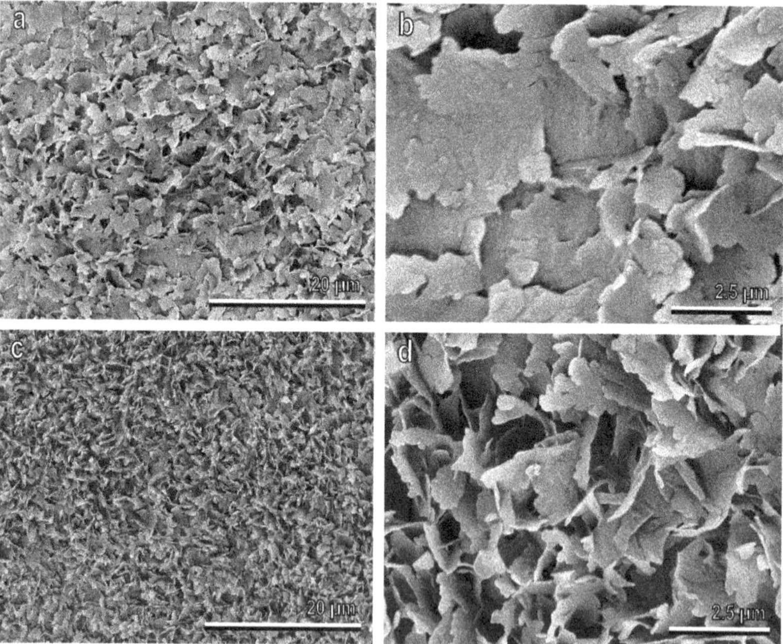

Figure 3. Olive (*Olea europaea*) fruit surfaces of the cultivars Manzanilla (**a,b**) and Picholine (**c,d**) in the cryo-scanning electron microscope (SEM). In (**b**) and (**d**), details of the epicuticular wax (EW) coverage composed of flat projections (platelets) with irregular sinuate margins, are shown. The platelets are attached to the fruit surface by their narrow side protruding from the surface at different angles (compare (**b**) with (**c**)).

The olive fruit surface in all tested cultivars is hydrophobic, with a water contact angle statistically differing among the cultivars ($F = 13.86$, *d.f.* $= 10, 95$, $p < 0.001$), ranging from 102.3° in the Manzanilla cultivar to 147.6° in the Picholine cultivar (Table 2).

Table 2. Contact angles of water on olive fruit surface of different cultivars.

Cultivars	Contact Angle (°)	n
Arbequina	135.18 ± 5.79 [abc]	10
Carolea	144.18 ± 3.07 [a]	11
Dolce Agogia	138.16 ± 5.59 [ab]	9
Frantoio	120.40 ± 4.41 [bd]	10
Kalamata	140.48 ± 3.50 [ab]	10
Leccino	105.62 ± 5.95 [d]	11
Manzanilla	102.30 ± 3.72 [d]	10
Nostrale	106.88 ± 3.67 [d]	8
Pendolino	115.02 ± 5.05 [cd]	8
Picholine	147.62 ± 4.72 [a]	10
San Felice	138.59 ± 3.92 [ab]	9

Data are presented as mean ± s.e.m. Different letters show statistical differences at $p < 0.05$, one-way ANOVA.

3.3. Attachment Ability of Bactrocera oleae Females to the Olive Fruit Surface of Different Cultivars and to Hydrophilic and Hydrophobic Glass

The friction experiments with *B. oleae* females on the ripe fruit surface of different cultivars of *O. europaea* showed that the safety factor (force divided by the weight of insect individual) obtained from the friction force varied significantly depending on the olive cultivar ($F = 7.85$, *d.f.* $= 14, 10, 140$, $p < 0.0001$). Among the tested cultivars, the highest safety factor was recorded on Manzanilla cultivar, together with Nostrale di Rigali, Leccino, Frantoio and San Felice, while the lowest safety factor was recorded on Carolea, together with Picholine, Arbequina, Kalamata and Pendolino. In Dolce Agogia, it was intermediate (Figure 4a).

The safety factor obtained from the pull off force also varied significantly depending on the olive cultivar ($F = 18.67$, *d.f.* $= 14, 10, 140$, $p < 0.0001$). Among the tested olive surfaces from different cultivars, the highest safety factor was recorded on the Manzanilla cultivar, together with Nostrale di Rigali, Leccino and Frantoio, while the lowest safety factor was recorded on Carolea, together with Kalamata, Dolce Agogia and Picholine. In Arbequina, Pendolino and San Felice, it was intermediate (Figure 4b).

The safety factor values obtained on glass at the beginning (friction: 133.4 ± 9.7; pull off: 39.7 ± 2.0) and at the end (friction: 162.3 ± 20.2; pull off: 35.9 ± 2.5) of experiments with each set of olives were not significantly different for both friction ($t = 1.66$, *d.f.* $= 14$, $p = 0.1193$) and pull off forces ($t = 1.64$, *d.f.* $= 14$, $p = 0.1223$).

In the experiments, aiming to compare the attachment ability (friction and pull off forces) of *B. oleae* females on the fruit surface with intact EWs, and with mechanically-removed waxes (dewaxed) (Figure 5), the safety factor was significantly higher on dewaxed olives than on intact olives for both friction ($t = 3.01$, *d.f.* $= 12$, $p = 0.0108$) and pull off forces ($t = 10.14$, *d.f.* $= 14$, $p < 0.0001$) (Figure 5). The contact angle of the intact olive surface was significantly higher than that of the dewaxed olive surface of the same cultivar (140.5 ± 3.5° and 107.8 ± 0.8°, respectively; $t = 10.58$, *d.f.* $= 18$, $p < 0.0001$).

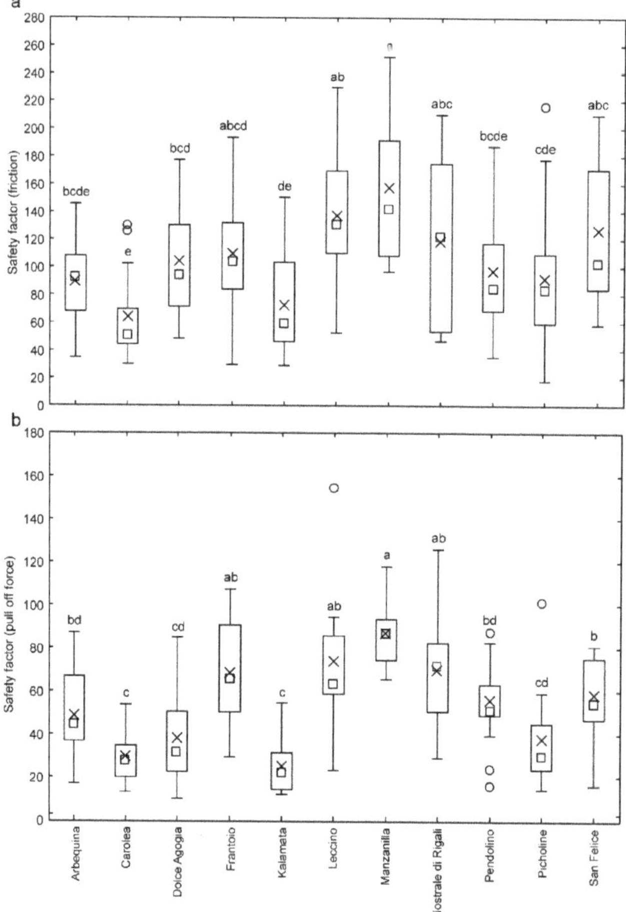

Figure 4. Safety factor (force divided by the insect weight) related to friction force (**a**) and pull off force (**b**) of *Bactrocera oleae* females on the fruit surface of different cultivars of *Olea europaea*. Boxplots show the interquartile range and the median, whiskers indicate 1.5 x interquartile range, "X" shows the arithmetic mean and circles show outliers. Boxplots with different lower-case letters are significantly different at $p < 0.05$, one-way repeated measures analysis of variance (ANOVA), Tukey's HSD post-hoc test.

In the traction experiments on hydrophilic glass (contact angle: 19.94 ± 0.41°) and hydrophobic glass (contact angle: 112.62 ± 1.50°), the safety factor varied significantly depending on the type of the tested surface ($t = 11.26$, *d.f.* $= 9$, $p < 0.0001$). It was higher on hydrophilic glass than on hydrophobic glass (Figure 6).

The normalized safety factor (safety factor normalized on that obtained on glass, to reduce the variability caused by individual insects) related to friction force (Figure 7a) and pull off force (Figure 7b) of *B. oleae* females obtained in our force experiments on the olive fruit surface of the different cultivars of *O. europaea* was negatively correlated with the water contact angle on the olive surfaces (friction force: $F = 9.87$, *d.f.* $= 1, 9$, $p = 0.0119$; pull off force: $F = 35.08$, *d.f.* $= 1, 9$; $p = 0.0002$).

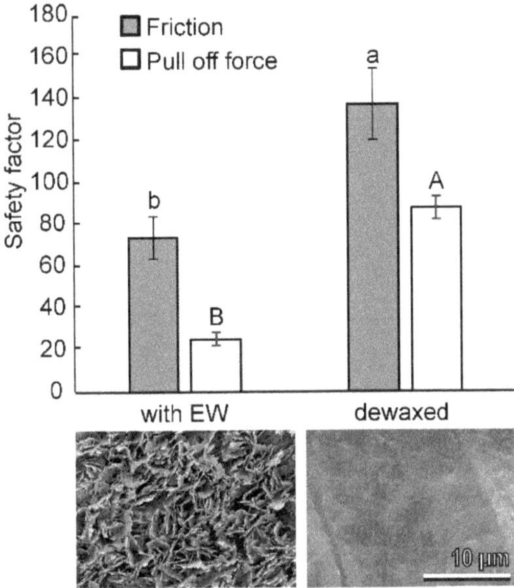

Figure 5. Safety factor (force divided by the insect weight) based on measured friction and pull off forces of *Bactrocera oleae* females on the olive fruit surface of the cultivar Kalamata tested intact (with EW) and with mechanically removed waxes (dewaxed). Bars with different upper-case letters and lower case letters are significantly different at $p < 0.05$, respectively, *t*-test for dependent samples. Images show the corresponding olive fruit surfaces in cryo-SEM.

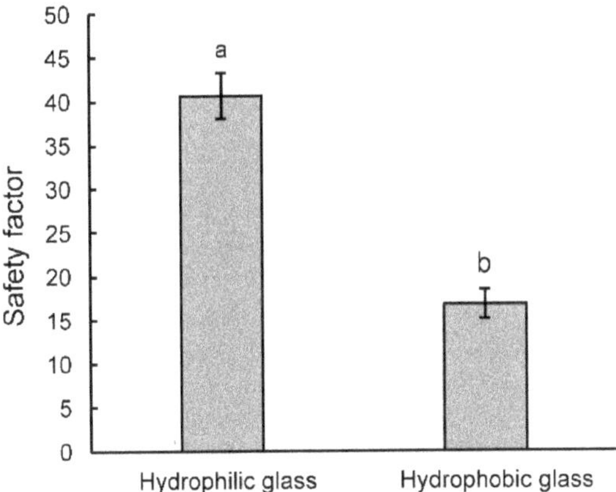

Figure 6. Safety factor (force divided by the insect weight) based on measured traction forces of *Bactrocera oleae* females obtained on hydrophobic and hydrophilic glass. Bars with different letters are significantly different at $p < 0.05$, *t*-test for dependent samples.

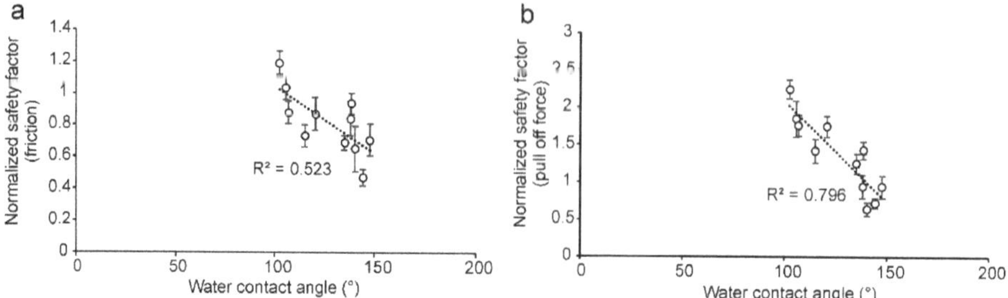

Figure 7. Relationship between the water contact angle on olive fruit surfaces of different cultivars of *Olea europaea* and the normalized safety factors relative to friction force (**a**) and pull off forces (**b**) of *Bactrocera oleae* females.

4. Discussion

4.1. EWs Effect on the Attachment Ability of Bactrocera oleae *to Olive Fruit Surfaces*

The importance of 3D EWs in decreasing surface wettability is known [2,8], and their role in reducing insect attachment to the plant surface has been studied in many insect and plant species using different experimental approaches [27–31] (see also reviews in [10–12]). In our recent investigation, testing the attachment ability of females of another Tephritidae, the Mediterranean fruit fly *Ceratitis capitata* Wiedemann (Diptera: Tephritidae) to fruits of various host plants characterized by different surface morphology (smooth, hairy, waxy) [32], a strong reduction in attachment ability was recorded on *Prunus domestica*, compared with other tested fruit surfaces, due to the dense and regular 3D EW coverage. This last in *P. domestica* is composed of numerous, very short, thin-walled tubules oriented at different angles to the surface, making the fruit surface superhydrophobic. The comparison between the attachment ability to waxy surfaces of *C. capitata*, which is a typical poliphagous species infesting in more than 300 plant species their fruits and nuts [33,34] having different physico-chemical properties, and the olive fruit fly, which is a monophagous species feeding exclusively on wild and cultivated olives [13] typically characterized by hydrophobic surface, is particularly interesting, in relation to possible adaptations to the trophic niche. Both insect species show morphologically similar attachment organs represented by pretarsal paired pulvilli covered by hundreds of capitate "tenent setae" with different terminal plate shapes in the proximal and distal portions of the pulvillus, which is a typical condition described in other brachyceran families, such as Calliphoridae [35–40] and Syrphidae [38,41]. As far as their attachment ability is concerned, the experiments demonstrated that fruit EW plays an important role in reducing attachment, not only in *C. capitata*, but also in *B. oleae*, since the attachment ability of this insect to a dewaxed olive (Kalamata) (contact angle of 107.8 ± 0.8°) was higher than that to the intact olive of the same cultivar (contact angle of 140.5 ± 3.5°).

In our experiments, the safety factor values of *B. oleae* obtained on glass at the beginning and at the end of each set of tests with olives were not significantly different. Similarly, in force experiments with the beetle *Chrysolina fastuosa* (Scopoli) (Coleoptera: Chrysomelidae) [42], with the aphid *Acyrthosiphon pisum* Harris (Hemiptera: Aphididae) [43] and with the bug *Nezara viridula* L. (Hemiptera: Pentatomidae) [44] on plant species with EWs, no influence of the waxy surfaces on the subsequent insect attachment ability was observed, or recovery of the attachment ability occurred rather quickly. Therefore, we can exclude that one of the main reasons of the reduction in *B. oleae* attachment to the olive surface could be linked to contamination of insect adhesive pads with broken fragments of wax projections, as demonstrated previously for several insect species with the hairy type of pads, such as *Adalia bipunctata* (L.) (Coleoptera: Coccinellidae) [45] or *C. fastuosa* [46]. The reduction of insect adhesion of *B. oleae* could be linked better to the roughness hypothesis [42,47], stating that wax projections reduce insect adhesion owing to the surface microroughness, similar to the effect of microrough polishing paper with 0.3–1.0 µm asperity sizes minimizing insect pad contact areas due to small surface irregularities. Such a reduction has been demonstrated in many different insect species [44,48–55], among which the Mediterranean fruit fly *C. capitata* [32].

Our experiments with smooth, artificial surfaces also confirmed an important role of surface wettability in affecting *B. oleae* attachment. Indeed, a reduction of traction force of *B. oleae* on a hydrophobic surface compared with a hydrophylic one was clearly visible: the safety factor was higher on hydrophilic glass (contact angle = 19.94 ± 0.41°) than on hydrophobic glass (contact angle = 112.62 ± 1.50°). A significant decrease in the attachment force on artificial surfaces with an increasing surface contact angle has been reported also in another Diptera species, *C. capitata* [32], and in different other insect species belonging to Coleoptera, such as *Gastrophysa viridula* (De Geer) (Coleoptera, Chrysomelidae) [22], *Cylas puncticolis* Boheman (Coleoptera, Brentidae) [56], *Cryptolaemus montrouzieri* Mulsant, (Coleoptera, Coccinellidae) [57] and *Coccinella septempunctata* L. (Coleoptera, Coccinellidae) females [24,58] as well as to Heteroptera, such as *N. viridula* [55]. These results are explained by the reduction of the attachment force on microstructured substrates owing to the reduced role of the adhesive fluid in generation of capillary forces, by either too low or too strong affinity of the insect adhesive fluid to the microstructured substrates [59]. However, in the beetle *Galerucella nymphaeae* (L.) (Coleoptera: Chrysomelidae) [60], insects showed the highest forces on rather smooth surfaces with a water contact angle of 83° (similar to that of the host plant), while hydrophilic (contact angle of 6° and 26°) and hydrophobic (contact angle of 109°) surfaces caused a reduction of their attachment ability. This suggested that the contact angle of the host plant might rule the attachment behavior of species on different surfaces in relation with adaptation [60]. In this regard, as above reported, the comparison between the attachment ability of two Tephritidae species, such as the polyphagous *C. capitata* [32] and the monophagous *B. oleae*, is particularly interesting. However, the similar results regarding the attachment ability to smooth hydrophilic and hydrophobic surfaces (the same glass surfaces were tested) obtained in these two species do not confirm the hypothesis (at least in Tephritidae) that the effect of the substrate chemistry on insect attachment could depend on insect species and its level of specialization. Further investigations on the tarsal adhesive organs and the attachment ability to hydrophilic or hydrophobic surfaces of other representative insect species, which are more or less specialized, could further clarify this aspect.

4.2. Different Attachment Ability of Bactrocera oleae *to the Olive Fruits of Different Cultivars of* Olea europaea

EWs appear in many different morphological forms (see review in [1]), whose structure is strictly depending on chemical composition [1,2]. A high variability of olive EW morphology in different cultivars of *O. europaea* has been reported by Lanza and Di Serio [61]. The EW of ripe olive is mainly composed of triterpenic acids, alkanes, alcohols, aldehydes, alkyl esters, benzyl esters, triacylglycerols, and fatty acids [62–67]. A chemical analysis of EWs of olives in different cultivars has revealed the presence of various fractions consisting of chemicals from the above reported compound families, whose proportions depended strongly upon the olive cultivar [21,67]. Our data on the attachment ability of *B. oleae* females to the ripe fruit surface of different cultivars of *O. europaea* revealed that both friction force (force preventing sliding of two contacting bodies) and pull off force (force resisting separation of two contacting bodies) varied significantly depending on the olive cultivar and these effects were similar for both above forces. Moreover, from our data it emerges that (1) the olive surface in different cultivars is characterized by different mean values of water contact angles ranging from 102.30° (Manzanilla) to 147.62° (Picholine) and (2) there is a negative correlation between *B. oleae* attachment ability and the olive surface contact angle mean value.

The EW coverage of ripe olives in different cultivars of *O. europaea* examined in the present study is represented by complex 3D structures in the form of irregular platelets. These EWs appear much more similar in the different cultivars than those described by Lanza and Di Serio [61], but this could be due to the different analyzed cultivars (only one was common in both investigations). For the only common Kalamata cultivar, our results on the micromorphology of the EW coverage are on the whole in line with the previous study [61] and some differences could be explained by the different stage of olive ripening used (ripe *vs* green) or different fixation method employed (cryo fixation vs 3–5 h in the oven at 30 °C). However, different cultivars in our study showed some variability in the orientation of the EW platelets. Most of them were oriented perpendicularly to the olive surfaces (e.g., in Picholine), but in some cultivars (especially those revealing a higher wettability by water, such as Manzanilla), wax platelets were oriented at rather shallow angles. In the first case, wax projections created more distinct surface microroughness, responsible for higher hydrophobicity of these surfaces. On fruits of these cultivars, the tested insects showed the poorest attachment (the lowest safety factors for both friction and pull off forces). In the second case, more flattened microrough fruit surfaces having lower hydrophobic properties represented more suitable substrates for the attachment of the olive fruit fly (significantly higher safety factors for both friction and pull off forces). The different angle of wax projection orientation might influence not only the wettability of the surface by different fluids, but also the amount of the direct real contact between terminal contact elements of the fly tenent setae and olive surface. This latter effect might be explained by different relative relationships between dimensions of terminal contact elements of fly tenent setae and substrate asperities. Vertically oriented wax projections generate very small asperities of high amplitude, whereas flattened wax projections produce larger asperities with relatively low amplitude. As previously demonstrated in different theoretical considerations, the first type of substrate will much stronger reduce real contact area with the terminal contact elements of the fly setae and in turn much stronger reduce fly adhesion than the second type of substrate [49,51,54].

Considering that insect attachment depends not only on the presence of 3D wax, but also on the projection size, density of the EW coverage or distribution of individual projections, as shown for the ladybird beetle *C. montrouzieri* walking on *Pisum sativum* plants with wild-type waxes and with reduced waxes caused by mutation [68] and for the ladybird beetle *C. septempunctata* on bio-inspired wax surfaces formed by four alkanes of varying chain length [69], we suggest that a different pattern of the fruit EWs in the different cultivars of *O. europaea* creates a different microroughness and produces a different level of wettability and different amount of direct real contact area between terminal contact elements of the fly setae and EW projections, thus effecting attachment ability of *B. oleae* females to the olive surface, as demonstrated in the present study.

4.3. Attachment Ability of Bactrocera oleae *and Cultivar Susceptibility to the Olive Fruit Fly*

The factors underlining the tolerance of the different cultivars of *O. europaea* to the olive fruit fly are complex [70] and may be due to different factors, such as mechanical barriers, chemical or morphological features and their combination. The role of EWs in reducing *B. oleae* oviposition has been demonstrated in previous investigations [18,19], even if the susceptibility to the fly of the different cultivars in relation with the olive EW coverage still needs to be clarified. According to Olive Germplasm database [71], an increasing level of susceptibility to *B. oleae* can be attributed to the cultivars tested in the present study, in the following order (from the lower to the higher): Kalamata, Picholine, Pendolino, Nostrale di Rigali, Leccino, San Felice, Frantoio and Manzanilla. This trend fits well to the different levels of *B. oleae* attachment ability to the olive fruit recorded in the present investigation. Such a correspondence is less clear for Arbequina, Carolea and Dolce Agogia (high susceptibility and low *B. oleae* attachment ability), probably owing to other factors acting together with the attachment ability of the fly and affecting the resistance properties of different cultivars. Further studies are necessary to better clarify these aspects.

5. Conclusions

The research efforts aimed to investigate the interaction between the olive and its key enemy are relatively few compared with the economic impact of *B. oleae* [17]. In this context, studies clarifying the attachment ability of *B. oleae* in relation to the different physico-chemical features of the EW in different *O. europaea* cultivars could contribute to deepen the knowledge about this important insect pest, thus helping to develop control methods alternative to the use of pesticide harmful for human health. In particular, starting from the results of the present investigation highlighting that *B. oleae* friction force and pull off force on the olive fruit surface varied significantly depending on the olive cultivar, in consideration that EW morphology and chemical composition can change during fruit development [21,62,72,73] and could be affected by environmental abiotic factors potentially changing the surface micro- and nanostructure that in turn influence the degree of porosity, wettability of surface as well as the ability of the fly terminal contact elements to form real contact area with these surfaces [10], further investigations on the mechanical ecology of the olive fruit fly adhesion to the olive fruit surface are advisable.

Author Contributions: The study was designed by all the authors. S.G. and E.G. performed the cryo-scanning electron microscopy investigations. G.S., M.R. and S.P. performed the force experiments. E.G. and M.R. characterized the tested surfaces. The manuscript was written by G.S., M.R. and E.G. All authors discussed the analysis and interpretation of the results and participated in the final editing of the manuscript. All authors have read and agreed to the published version of the manuscript.

Funding: This research was funded by the European Cooperation in Science and Technology, EMBA COST Action CA15216, STSM grant (COST-STSM–CA15216-44544).

Acknowledgments: We are very grateful to Daniela Farinelli for giving us access to the germplasm collection of the Department of Agricultural, Food and Environmental Science of the University of Perugia.

Conflicts of Interest: The authors declare no conflict of interest. The funders had no role in the design of the study; in the collection, analyses, or interpretation of data; in the writing of the manuscript, or in the decision to publish the results.

References

1. Barthlott, W.; Neinhuis, C.; Cutler, D.; Ditsch, F.; Meusel, I.; Theisen, I.; Wilhelmi, H. Classification and terminology of plant epicuticular waxes. *Bot. J. Linn. Soc.* **1998**, *126*, 237–260. [CrossRef]

2. Bargel, H.; Koch, K.; Cerman, Z.; Neinhuis, C. Structure–function relationships of the plant cuticle and cuticular waxes—A smart material? *Funct. Plant Biol.* **2006**, *33*, 893–910. [CrossRef]

3. Jetter, R.; Kunst, L.; Samuels, A.L. Composition of plant cuticular waxes. In *Biology of the Plant Cuticle*; Riederer, M., Müller, C., Eds.; Blackwell: Oxford, UK, 2006; pp. 145–181.

4. Jeffree, C.E. The cuticle, epicuticular waxes and trichomes of plants, with reference to their structure, function and evolution. In *Insects and the Plant Surface*; Juniper, B., Southwood, T.R.E., Eds.; Edward Arnold Publishers: London, UK, 1986; pp. 23–63.

5. Riederer, M.; Schreiber, L. Protecting against water loss: Analysis of the barrier properties of plant cuticles. *J. Exp. Bot.* **2001**, *52*, 2023–2032. [CrossRef] [PubMed]

6. Barnes, J.D.; Cardoso-Vilhena, J. Interactions between electromagnetic radiation and the plant cuticle. In *Plant Cuticles, an Integrated Approach*; Kerstiens, G., Ed.; Bios Scientific Publisher: Oxford, UK, 1996; pp. 157–170.

7. Barthlott, W.; Neinhuis, C. Purity of the sacred lotus or escape from contamination in biological surfaces. *Planta* **1997**, *202*, 1–7. [CrossRef]

8. Fürstner, R.; Barthlott, W.; Neinhuis, C.; Walzel, P. Wetting and self cleaning properties of artificial superhydrophobic surfaces. *Langmuir* **2005**, *21*, 956–961. [CrossRef]

9. Garcia, S.; Garcia, C.; Heinzen, H.; Moyna, P. Chemical basis of the resistance of barley seeds to pathogenic fungi. *Phytochemistry* **1997**, *44*, 415–418. [CrossRef]

10. Gorb, S.N.; Gorb, E. Anti-adhesive effects of plant wax coverage on insect attachment. *J. Exp. Bot.* **2017**, *68*, 5323–5337. [CrossRef]

11. Eigenbrode, S.D. Plant surface waxes and insect behaviour. In *Plant Cuticles—An Integrated Functional Approach*; Kerstiens, G., Ed.; BIOS Scientific Publishers: Oxford, UK, 1996; pp. 201–222.

12. Müller, C. Plant-insect interactions on cuticular surfaces. In *Biology of the Plant Cuticle*; Riederer, M., Müller, C., Eds.; Blackwell: Oxford, UK, 2006; pp. 398–422.

13. Daane, K.M.; Johnson, M.W. Olive Fruit Fly: Managing an Ancient Pest in Modern Times. *Annu. Rev. Entomol.* **2010**, *55*, 151–169. [CrossRef]

14. Malheiro, R.; Casal, S.; Baptista, P.; Pereira, J.A. A review of *Bactrocera oleae* (Rossi) impact in olive products: From the tree to the table. *Trends Food Sci. Technol.* **2015**, *44*, 226–242. [CrossRef]

15. Neuenschwander, P.; Michelakis, S. Olive Fruit Drop Caused by *Dacus oleae* (Gmel) (Dipt. Tephritidae). *Z. Angew. Entomol.* **1981**, *91*, 193–205. [CrossRef]

16. Barrios, G.; Mateu, J.; Ninot i Cort, A.; Romero, A.; Vichi, S. Sensibilidad varietal del olivo a *Bactrocera oleae* y su incidencia en la Gestión Integrada de Plagas. *Phytoma España Rev. Prof. Sanid. Veg.* **2015**, *268*, 21–28.

17. Grasso, F.; Coppola, M.; Carbone, F.; Baldoni, L.; Alagna, F.; Perrotta, G.; PeÂrez-Pulido, A.J.; Garonna, A.; Facella, P.; Daddiego, L.; et al. The transcriptional response to the olive fruit fly (*Bactrocera oleae*) reveals extended differences between tolerant and susceptible olive (*Olea europaea* L.) varieties. *PLoS ONE* **2017**, *12*, e0183050. [CrossRef]

18. Neuenschwander, P.; Michelakis, S.; Holloway, P.; Berchtol, W. Factors affecting the susceptibility of fruits of different olive varieties to attack by *Dacus oleae* (Gmel.) (Dipt, Tephritidae). *Z. Angew. Entomol.* **1985**, *100*, 174–188. [CrossRef]

19. Kombargi, W.S.; Michelakis, S.E.; Petrakis, C.A. Effect of olive surface waxes on oviposition by *Bactrocera oleae* (Diptera: Tephritidae). *J. Econ. Entomol.* **1998**, *91*, 993–998. [CrossRef]

20. Rizzo, R.; Caleca, V.; Lombardo, A. Relation of fruit color, elongation, hardness, and volume to the infestation of olive cultivars by the olive fruit fly, *Bactrocera oleae*. *Entomol. Exp. Appl.* **2012**, *145*, 15–22. [CrossRef]

21. Vichi, S.; Cortés-Francisco, N.; Caixach, J.; Barrios, G.; Mateu, J.; Ninot, A. Epicuticular wax in developing olives (*Olea europaea*) is highly dependent upon cultivar and fruit ripeness. *J. Agric. Food Chem.* **2016**, *64*, 5985–5994. [CrossRef]

22. Gorb, E.V.; Gorb, S.N. Functional surfaces in the pitcher of the carnivorous plant *Nepenthes alata*: A cryo-SEM approach. In *Functional Surfaces in Biology—Adhesion Related Phenomena*; Gorb, S.N., Ed.; Springer: Dordrecht, The Netherlands; Heidelberg, Germany; London, UK; New York, NY, USA, 2009; Volume 2, pp. 205–238.

23. Schneider, C.A.; Rasband, W.S.; Eliceiri, K.W. NIH Image to ImageJ: 25 years of image analysis. *Nat. Methods* **2012**, *9*, 671–675. [CrossRef]

24. Gorb, E.V.; Hosoda, N.; Miksch, C.; Gorb, S.N. Slippery pores: Anti-adhesive effect of nanoporous substrates on the beetle attachment system. *J. R. Soc. Interface* **2010**, *7*, 1571–1579. [CrossRef]

25. StatSoft Inc. *Statistica (Data Analysis Software System)*; Version 6; StatSoft Italia S.R.L: Vigonza (PD), Italy, 2001.

26. Sokal, R.R.; Rohlf, F.J. *Biometry: The Principles and Practice of Statistics in Biological Research*; W.H. Freeman and Company: New York, NY, USA, 1998; p. 887.

27. Anstey, T.H.; Moore, J.F. Inheritance of glossy foliage and cream petals in Green Sprouting Broccoli. *J. Hered.* **1954**, *45*, 39–41. [CrossRef]

28. Edwards, P.B. Do waxes of juvenile *Eucalyptus* leaves provide protection from grazing insects? *Aust. J. Ecol.* **1982**, *7*, 347–352. [CrossRef]

29. Bodnaryk, R.P. Leaf epicuticular wax, an antixenotic factor in Brassicaceae that affects the rate and pattern of feeding in flea beetles, *Phyllotreta cruciferae* (Goeze). *Can. J. Plant Sci.* **1992**, *72*, 1295–1303. [CrossRef]

30. Brennan, E.B.; Weinbaum, S.A. Effect of epicuticular wax on adhesion of psyllids to glaucous juvenile and glossy adult leaves of *Eucalyptus globulus* Labillardière. *Aust. J. Entomol.* **2001**, *40*, 270–277. [CrossRef]

31. Chang, G.C.; Neufeld, J.; Eigenbrode, S.D. Leaf surface wax and plant morphology of peas influence insect density. *Entomol. Exp. Appl.* **2006**, *119*, 197–205. [CrossRef]

32. Salerno, G.; Rebora, M.; Piersanti, S.; Gorb, E.V.; Gorb, S.N. Mechanical ecology of fruit-insect interaction in the adult Mediterranean fruit fly *Ceratitis capitata* (Diptera: Tephritidae). *Zoology* **2020**, *139*, 1257482. [CrossRef] [PubMed]

33. Liquido, N.J.; Shinoda, L.A.; Cunningham, R.T. Host plants of the Mediterranean fruit fly (Diptera: Tephritidae): An annotated world review. *Ann. Entomol. Soc. Am.* **1991**, *77*, 1–52.

34. White, I.M.; Elson-Harris, M.M. *Fruit Flies of Economic Significance: Their Identification and Bionomics*; CAB International: Wallingford, CT, USA; Oxon, UK, 1992.

35. Bauchhenß, E.; Renner, M. Pulvillus of *Calliphora erythrocephala* Meig (Diptera; Calliphoridae). *Int. J. Insect Morphol. Embryol.* **1977**, *6*, 225–227. [CrossRef]

36. Bauchhenß, E. Die Pulvillen von *Calliphora erythrocephala* Meig. (Diptera, Brachycera) als Adhäsionsorgane. *Zoomorphologie* **1979**, *93*, 99–123. [CrossRef]

37. Walker, G.; Yule, A.B.; Ratcliffe, J. The adhesive organ of the blowfly, *Calliphora vomitoria*: A functional approach (Diptera: Calliphoridae). *J. Zool.* **1985**, *205*, 297–307. [CrossRef]

38. Gorb, S.N. The design of the fly adhesive pad: Distal tenent setae are adapted to the delivery of an adhesive secretion. *Proc. R. Soc. Lond. Ser. B Biol. Sci.* **1998**, *265*, 747–752. [CrossRef]

39. Niederegger, S.; Gorb, S.N.; Jiao, Y. Contact behaviour of tenent setae in attachment pads of the blowfly *Calliphora vicina* (Diptera, Calliphoridae). *J. Comp. Physiol. A* **2002**, *187*, 961–970. [CrossRef]

40. Gorb, S.N.; Schuppert, J.; Walther, P.; Schwarz, H. Contact behaviour of setal tips in the hairy attachment system of the fly *Calliphora vicina* (Diptera, Calliphoridae): A cryo-SEM approach. *Zoology* **2012**, *115*, 142–150. [CrossRef] [PubMed]

41. Gorb, S.N.; Gorb, E.V.; Kastner, V. Scale effects on the attachment pads and friction forces in syrphid flies (Diptera, Syrphidae). *J. Exp. Biol.* **2001**, *204*, 1421–1431. [PubMed]

42. Gorb, E.V.; Gorb, S.N. Attachment ability of the beetle *Chrysolina fastuosa* on various plant surfaces. *Entomol. Exp. Appl.* **2002**, *105*, 13–28. [CrossRef]

43. Friedemann, K.; Kunert, G.; Gorb, E.; Gorb, S.N.; Beutel, R.G. Attachment forces of pea aphids (*Acyrthosiphon pisum*) on different legume species. *Ecol. Entomol.* **2015**, *40*, 732–740. [CrossRef]

44. Salerno, G.; Rebora, M.; Gorb, E.; Gorb, S. Attachment ability of the polyphagous bug *Nezara viridula* (Heteroptera: Pentatomidae) to different host plant surfaces. *Sci. Rep.* **2018**, *8*, 10975. [CrossRef]

45. Gorb, E.; Haas, K.; Henrich, A.; Enders, S.; Barbakadze, N.; Gorb, S. Composite structure of the crystalline epicuticular wax layer of the slippery zone in the pitchers of the carnivorous plant *Nepenthes alata* and its effect on insect attachment. *J. Exp. Biol.* **2005**, *208*, 4651–4662. [CrossRef]

46. Gorb, E.V.; Gorb, S.N. Do plant waxes make insect attachment structures dirty? Experimental evidence for the contamination hypothesis. In *Ecology and Biomechanics: A Mechanical Approach to the Ecology of Animals and Plants*; Herrel, A., Speck, T., Rowe, N.P., Eds.; CRC Press: Boca Raton, FL, USA, 2006; pp. 147–162.

47. Scholz, I.; Bückins, L.; Dolge, L.; Erlinghagen, T.; Weth, A.; Hischen, F.; Mayer, J.; Hoffmann, S.; Riederer, M.; Riedel, M.; et al. Slippery surfaces of pitcher plants: *Nepenthes* wax crystals minimize insect attachment via microscopic surface roughness. *J. Exp. Biol.* **2010**, *213*, 1115–1125. [CrossRef]

48. Gorb, S.N. *Attachment Devices of Insect Cuticle*; Kluwer Academic: Dordrecht, The Netherlands, 2001.

49. Peressadko, A.; Gorb, S.N. Surface profile and friction force generated by insects. In *First International Industrial Conference Bionik 2004*; Boblan, I., Bannasch, R., Eds.; VDI Verlag: Düsseldorf, Germany, 2004; pp. 257–263.

50. Voigt, D.; Schuppert, J.M.; Dattinger, S.; Gorb, S.N. Sexual dimorphism in the attachment ability of the Colorado potato beetle *Leptinotarsa decemlineata* (Coleoptera: Chrysomelidae) to rough substrates. *J. Insect. Physiol.* **2008**, *54*, 765–776. [CrossRef]

51. Wolff, J.O.; Gorb, S.N. Surface roughness effects on attachment ability of the spider *Philodromus dispar* (Araneae, Philodromidae). *J. Exp. Biol.* **2012**, *215*, 179–184. [CrossRef]

52. Zhou, Y.; Robinson, A.; Steiner, U.; Federle, W. Insect adhesion on rough surfaces: Analysis of adhesive contact of smooth and hairy pads on transparent microstructured substrates. *J. R. Soc. Interface* **2014**, *11*, 20140499. [CrossRef]

53. Zurek, D.B.; Gorb, S.N.; Voigt, D. Changes in tarsal morphology and attachment ability to rough surfaces during ontogenesis in the beetle *Gastrophysa viridula* (Coleoptera, Chrysomelidae). *Arthropod Struct. Dev.* **2017**, *46*, 130–137. [CrossRef]

54. Kovalev, A.; Filippov, A.E.; Gorb, S.N. Critical roughness in animal hairy adhesive pads: A numerical modeling approach. *Bioinspiration Biomim.* **2018**, *13*, 66004. [CrossRef] [PubMed]

55. Salerno, G.; Rebora, M.; Gorb, E.V.; Kovalev, A.; Gorb, S.N. Attachment ability of the southern green stink bug *Nezara viridula* (Heteroptera: Pentatomidae). *J. Comp. Physiol. A* **2017**, *203*, 1–11. [CrossRef] [PubMed]

56. Lüken, D.; Voigt, D.; Gorb, S.N.; Zebitz, C.P.W. Die Tarsenmorphologie und die Haftfähigkeit des Schwarzen Batatenkäfers *Cylas puncticollis* (Boheman) auf glatten Oberflächen mit unterschiedlichen physiko-chemischen Eigenschaften. *Mitt Dtsch. Ges Allg. Angew. Entomol.* **2009**, *17*, 109–113.

57. Gorb, E.V.; Gorb, S.N. Attachment ability of females and males of the ladybird beetle *Cryptolaemus montrouzieri* to different artificial surfaces. *J. Insect Physiol.* **2020**, *121*, 104011. [CrossRef]

58. Hosoda, N.; Gorb, S.N. Underwater locomotion in a terrestrial beetle: Combination of surface de-wetting and capillary forces. *Proc. R. Soc. Lond. Ser. B Biol. Sci.* **2012**, *279*, 4236–4242. [CrossRef]

59. England, M.W.; Sato, T.; Yagihashi, M.; Hozumi, A.; Gorb, S.N.; Gorb, E.V. Surface roughness rather than surface chemistry essentially affects insect adhesion. *Beilstein J. Nanotechnol.* **2016**, *7*, 1471–1479. [CrossRef]

60. Grohmann, C.; Blankenstein, A.; Koops, S.; Gorb, S.N. Attachment of *Galerucella nymphaeae* (Coleoptera, Chrysomelidae) to surfaces with different surface energy. *J. Exp. Biol.* **2014**, *217*, 4213–4220. [CrossRef]

61. Lanza, B.; Di Serio, M.G. SEM characterization of olive (*Olea europaea* L.) fruit epicuticular waxes and epicarp. *Sci. Hortic.* **2015**, *191*, 49–56. [CrossRef]

62. Bianchi, G.; Murelli, C.; Vlahov, G. Surface waxes from olive fruits. *Phytochemistry* **1992**, *31*, 3503–3506. [CrossRef]

63. Bianchi, G.; Pozzi, N.; Vlahov, G. Pentacyclic triterpene acids in olives. *Phytochemistry* **1994**, *37*, 205–207. [CrossRef]

64. Guinda, A.; Rada, M.; Delgado, T.; Gutiérrez-Adánez, P.; Castellano, J.M. Pentacylic triterpenoids from olive fruit and leaf. *J. Agric. Food Chem.* **2010**, *58*, 9685–9691. [CrossRef] [PubMed]

65. Vlahov, G.; Rinaldi, G.; Del Re, P.; Giuliani, A.A. 13C nuclear magnetic resonance spectroscopy for determining the different components of epicuticular waxes of olive fruit (*Olea europea*) Dritta cultivar. *Anal. Chim. Acta* **2008**, *624*, 184–194. [CrossRef] [PubMed]

66. Vichi, S.; Cortés-Francisco, N.; Romero, A.; Caixach, J. Direct chemical profiling of olive (*Olea europaea*) fruit epicuticular waxes by direct electrospray-Ultrahigh resolution mass spectrometry. *J. Mass Spectrom.* **2015**, *50*, 558–566. [CrossRef]

67. Diarte, C.; Lai, P.-H.; Huang, H.; Romero, A.; Casero, T.; Gatius, F.; Graell, J.; Medina, V.; East, A.; Riederer, M.; et al. Insights Into Olive Fruit Surface Functions: A Comparison of Cuticular Composition, Water Permeability, and Surface Topography in Nine Cultivars During Maturation. *Front. Plant Sci.* **2019**, *10*, 1484. [CrossRef]

68. Gorb, E.; Voigt, D.; Eigenbrode, S.D.; Gorb, S. Attachment force of the beetle *Cryptolaemus montrouzieri* (Coleoptera, Coccinellidae) on leaflet surfaces of mutants of the pea *Pisum sativum* (Fabaceae) with regular and reduced wax coverage. *Arthropod–Plant Interact.* **2008**, *2*, 247–259. [CrossRef]

69. Gorb, E.; Böhm, S.; Jacky, N.; Maier, L.P.; Dening, K.; Pechook, S.; Pokroy, B.; Gorb, S. Insect attachment on crystalline bioinspired wax surfaces formed by alkanes of varying chain lengths. *Beilstein J. Nanotechnol.* **2014**, *5*, 1031–1041. [CrossRef]

70. Corrado, G.; Garonna, A.; Cabanàs, C.G.-L.; Gregoriou, M.; Martelli, G.P.; Mathiopoulos, K.D.; Saponari, J.M.-B.M.; Tsoumani, K.T.; Rao, R. Host Response to Biotic Stresses. In *The Olive Tree Genome*; Rugini, E., Baldoni, L., Muleo, R., Sebastiani, L., Eds.; Springer International Publishing: Cham, Germany, 2016; pp. 75–98.

71. Bartolini, G. Olive Germplasm (*Olea europaea* L.). Istituto per la Valorizzazione del Legno e delle Specie Arboree (IVALSA), Trees and Timber Institute. Available online: www.oleadb.it/olivodb.html (accessed on 20 January 2020). [CrossRef]

72. Yeats, T.H.; Rose, J.K.C. The formation and function of plant cuticles. *Plant Physiol.* **2013**, *163*, 5–20. [CrossRef]

73. Trivedi, P.; Nguyen, N.; Hykkerud, A.L.; Häggman, H.; Martinussen, I.; Jaakola, L.; Karppinen, K. Developmental and environmental regulation of cuticular wax biosynthesis in fleshy fruits. *Front. Plant Sci.* **2019**, *10*, 431. [CrossRef]

MDPI

St. Alban-Anlage 66

4052 Basel

Switzerland

www.mdpi.com

Insects Editorial Office

E-mail: insects@mdpi.com

www.mdpi.com/journal/insects